UNIVERSITY *of* LIMERICK

Telephone 061 333644 Extension 2158

**Please return by latest date shown**

**PLEASE NOTE:** This item is subject to recall after two weeks
if required by another reader.

D1348563

4371119

# MONOGRAPHS ON
# STATISTICS AND APPLIED PROBABILITY

## General Editors

### D.R. Cox, V. Isham, N. Keiding, N. Reid and H. Tong

(Full details concerning this series are available from the Publishers).

# Retrial Queues

G.I. Falin

*Department of Probability Theory*
*Mechanics–Mathematics Faculty*
*Moscow State University*
*Moscow, Russia*

and

J.G.C. Templeton

*Department of Mechanical and Industrial Engineering*
*University of Toronto*
*Toronto, Canada*

**CHAPMAN & HALL**

London · Weinheim · New York · Tokyo · Melbourne · Madras

Published by Chapman & Hall, 2–6 Boundary Row, London SE1 8HN, UK

Chapman & Hall, 2–6 Boundary Row, London SE1 8HN, UK

Chapman & Hall GmbH, Pappelallee 3, 69469 Weinheim, Germany

Chapman & Hall USA., 115 Fifth Avenue, New York, NY 10003, USA

Chapman & Hall Japan, ITP-Japan, Kyowa Building, 3F, 2-2-1 Hirakawacho, Chiyoda-ku, Tokyo 102, Japan

Chapman & Hall Australia, 102 Dodds Street, South Melbourne, Victoria 3205, Australia

Chapman & Hall India, R. Seshadri, 32 Second Main Road, CIT East, Madras 600 035, India

First edition 1997

© 1997 Chapman & Hall

Printed in Great Britain by St Edmundsbury Press, Bury St Edmunds, Suffolk.

ISBN  0  412  78550 1

A catalogue record for this book is available from the British Library

∞ Printed on permanent acid-free text paper, manufactured in accordance with ANSI/NISO Z39.48 - 1992 and ANSI/NISO Z39.48 - 1984 (Permanence of Paper).

# Contents

# Preface

It is well known that a telephone subscriber who obtains a busy signal usually repeats the call until the required connection is made. As a result, the flow of calls circulating in a telephone network consists of two parts: the flow of primary calls, which reflects the real wishes of the telephone subscribers, and the flow of repeated calls, which is the consequence of the lack of success of previous attempts. The standard models of telephone systems, queueing systems with losses, do not take into consideration this structure of the real flow of calls and therefore cannot be applied in solving a number of practically important problems.

A new class of queueing systems, systems with repeated calls (or retrial queues, queues with returning customers, repeated orders, etc.) has been introduced for their analysis. This class of queues is characterized by the following feature: a customer arriving when all servers accessible for him are busy leaves the service area but after some random time repeats his demand. This feature plays a special role in several computer and communication networks as well. Other applications include stacked aircraft waiting to land, and queues of retail shoppers who may leave a long waiting line hoping to return later when the line may be shorter, so that the area of possible applications of such systems is wide.

Retrial queues can be regarded as networks with reservicing after blocking. In the simplest cases, which are also the most important, this network contains two nodes: the main node where blocking is possible and a delay node for repeated trials. As for other networks with blocking, the investigation of such systems presents great analytical difficulties. Detailed results exist for some special retrial queues, while for many others there is extremely poor information. Nevertheless, the main features of the theory of retrial queueing systems as an independent part of queueing theory are quite clearly drawn. In particular, the nature of results obtained, methods of analysis and areas of application allow us to divide retrial queues into two large groups in a natural way: single-channel systems and multichannel (fully available) systems. This division is of course arbitrary; for example, from a mathematical point of view, it is natural to analyse single-channel queues with buffers as multichannel queues. As special parts of the theory we must mention queueing networks with repeated calls, statistics of systems and engineering aspects with applications.

The first mathematical results about retrial queues were published in the 1950s and at present a full bibliography consists of several hundred items. Research papers on retrial queues and their applications are published in mathematical and statistical journals such as *Journal of the Royal Statistical Society, Journal of Applied Probability, Advances in Applied Probability, Probability Theory and Its Applications*, etc., OR journals such as *Queueing Systems, ORSA Journal on Computing, Acta Informatica, OR Spectrum*, etc., telecommunication journals such as *The Bell System Technical Journal, Philips Telecommunication Review, IEEE Transactions on Information Theory, Transactions of the Institute of Electronics and Communication Engineers of Japan, Electronics and Communications in Japan, Budavox Telecommunication Review, Australian Telecommunication Research, Ericsson Technics*, and proceedings of conferences (we should specifically mention Proceedings of regular International Teletraffic Congresses). Some textbooks and monographs on queueing theory and teletraffic theory include sections devoted to retrial queues, where simple results concerning this type of system are stated. Some results have appeared only in Ph.D. and Master's Theses, technical reports of universities and research centres. A significant contribution to the theory of retrial queues has been and is being made by Russian scientists. Since much of their work appears in Russian language journals which are not available in English translation, it is difficult for unilingual English readers to keep up with this very active research area.

At present, the theory of retrial queues is recognized as an important part of queueing theory and teletraffic theory. As noted by L.Kosten in his monograph *Stochastic Theory of Service Systems*, Pergamon Press, 1973: 'any theoretical result that does not take into consideration this repetition effect should be considered suspect'.

Results and proofs have not simply been taken from existing literature. Many known results have been given new derivations using general methods of wide applicability. As a result of these changes, these is a fairly close resemblance among many of the proofs in the book. In some cases, similar proofs have been given at length in order to show the similarity. In other cases, where the proof of one result is particularly close to the proof of another result, the second proof is replaced by a reference to the first proof.

The authors wish to acknowledge with thanks all those who contributed to the preparation of this book. Both authors thank the Natural Sciences and Engineering Research Council of Canada for financial support through a research grant, and thank Ivan Falin for his help with LaTeX typing. Guennadi Falin gives thanks for hospitality and support to Universidad Complutense (Madrid), where he gave a course on Retrial Queues during the second term of the 1991–92 academic year, and to the University of Toronto for access to its libraries and other facilities during several visits. Jim Templeton thanks the University of Toronto for continued support during his retirement, in the form of office space and access to services and facilities.

*Guennadi Falin*
*James G. C. Templeton*
February, 1997

# CHAPTER 1

# The main single-server model

## 1.1 Description of the main model of $M/G/1$ type

Consider a single server queueing system in which customers arrive in a Poisson process with rate $\lambda$. These customers are identified as primary calls. If the server is free at the time of a primary call arrival, the arriving call begins to be served immediately and leaves the system after service completion. Otherwise, if the server is busy, the arriving customer becomes a source of repeated calls (a customer in orbit, a customer in pool, etc.). The pool of sources of repeated calls may be viewed as a sort of queue. Every such source produces a Poisson process of repeated calls with intensity $\mu$. If an incoming repeated call finds the server free, it is served and leaves the system after service, while the source which produced this repeated call disappears. Otherwise, the system state does not change.

The service time distribution function is $B(x)$ for both primary calls and repeated calls.

The input flow of primary calls, intervals between repetitions, and service times are mutually independent.

The queueing process evolves in the following manner. Suppose that the $(i-1)$th call completes its service at epoch $\eta_{i-1}$ (the calls are numbered in the order of service) and the server becomes free. Even if there are some customers in the system who want to get service they cannot occupy the server immediately, because of their ignorance of the server state. Therefore the next, $i$th, call enters service only after some time interval $R_i$ during which the server is free while there may be waiting customers. If the number of sources of repeated calls at the time $\eta_{i-1}$, $N_{i-1}$, is equal to $n$, then the random variable $R_i$ has an exponential distribution with parameter $\lambda+n\mu$. The $i$th call is a primary call with probability $\frac{\lambda}{\lambda+n\mu}$ and it is a repeated call with probability $\frac{n\mu}{\lambda+n\mu}$. At epoch $\xi_i = \eta_{i-1} + R_i$ the $i$th call's service starts and continues during a time $S_i$ (i.e. $S_i$ is the service time of the $i$th call). All primary calls arriving during the service time form sources of repeated calls. Repeated calls which arrive during this time interval do not influence the process. Then, at epoch $\eta_i = \xi_i + S_i$ the $i$th call completes service and the server becomes free again.

At time $t$ let $N(t)$ be the number of sources of repeated calls (which may be viewed as a sort of queue) and $C(t)$ the number of busy servers (for single-server queues $C(t) = 1$ or $0$ according as the server is busy or free). The process $(C(t), N(t))$ which describes the number of customers in the system is the simplest and simultaneously the most important process associated with the above queueing system.

If the service time distribution is not exponential then the process $(C(t), N(t))$ is not Markov. In this case we introduce a supplementary variable: if $C(t) = 1$, we define $\xi(t)$ as the elapsed service time of the call being served.

Let $\beta(s) = \int_0^\infty e^{-sx} dB(x)$ be the Laplace-Stieltjes transform of the service time distribution function $B(x)$, $\beta_k = (-1)^k \beta^{(k)}(0)$ be the $k$th moment of the service time about the origin, $\rho = \lambda\beta_1$ the system load due to primary calls, $b(x) = B'(x)/(1 - B(x))$ be the instantaneous service intensity given that the elapsed service time is equal to $x$, $k(z) = \beta(\lambda - \lambda z)$. It is easy to see that $k(z) = \sum_{n=0}^\infty k_n z^n$ , where

$$k_n = \int_0^\infty \frac{(\lambda x)^n}{n!} e^{-\lambda x} dB(x)$$

is the distribution of the number of primary calls which arrive during the service time of a call.

## 1.2 Joint distribution of the server state and the queue length in the steady state

In this section we carry out the simplest and at the same time the most important (from an applied point of view) analysis of the system. Namely, we investigate the joint distribution of the server state and the queue length in the steady state. As we will show later on the stationary regime exists if and only if $\rho < 1$, so the condition $\rho < 1$ is assumed to hold from now on.

*1.2.1   M/M/1 case*

First consider the important particular case of exponentially distributed service times when $B(x) = 1 - e^{-\nu x}$.

**Theorem 1.1** *For an M/M/1 retrial queue in the steady state the joint distribution of server state $C(t)$ and queue length $N(t)$, $p_{in} = \mathrm{P}\{C(t) = i, N(t) = n\}$, is given by*

$$p_{0n} = \frac{\rho^n}{n!\mu^n} \prod_{i=0}^{n-1} (\lambda + i\mu) \cdot (1 - \rho)^{\frac{\lambda}{\mu}+1}, \qquad (1.1)$$

$$p_{1n} = \frac{\rho^{n+1}}{n!\mu^n} \prod_{i=1}^{n} (\lambda + i\mu) \cdot (1 - \rho)^{\frac{\lambda}{\mu}+1}. \qquad (1.2)$$

*The corresponding partial generating functions are given by*

$$p_0(z) \equiv \sum_{n=0}^{\infty} z^n p_{0n} = (1 - \rho) \left( \frac{1 - \rho}{1 - \rho z} \right)^{\frac{\lambda}{\mu}}, \qquad (1.3)$$

$$p_1(z) \equiv \sum_{n=0}^{\infty} z^n p_{1n} = \rho \left( \frac{1 - \rho}{1 - \rho z} \right)^{\frac{\lambda}{\mu}+1}. \qquad (1.4)$$

*Proof.* In the case of exponentially distributed service times process $(C(t), N(t))$ is a Markov process with the state space $\{0; 1\} \times Z_+$, where $Z_+$ is the set of nonnegative integers.

From a state $(0, n)$ only transitions into the following states are possible:

1. $(1, n)$ with rate $\lambda$;
2. $(1, n - 1)$ with rate $n\mu$.

The first transition is due to arrival of a primary call and the second is due to arrival of a repeated call. Since the state $(0, n)$ means that the server is free, there is no transition corresponding to the service completion.

Reaching state $(0, n)$ is possible only from state $(1, n)$ with rate $\nu$.

From a state $(1, n)$ only transitions into the following states are possible:

1. $(1, n + 1)$ with rate $\lambda$;
2. $(0, n)$ with rate $\nu$.

The first transition is due to arrival of a primary call and the second is due to a service completion. Since in state $(1, n)$ the server is busy, arrivals of repeated calls do not influence the process. Thus there is no transition corresponding to arrival of a repeated call.

Reaching state $(1, n)$ is possible only from the states:

1. $(0, n)$ with rate $\lambda$;
2. $(0, n + 1)$ with rate $(n + 1)\mu$;
3. $(1, n - 1)$ with rate $\lambda$.

Thus the set of statistical equilibrium equations for the probabilities $p_{0n}$, $p_{1n}$ is

$$(\lambda + n\mu)p_{0n} = \nu p_{1n}, \tag{1.5}$$

$$(\lambda + \nu)p_{1n} = \lambda p_{0n} + (n + 1)\mu p_{0,n+1} + \lambda p_{1,n-1} \tag{1.6}$$

We give two different ways of solving these equations.

The first one uses partial generating functions

$$p_0(z) \equiv \sum_{n=0}^{\infty} z^n p_{0n}, \; p_1(z) \equiv \sum_{n=0}^{\infty} z^n p_{1n}.$$

For them the equations (1.5), (1.6) become:

$$\lambda p_0(z) + \mu z p_0'(z) = \nu p_1(z), \tag{1.7}$$

$$(\nu + \lambda - \lambda z)p_1(z) = \lambda p_0(z) + \mu p_0'(z).$$

Eliminating $p_1(z)$ we get the following differential equation for $p_0(z)$:

$$p_0'(z) = \frac{\lambda \rho}{\mu(1 - \rho z)}p_0(z)$$

with solution

$$p_0(z) = \frac{\text{Const}}{(1 - \rho z)^{\frac{\lambda}{\mu}}}. \tag{1.8}$$

Now from (1.7) we have:

$$p_1(z) = \rho p_0(z) + \frac{\mu z}{\nu}p_0'(z) = \rho p_0(z) + \frac{\rho^2 z}{1 - \rho z}p_0(z)$$

$$= \frac{\rho}{1 - \rho z}p_0(z) = \frac{\rho \cdot \text{Const}}{(1 - \rho z)^{\frac{\lambda}{\mu}+1}}. \tag{1.9}$$

The constant can be found with the help of the normalizing condition

$$\sum_{n=0}^{\infty}(p_{0n} + p_{1n}) = p_0(1) + p_1(1) = 1,$$

which implies that

$$\text{Const} = (1 - \rho)^{\frac{\lambda}{\mu}+1}. \tag{1.10}$$

From (1.8)–(1.10) we immediately have (1.3) and (1.4).

To get formulas (1.1) and (1.2) it is sufficient to expand functions $(1 - \rho z)^{-\frac{\lambda}{\mu}}$ and $(1 - \rho z)^{-\frac{\lambda}{\mu}-1}$ in power series with the help of the classical binomial formula:

$$(1 + x)^m = \sum_{n=0}^{\infty} \frac{x^n}{n!} \prod_{i=0}^{n-1} (m - i). \tag{1.11}$$

The second way of solving the equations (1.5), (1.6) consists of the following.

With the help of equation (1.5) eliminate probabilities $p_{1n}$ from equation (1.6) and rewrite the resulting equation as

$$(n+1)\mu\nu p_{0,n+1} - \lambda(\lambda + n\mu)p_{0n} = n\mu\nu p_{0n} - \lambda(\lambda + (n-1)\mu)p_{0,n-1}.$$

This implies that

$$n\mu\nu p_{0n} - \lambda(\lambda + (n-1)\mu)p_{0,n-1} = 0,$$

i.e.

$$
\begin{aligned}
p_{0n} &= \frac{\lambda(\lambda + (n-1)\mu)}{n\mu\nu} p_{0,n-1} \\
&= \frac{\rho^n}{n!\mu^n} \prod_{i=0}^{n-1} (\lambda + i\mu) p_{00}.
\end{aligned}
$$

Now from equation (1.5) we have

$$
\begin{aligned}
p_{1n} &= \frac{\rho^n}{n!\mu^n\nu} \prod_{i=0}^{n} (\lambda + i\mu) p_{00} \\
&= \frac{\rho^{n+1}}{n!\mu^n} \prod_{i=1}^{n} (\lambda + i\mu) p_{00}.
\end{aligned}
$$

The probability $p_{00}$ may be found with the help of the normalizing condition $\sum_{n=0}^{\infty} p_{0n} + \sum_{n=0}^{\infty} p_{1n} = 1$:

$$p_{00}^{-1} = \sum_{n=0}^{\infty} \frac{\rho^n}{n!\mu^n} \prod_{i=0}^{n-1} (\lambda + i\mu) + \sum_{n=0}^{\infty} \frac{\rho^{n+1}}{n!\mu^n} \prod_{i=1}^{n} (\lambda + i\mu).$$

The sums in the right-hand side of this formula can be reduced to binomial series (1.11):

$$\sum_{n=0}^{\infty} \frac{\rho^n}{n!\mu^n} \prod_{i=0}^{n-1} (\lambda + i\mu) = \sum_{n=0}^{\infty} \frac{\rho^n}{n!} \prod_{i=0}^{n-1} \left( \frac{\lambda}{\mu} + i \right)$$

$$= \sum_{n=0}^{\infty} \frac{(-\rho)^n}{n!} \prod_{i=0}^{n-1} (-\frac{\lambda}{\mu} - i)$$

$$= (1 - \rho)^{-\frac{\lambda}{\mu}};$$

$$\sum_{n=0}^{\infty} \frac{\rho^{n+1}}{n!\mu^n} \prod_{i=1}^{n} (\lambda + i\mu) = \rho \sum_{n=0}^{\infty} \frac{(-\rho)^n}{n!} \prod_{j=0}^{n-1} (-\frac{\lambda}{\mu} - 1 - j)$$

$$= \rho(1 - \rho)^{-\frac{\lambda}{\mu} - 1}.$$

Thus,

$$p_{00}^{-1} = (1 - \rho)^{-\frac{\lambda}{\mu}} + \rho(1 - \rho)^{-\frac{\lambda}{\mu} - 1} = (1 - \rho)^{-\frac{\lambda}{\mu} - 1},$$

which implies equations (1.1) and (1.2). Formulas (1.3) and (1.4) can be obtained with the help of binomial series.  □

Now we can get various performance characteristics of the system in the steady state:

(a) The stationary distribution of the number of sources of repeated calls $q_n = P\{N(t) = n\}$ has the generating function

$$p(z) = p_0(z) + p_1(z) = (1 + \rho - \rho z) \left(\frac{1 - \rho}{1 - \rho z}\right)^{\frac{\lambda}{\mu} + 1}. \qquad (1.12)$$

This implies that factorial moments $\Phi_n = E(N(t))_n$ of the queue length are given by:

$$\Phi_n = \frac{\rho^{n+1}}{(1 - \rho)^n} \frac{(\lambda + \mu) \dots (\lambda + (n-1)\mu)(\nu + n\mu)}{\mu^n}. \qquad (1.13)$$

In particular, the mean $EN(t) = \Phi_1$ and variance $\text{Var}N(t) = \Phi_2 + \Phi_1 - \Phi_1^2$ of the random variable $N(t)$ are given by:

$$EN(t) = \frac{\rho(\lambda + \rho\mu)}{(1 - \rho)\mu},$$

$$\text{Var}N(t) = \frac{\rho(\lambda + \rho\mu + \rho^2\mu - \rho^3\mu)}{(1 - \rho)^2\mu}.$$

(b) The stationary distribution of the number of customers in the system $Q_n = P\{K(t) = n\}$ has the generating function

$$Q(z) = p_0(z) + zp_1(z) = \left(\frac{1 - \rho}{1 - \rho z}\right)^{\frac{\lambda}{\mu} + 1}.$$

In particular, the mean and variance of the number of customers in the system are given by

$$EK(t) = \frac{\rho(\lambda + \mu)}{(1 - \rho)\mu},$$

$$VarK(t) = \frac{\rho(\lambda + \mu)}{(1 - \rho)^2\mu}.$$

(c) The blocking probability $p_1$, i.e. probability that the server is busy, is given by:

$$p_1 = p_1(1) = \rho.$$

The state of the server can be described in more detail. Namely, let $M(t)$ be the total number of arrivals (both primary and repeated) since the last departure time. This means that if $M(t) = 0$ then the server is free at time $t$. But if $M(t) = m \geq 1$, then the server is occupied at time $t$ and during the elapsed service time there were exactly $m - 1$ unsuccessful attempts to get service. The process $(M(t), N(t))$ is a Markov process with two-dimensional integer lattice $Z_+^2$ as the state space. The rates of transitions of the process $((M(t), N(t))$ are

1. if $m = 0$ then

$$q_{(m,n)(i,j)} = \begin{cases} \lambda, & \text{if } (i,j) = (1,n) \\ n\mu, & \text{if } (i,j) = (1, n-1) \\ -(\lambda + n\mu), & \text{if } (i,j) = (0,n) \\ 0, & \text{otherwise} \end{cases}$$

2. if $m \geq 1$ then

$$q_{(m,n)(i,j)} = \begin{cases} \lambda, & \text{if } (i,j) = (m+1, n+1) \\ n\mu, & \text{if } (i,j) = (m+1, n) \\ \nu, & \text{if } (i,j) = (0,n) \\ -(\lambda + \nu + n\mu), & \text{if } (i,j) = (m,n) \\ 0. & \text{otherwise} \end{cases}$$

Note also that $C(t) = \delta(M(t))$, where $\delta(n)$ is the indicator function of positive integers, and thus the process $(C(t), N(t))$ can be thought of as a function of the process $(M(t), N(t))$.

Let $s_{mn} = P(M(t) = m, N(t) = n)$ be the stationary distribution of the process $(M(t), N(t))$. The Kolmogorov equations for the probabilities $s_{mn}$ are

$$(\lambda + n\mu)s_{0n} = \nu \sum_{m=1}^{\infty} s_{mn} \tag{1.14}$$

$$(\lambda + \nu + n\mu)s_{mn} = (\lambda s_{m-1,n-1} + n\mu s_{m-1,n})(1 - \delta_{m,1})$$
$$+ (\lambda s_{0n} + (n+1)\mu s_{0,n+1})\delta_{m,1},$$
$$\text{if } m \geq 1, \tag{1.15}$$

where $\delta_{m,n}$ is Kronecker's delta. Note that since $C(t) = \delta(M(t))$, we have:

$$s_{0n} = p_{0n}, \quad \sum_{m=1}^{\infty} s_{mn} = p_{1n},$$

where $\{p_{in}\}$ is the stationary distribution of process $(C(t), N(t))$ which is given by formulas (1.1), (1.2).

To solve equations (1.14), (1.15) we introduce a generating function

$$s(x, z) = \sum_{m=0}^{\infty} x^m \sum_{n=0}^{\infty} z^n s_{mn} = \mathrm{E}\left(x^{M(t)} z^{N(t)}\right).$$

For the function $s(x, z)$ equations (1.14), (1.15) become:

$$\mu z(1-x)s'_z(x, z) = (\lambda xz - \lambda - \nu)s(x, z) + (\nu + \lambda x - \lambda xz)p(z), \tag{1.16}$$

where $p(z)$ is given by formula (1.12).

Later on we will need factorial moments

$$F(m, n) = \mathrm{E}\left((M(t))_m(N(t))_n\right)$$
$$= \frac{\partial^{m+n} s(1, 1)}{(\partial x)^m (\partial z)^n}$$

rather than the distribution $\{s_{mn}\}$. With this goal we differentiate equation (1.16) $m$ times with respect to $x$ and $n$ times with respect to $z$ at the point $x = 1$, $z = 1$. After some algebra we get:

$$\nu F(m, n) = m\mu F(m-1, n+1) + m(\lambda + n\mu)F(m-1, n)$$
$$+ \lambda nm F(m-1, n-1) + \lambda n F(m, n-1)$$
$$+ \psi(m, n), \tag{1.17}$$

where

$$\psi(m, n) = \begin{cases} \nu \Phi_n - \lambda n \Phi_{n-1}, & \text{if } m = 0, \\ -\lambda n \Phi_{n-1}, & \text{if } m = 1, \\ 0, & \text{if } m \geq 2, \end{cases} \tag{1.18}$$

and the factorial moments of the queue length, $\Phi_n$, are given by (1.13). Since $F(0, n) = \Phi_n$, from equations (1.17), (1.18) we can find recursively all moments $F(m, n)$. In particular,

$$\nu F(1, 0) = \mathrm{E}M(t) = \mu \Phi_1 + \lambda,$$

$$\nu^2 F(1,1) = \mathrm{E}\{M(t)N(t)\} = \nu\mu\Phi_2 + (\lambda\nu + \mu\nu + \lambda\mu)\Phi_1 + \lambda^2,$$

$$\nu^3 F(1,2) = \mathrm{E}\{M(t)N(t)(N(t)-1)\}$$
$$= \nu^2\mu\Phi_3 + (\lambda\nu^2 + 2\mu\nu^2 + 2\lambda\mu\nu)\Phi_2$$
$$+ \ 2\lambda(\lambda\nu + \mu\nu + \lambda\mu)\Phi_1 + 2\lambda^3,$$

$$\nu^2 F(2,0) = \mathrm{E}\{M(t)(M(t)-1)\}$$
$$= 2\mu^2\Phi_2 + 2\mu(2\lambda + \mu + \rho\mu)\Phi_1 + 2\lambda\rho(\nu+\mu).$$

These formulas can be rewritten in terms of the central moments:

$$\mathrm{E}M(t) = \frac{\rho}{1-\rho}\left(1 + \rho\frac{\mu}{\nu}\right), \tag{1.19}$$

$$\mathrm{Cov}(M(t), N(t)) = \frac{\rho^2}{(1-\rho)^2}\left(2 - \rho + (1 + 2\rho - 2\rho^2)\frac{\mu}{\nu}\right), \tag{1.20}$$

$$\mathrm{Var}M(t) = \frac{\rho}{(1-\rho)^2}\left(1 + (5-\rho)\rho\frac{\mu}{\nu} + (2 + 4\rho - 3\rho^2)\rho\left(\frac{\mu}{\nu}\right)^2\right), \tag{1.21}$$

$$\mathrm{Cov}\{M(t), N(t)(N(t)-1)\} = \frac{2\rho^3}{(1-\rho)^3\mu\nu}((2\rho - \rho^2)\nu^2$$
$$+\mu\nu(3 + 2\rho^2 - 2\rho^3) + \mu^2(3 + 2\rho - 4\rho^2 + \rho^3)). \tag{1.22}$$

### 1.2.2 M/G/1 case

Consider now the case of a general distribution function $B(x)$ of the service times.

**Theorem 1.2** *For an* $M/G/1$ *retrial queue in the steady state the joint distribution of the server state and queue length*

$$p_{0n} = \mathrm{P}\{C(t) = 0, N(t) = n\},$$

$$p_{1n}(x) = \frac{d}{dx}\mathrm{P}\{C(t) = 1, \xi(t) < x, N(t) = n\},$$

*has partial generating functions*

$$p_0(z) \equiv \sum_{n=0}^{\infty} z^n p_{0n}$$

$$= (1-\rho)\exp\left\{\frac{\lambda}{\mu}\int_1^z \frac{1 - k(u)}{k(u) - u}du\right\}, \tag{1.23}$$

$$p_1(z,x) \equiv \sum_{n=0}^{\infty} z^n p_{1n}(x)$$

$$= \lambda \frac{1-z}{k(z)-z} p_0(z)[1-B(x)]e^{-(\lambda-\lambda z)x}. \quad (1.24)$$

*If in the case $C(t) = 1$ we neglect the elapsed service time $\xi(t)$, then for the probabilities $p_{1n} = P\{C(t) = 1, N(t) = n\}$ we have*

$$p_1(z) \equiv \sum_{n=0}^{\infty} z^n p_{1n} = \frac{1-k(z)}{k(z)-z} p_0(z). \quad (1.25)$$

*Proof.* In a general way we obtain the equations of statistical equilibrium:

$$(\lambda + n\mu)p_{0n} = \int_0^{\infty} p_{1n}(x)b(x)dx,$$

$$p_{1n}'(x) = -(\lambda + b(x))p_{1n}(x) + \lambda p_{1,n-1}(x),$$

$$p_{1n}(0) = \lambda p_{0n} + (n+1)\mu p_{0,n+1}.$$

For generating functions $p_0(z)$ and $p_1(z,x)$ these equations give:

$$\lambda p_0(z) + \mu z \frac{dp_0(z)}{dz} = \int_0^{+\infty} p_1(z,x)b(x)dx, \quad (1.26)$$

$$\frac{\partial p_1(z,x)}{\partial x} = -(\lambda - \lambda z + b(x))p_1(z,x), \quad (1.27)$$

$$p_1(z,0) = \lambda p_0(z) + \mu \frac{dp_0(z)}{dz}. \quad (1.28)$$

From (1.27) we find that $p_1(z,x)$ depends upon $x$ as follows:

$$p_1(z,x) = p_1(z,0)[1-B(x)]e^{-(\lambda-\lambda z)x}. \quad (1.29)$$

With the help of (1.29), equation (1.26) can be rewritten as

$$\lambda p_0(z) + \mu z \frac{dp_0(z)}{dz} = k(z)p_1(z,0). \quad (1.30)$$

Eliminating $p_0'(z)$ from (1.30) and (1.28) we get:

$$p_1(z,0) = \lambda \frac{1-z}{k(z)-z} p_0(z),$$

so that from (1.29) we have equation (1.24).

Integrating (1.24) with respect to $x$ and using well known formula

$$\int_0^{\infty} e^{-sx}(1-B(x))dx = \frac{1-\beta(s)}{s}$$

we have equation (1.25).

Note that (1.25) yields that $p_1(1) = \rho p_0(1)/(1 - \rho)$. Using the normalizing condition $p_0(1) + p_1(1) = 1$ we get:

$$p_0(1) = 1 - \rho, \ p_1(1) = \rho.$$

In order to find still unknown function $p_0(z)$, eliminate $p_1(z, 0)$ from (1.30) and (1.28):

$$\mu[k(z) - z]\frac{dp_0(z)}{dz} = \lambda[1 - k(z)]p_0(z). \tag{1.31}$$

Consider the coefficient $f(z) \equiv k(z) - z$. Note that:

(a) $f(1) = \beta(0) - 1 = 1 - 1 = 0$;

(b) $f'(z) = -\lambda\beta'(\lambda - \lambda z) - 1$ and thus $f'(1) = -\lambda\beta'(0) - 1 = \rho - 1 < 0$;

(c) $f''(z) = \lambda^2\beta''(\lambda - \lambda z) \geq 0$.

Therefore the function $f(z)$ is decreasing on the interval $[0, 1]$, $z = 1$ is the only zero there and for $z \in [0, 1)$ the function is positive, i.e. (as $\rho < 1$) for $z \in [0, 1)$ we have:

$$z < k(z) \leq 1.$$

Besides,

$$\lim_{z \to 1-0} \frac{1 - k(z)}{k(z) - z} = \frac{k'(1)}{1 - k'(1)} = \frac{\rho}{1 - \rho} < \infty,$$

i.e. the function $(1 - k(z))/(k(z) - z)$ can be defined at the point $z = 1$ as $\rho/(1 - \rho)$. This means that for $z \in [0; 1]$ we can rewrite equation (1.31) as

$$\frac{dp_0(z)}{dz} = \frac{\lambda}{\mu}\frac{1 - k(z)}{k(z) - z}p_0(z).$$

Since $p_0(1) = 1 - \rho$ is known, we can solve this simple differential equation, which yields (1.23) and completes the proof. $\square$

With the help of generating functions $p_0(z)$, $p_1(z)$ we can get various performance characteristics of the system:

(a) The distribution of the number of sources of repeated calls $q_n = P\{N(t) = n\}$ has generating function:

$$\begin{aligned}
p(z) &= p_0(z) + p_1(z) \\
&= (1 - \rho)\frac{1 - z}{k(z) - z}\exp\left\{\frac{\lambda}{\mu}\int_1^z \frac{1 - k(u)}{k(u) - u}du\right\}. \tag{1.32}
\end{aligned}$$

In particular, the mean $EN(t) = p'(1)$ and the variance $\mathrm{Var}N(t) = p''(1) + p'(1) - [p'(1)]^2$ of the queue length are given by:

$$EN(t) = \frac{\lambda^2}{1-\rho}\left(\frac{\beta_1}{\mu} + \frac{\beta_2}{2}\right), \tag{1.33}$$

$$\mathrm{Var}N(t) = \frac{\lambda^3\beta_3}{3(1-\rho)} + \frac{\lambda^3\beta_2}{2\mu(1-\rho)^2} \tag{1.34}$$

$$+ \frac{\lambda^4\beta_2^2}{4(1-\rho)^2} + \frac{\lambda\rho}{\mu(1-\rho)} + \frac{\lambda^2\beta_2}{2(1-\rho)}. \tag{1.35}$$

(b) The distribution of the number of customers in the system $Q_n = P\{K(t) = n\}$ has the generating function:

$$Q(z) = p_0(z) + zp_1(z)$$

$$= (1-\rho)\frac{(1-z)k(z)}{k(z)-z}\exp\left\{\frac{\lambda}{\mu}\int_1^z\frac{1-k(u)}{k(u)-u}du\right\} \tag{1.36}$$

In particular,

$$EK(t) = \rho + \frac{\lambda^2}{1-\rho}\left(\frac{\beta_1}{\mu} + \frac{\beta_2}{2}\right), \tag{1.37}$$

$$\mathrm{Var}K(t) = \rho - \rho^2 + \frac{\lambda^3\beta_3}{3(1-\rho)} + \frac{\lambda^4\beta_2^2}{4(1-\rho)^2} + \frac{\lambda^2\beta_2}{2(1-\rho)}$$

$$+ \lambda^2\beta_2 + \frac{\lambda^3\beta_2}{2\mu(1-\rho)^2} + \frac{\lambda\rho}{\mu(1-\rho)}. \tag{1.38}$$

(c) The blocking probability $p_1$ is given by

$$p_1 = p_1(1) = \rho.$$

### 1.2.3 Some comments about the stationary distribution of the server state

Note that the stationary distribution of the server state

$$p_0 = 1 - \rho, \quad p_1 = \rho$$

depends on the service time distribution $B(x)$ only through its mean $\beta_1$ and does not depend on the rate of retrial $\mu$. In fact, this result is quite obvious if we take into account that for any single server queue the probability $p_1$ is equal to the mean number of busy servers, which, in turn, is equal to the intensity of carried traffic. Since the customers are not lost, the carried traffic is equal to the offered traffic, which is obviously equal to $\lambda \beta_1 = \rho$.

The stationary distribution of the number of busy servers:

$$p_0 = 1 - \rho, \ p_1 = \rho,$$

can be rewritten as

$$p_0 = \frac{1}{1 + \rho'}, \ p_1 = \frac{\rho'}{1 + \rho'}$$

where $\rho' = \rho + \rho^2/(1 - \rho)$.

These equations show that the stationary distribution of the process $C(t)$ coincides with the distribution of the number of busy servers in the M/G/1/0 Erlang loss model with increased arrival rate $\lambda' = \lambda + \lambda\rho/(1 - \rho)$. By (1.33) the additional intensity equals $\lim_{\mu \to 0} \mu EN(t)$. But $\mu EN(t)$ is mean rate of flow of repeated calls. Thus as $\mu \to 0$ the additional load can be thought of as a load formed by sources of repeated calls.

But we would have a different situation if we put $\mu = 0$ in the initial process. In this case the retrial queue can be thought of as an Erlang loss model with the same arrival rate $\lambda$. So the stationary distribution of the process $C(t)$ is given by $p_0 = 1/(1 + \rho)$, $p_1 = \rho/(1 + \rho)$.

The statement 'when retrial rate is small enough the retrial queue can be thought of as equivalent to the corresponding system with losses' can be found in many papers on retrial queues. But as the above arguments show it is necessary to distinguish between cases $\mu \to 0$ and $\mu = 0$.

### 1.2.4 Stochastic decomposition

To emphasize the dependence of characteristics of the $M/G/1$ retrial queue on the rate of retrial $\mu$, denote the variables $C \equiv C(t)$ and $N \equiv N(t)$ as $C_\mu$ and $N_\mu$ respectively (we omit the argument $t$ since the steady state distributions of these random variables do not depend on time; however, it is important to remember that both these variables are considered at the same time $t$). Denote also by $C_\infty$ and $N_\infty$ the corresponding variables for the standard $M/G/1$ queueing system in the steady state. Thus, $C_\infty$ is equal to 0 or 1 according to the server is free or busy and $N_\infty$ is the number of customers in the queue excluding the customer in service. This system can be thought of as a limit case of the retrial queue with retrial rate equal to infinity and so this notation is consistent with the previous one. It is well known that usually the state of the standard M/G/1 queue is described by the single variable $K_\infty$ which represents the total number of customers in the system rather than by the vector $(C_\infty, N_\infty)$. However there is a very simple relation between $K_\infty$ and $(C_\infty, N_\infty)$ :

$$C_\infty = I(K_\infty > 0); \quad N_\infty = (K_\infty - 1)^+,$$

where $I(A)$ is the indicator of the event $A$, and $(a)^+ = \max(a, 0)$. This yields a very simple relation between the distributions of the random variable $K_\infty$ and the random vector $(C_\infty, N_\infty)$:

$$
\begin{aligned}
P(C_\infty = 0, N_\infty = 0) &= P(K_\infty = 0), \\
P(C_\infty = 0, N_\infty = n) &= 0, \text{ if } n \geq 1, \\
P(C_\infty = 1, N_\infty = n) &= P(K_\infty = n+1), \text{ if } n \geq 0.
\end{aligned}
$$

The distribution of the random variable $K_\infty$ in the steady state is given (in terms of the generating function) by the well known Pollaczek–Khintchine formula:

$$\mathrm{E}z^{K_\infty} = (1-\rho)\frac{1-z}{k(z)-z}k(z).$$

Correspondingly, in the steady state the distribution of the vector $(C_\infty, N_\infty)$ is given by the following partial generating functions

$$
\begin{aligned}
p_0^{(\infty)}(z) &\equiv \sum_{n=0}^{\infty} z^n P(C_\infty = 0, N_\infty = n) \\
&= (1-\rho), \tag{1.39}
\end{aligned}
$$

$$
\begin{aligned}
p_1^{(\infty)}(z) &\equiv \sum_{n=0}^{\infty} z^n P(C_\infty = 1, N_\infty = n) \\
&= (1-\rho)\frac{1-k(z)}{k(z)-z}. \tag{1.40}
\end{aligned}
$$

Now introduce a random variable $R_\mu$ with the generating function

$$\mathrm{E}z^{R_\mu} = \exp\left\{\frac{\lambda}{\mu}\int_1^z \frac{1-k(u)}{k(u)-u}du\right\}. \tag{1.41}$$

It is easy to see that the right-hand side of this equation is equal to

$$\frac{p_0(z)}{1-\rho} = \frac{\mathrm{E}(z^{N_\mu(t)}; C_\mu(t) = 0)}{P(C_\mu(t) = 0)} = \mathrm{E}(z^{N_\mu(t)}|C_\mu(t) = 0).$$

Thus it really is a generating function of a probability distribution and the distribution of the random variable $R_\mu$ coincides with the conditional distribution of the number of sources of repeated calls given that the server is free.

Comparing formulas for $p_0(z)$ and $p_1(z)$ given by Theorem 1.2 with formulas (1.39)–(1.41) we get the following result about stochastic decomposition of the vector $(C_\mu, N_\mu)$.

**Theorem 1.3** *The vector $(C_\mu, N_\mu)$ can be represented as a sum of two independent vectors. One of them is the vector $(C_\infty, N_\infty)$ and the other is $(0, R_\mu)$:*

$$(C_\mu, N_\mu) = (C_\infty, N_\infty) + (0, R_\mu). \qquad (1.42)$$

*In particular, the total number of customers in the M/G/1 retrial queue, $K_\mu$, can be represented as the sum of two independent random variables, one of which is the total number of customers in the standard M/G/1 queueing system, $K_\infty$, and the second is $R_\mu$:*

$$K_\mu = K_\infty + R_\mu. \qquad (1.43)$$

It is easy to show that in fact relation (1.43) is equivalent to (1.42). Indeed, since

$$
\begin{aligned}
Ez^{K_\mu} &= p_0(z) + zp_1(z), \\
Ez^{R_\mu} &= \frac{p_0(z)}{1-\rho}, \\
Ez^{K_\infty} &= (1-\rho)\frac{1-z}{k(z)-z}k(z),
\end{aligned}
$$

we can rewrite (1.43) as

$$p_0(z) + zp_1(z) = (1-\rho)\frac{1-z}{k(z)-z}k(z) \cdot \frac{p_0(z)}{1-\rho}$$

which yields that

$$p_1(z) = \frac{1-k(z)}{k(z)-z}p_0(z). \qquad (1.44)$$

Thus

$$
\begin{aligned}
p_1(z) &= (1-\rho)\frac{1-k(z)}{k(z)-z} \cdot \frac{p_0(z)}{1-\rho} \\
&= p_1^{(\infty)}(z) \cdot Ez^{R_\mu}, \\
p_0(z) &= (1-\rho) \cdot \frac{p_0(z)}{1-\rho}
\end{aligned}
$$

$$= p_0^{(\infty)}(z) \cdot \mathrm{E} z^{R_\mu}.$$

These relations are equivalent to (1.42).

As one can see from this analysis, the stochastic decomposition is equivalent to relation (1.44). This relation in turn is a consequence of relation (1.24), which describes the dependence of the function $p_1(z, x)$ on the variable $x$. Thus in order to establish the stochastic decomposition we need not have explicit formulas for the generating functions $p_0(z)$ and $p_1(z)$. As a matter of fact, it is sufficient to know the dependence of $p_1(z, x)$ upon $x$, i.e. to solve equation (1.27).

This remark allows us to show that the stochastic decomposition holds for retrial queues with more general structure of flow of repeated calls. Assume, for example, that the total rate of flow of repeated calls given that the number of customers in orbit equals $n$ is a general function $\mu_n$ ($\mu_0 = 0$). Then equations (1.26), (1.27), (1.28) become:

$$\lambda p_0(z) + \sum_{n=0}^{\infty} \mu_n z^n p_{0n} = \int_0^{+\infty} p_1(z, x) b(x) dx,$$

$$\frac{\partial p_1(z, x)}{\partial x} = -(\lambda - \lambda z + b(x)) p_1(z, x),$$

$$p_1(z, 0) = \lambda p_0(z) + \frac{1}{z} \sum_{n=0}^{\infty} \mu_n z^n p_{0n}$$

respectively. As in the basic case, from the second equation we find that $p_1(z, x)$ depends upon $x$ as follows:

$$p_1(z, x) = p_1(z, 0)[1 - B(x)] e^{-(\lambda - \lambda z) x}.$$

Besides, eliminating from the first and third equations the term $\sum_{n=0}^{\infty} \mu_n z^n p_{0n}$ we get:

$$p_1(z, 0) = \lambda \frac{1 - z}{k(z) - z} p_0(z).$$

Thus

$$p_1(z) \equiv \int_0^{\infty} p_1(z, x) dx = p_1(z, 0) \frac{1 - k(z)}{\lambda(1 - z)} = \frac{1 - k(z)}{k(z) - z} p_0(z),$$

which yields the stochastic decomposition.

Similar analysis shows that the stochastic decomposition holds in the case of general distribution of intervals between retrials or when the flow of repeated calls from a source is a Markovian Arrival Process (which allows us to include in the model dependence between inter-retrial times).

The above result on stochastic decomposition is very useful, in particular, for obtaining explicit formulas for moments of the random variable $N_\mu$ (it should be noted that technically it is not easy to obtain even the explicit formula (1.35) for $\text{Var}N(t)$).

To state our result we need some additional notations. Let $n$ and $j$ be nonnegative integers. Define a set $X_j(n)$ as a set of all vectors $\mathbf{x} = (x_1, ..., x_j) \in Z_+^j$ such that $x_1 > 0, ..., x_j > 0$ and $x_1 + ... + x_j = n$ (for example, $X_1(4) = \{4\}$, $X_2(4) = \{(1,3),(2,2),(3,1)\}$, $X_3(4) = \{(1,1,2),(1,2,1),(2,1,1)\}$). For such a vector $\mathbf{x}$ denote $\mathbf{x}! = x_1!...x_j!$. Besides let $\Phi_n^{(\infty)} \equiv \text{E}\left((N_\infty)_n\right)$ be the $n$th factorial moment of the queue length distribution in the standard $M/G/1$ queue in the steady state. This distribution is studied in queueing theory in full details and can be considered as standard (like Gaussian, gamma, etc.).

**Theorem 1.4** *The $n$th factorial moment of the number of sources of repeated calls, $\Phi_n$, is given by*

$$
\Phi_n = \Phi_n^{(\infty)} + \sum_{k=1}^{n} (n)_k \Phi_{n-k}^{(\infty)} \sum_{j=1}^{k} \frac{1}{j!} \left(\frac{\lambda}{\mu(1-\rho)}\right)^j
$$

$$
\times \sum_{\mathbf{x} \in X_j(k)} \frac{1}{\mathbf{x}!} \prod_{i=1}^{j} \left(\Phi_{x_i-1}^{(\infty)} - (1-\rho)\delta_{x_i,1}\right). \qquad (1.45)
$$

*Proof.* By Theorem 1.3 random variable $N_\mu$ can be represented as the sum of two independent random variables, $N_\infty$ and $R_\mu$. Making use of the Vandermonde binomial theorem we have from the independence of $N_\infty$ and $R_\mu$ that

$$
\Phi_n = \sum_{k=0}^{n} \binom{n}{k} \Phi_{n-k}^{(\infty)} \text{E}\left((R_\mu)_k\right). \qquad (1.46)
$$

To calculate the factorial moment $\text{E}\left((R_\mu)_k\right)$ of the random variable $R_\mu$ consider the logarithm of the generating function $\text{E}z^{R_\mu}$ which by (1.41) is given by the following formula:

$$
F(z) \equiv \ln \text{E}z^{R_\mu} = \frac{\lambda}{\mu} \int_1^z \frac{1-k(u)}{k(u)-u} du.
$$

This logarithm can be rewritten in terms of the generating function

$p^{(\infty)}(z)$ as follows:

$$F(z) = \frac{\lambda}{\mu(1-\rho)} \int_1^z p^{(\infty)}(u)du + \frac{\lambda}{\mu}(1-z).$$

Differentiating this equation at the point $z = 1$ we get:

$$F^{(n)}(1) = \frac{\lambda}{\mu(1-\rho)} \Phi_{n-1}^{(\infty)} - \frac{\lambda}{\mu}\delta_{n,1}, \qquad (1.47)$$

where $\delta_{n,1}$ is Kronecker's delta.

The variable $F^{(n)}(1)$ is known in probability theory as the $n$th factorial cumulant of the random variable $R_\mu$. Factorial moments $E((R_\mu)_k)$ can be expressed through the factorial cumulants with the help of the following formula (Shiryayev, A. N. (1984) *Probability*, Springer-Verlag):

$$E((R_\mu)_k) = k! \sum_{j=1}^k \frac{1}{j!} \sum_{\mathbf{x} \in X_j(k)} \frac{1}{\mathbf{x}!} \prod_{i=1}^j F^{(x_i)}(1), \ k \geq 1. \qquad (1.48)$$

From (1.46), (1.47) and (1.48) we get the main formula (1.45). $\quad\square$

## 1.3 The embedded Markov chain

Let $N_i = N(\eta_i)$ be the number of calls in orbit at the time $\eta_i$ of the $i$th departure. It is easy to see that

$$N_i = N_{i-1} - B_i + \nu_i, \qquad (1.49)$$

where $B_i$ is the number of sources which enter service at time $\xi_i$ (i.e. $B_i = 1$ if the $i$th call is a repeated call and $B_i = 0$ if the $i$th call is a primary call) and $\nu_i$ is the number of primary calls which arrive in the system during the service time $S_i$ of the $i$th call.

The Bernoulli random variable $B_i$ depends on the history of the system before time $\eta_{i-1}$ only through $N_{i-1}$ and its conditional distribution is given by

$$\left.\begin{array}{l} P\{B_i = 0 \mid N_{i-1} = n\} = \dfrac{\lambda}{\lambda + n\mu}, \\[2mm] P\{B_i = 1 \mid N_{i-1} = n\} = \dfrac{n\mu}{\lambda + n\mu}. \end{array}\right\} \qquad (1.50)$$

The random variable $\nu_i$ does not depend on events which have occurred before epoch $\xi_i$ and has distribution

$$k_n = P\{\nu_i = n\} = \int_0^\infty \frac{(\lambda x)^n}{n!} e^{-\lambda x} dB(x) \qquad (1.51)$$

with generating function

$$k(z) \equiv \sum_{n=0}^{\infty} k_n z^n = \beta(\lambda - \lambda z)$$

and mean value

$$\mathrm{E}\nu_i = \sum_{n=0}^{\infty} n k_n = \rho. \tag{1.52}$$

The above remarks imply that the sequence of random variables $\{N_i\}$ forms a Markov chain, which is the embedded chain for our queueing system. Its one-step transition probabilities $r_{mn} = \mathrm{P}\{N_i = n \mid N_{i-1} = m\}$ are given by the formula

$$r_{mn} = \frac{\lambda}{\lambda + m\mu} k_{n-m} + \frac{m\mu}{\lambda + m\mu} k_{n-m+1}. \tag{1.53}$$

Note that $r_{mn} \neq 0$ only for $m = 0, 1, ..., n + 1$.

To prove formula (1.53) assume that $N_{i-1} = m$, i.e. there are $m$ sources of repeated calls at time $\eta_{i-1}$ of a service completion. The next primary call will arrive into the system after a random interval of time $t'$ which has an exponential distribution with parameter $\lambda$, while the next repeated call will arrive into the system after a random interval of time $t''$ which has an exponential distribution with parameter $m\mu$. The $i$th call is a primary call if $t' < t''$ and it is a repeated call if $t'' < t'$. It is well known that if $t', t''$ are two independent exponentially distributed random variables with parameters $\lambda', \lambda''$ then

$$\mathrm{P}\{t' < t''\} = \frac{\lambda'}{\lambda' + \lambda''}, \ \mathrm{P}\{t'' < t'\} = \frac{\lambda''}{\lambda' + \lambda''}.$$

Thus with probability $\frac{\lambda}{\lambda + m\mu}$ the $i$th call is a primary call (and the number of sources does not change) and with probability $\frac{m\mu}{\lambda + m\mu}$ the $i$th call is a repeated call (and the number of sources decreases by 1). To have $n$ sources in the system at time $\eta_i$ we need $n - m$ new arrivals during the service time of the $i$th call in the first case (probability of this event is $k_{n-m}$) and $n - m + 1$ new arrivals during the service time of the $i$th call in the second case (probability of this event is $k_{n-m+1}$).

A more formal proof can be done with the help of recursive formula (1.49) and equations (1.50), (1.51):

$$\begin{aligned} r_{mn} &= \mathrm{P}\{N_i = n \mid N_{i-1} = m\} \\ &= \mathrm{P}\{N_{i-1} - B_i + \nu_i = n \mid N_{i-1} = m\} \end{aligned}$$

$$\begin{aligned}
&= \ \mathrm{P}\{\nu_i = n - m + B_i \mid N_{i-1} = m\} \\
&= \ \mathrm{P}\{\nu_i = n - m \mid N_{i-1} = m, B_i = 0\} \\
&\times \ \mathrm{P}\{B_i = 0 \mid N_{i-1} = m\} \\
&+ \ \mathrm{P}\{\nu_i = n - m + 1 \mid N_{i-1} = m, B_i = 1\} \\
&\times \ \mathrm{P}\{B_i = 1 \mid N_{i-1} = m\} \\
&= \ \mathrm{P}\{\nu_i = n - m\} \cdot \mathrm{P}\{B_i = 0 \mid N_{i-1} = m\} \\
&+ \ \mathrm{P}\{\nu_i = n - m + 1\} \cdot \mathrm{P}\{B_i = 1 \mid N_{i-1} = m\} \\
&= \ k_{n-m}\frac{\lambda}{\lambda + m\mu} + k_{n-m+1}\frac{m\mu}{\lambda + m\mu}.
\end{aligned}$$

### 1.3.1 Ergodicity of $\{N_i\}$

As usual, the first question to be investigated is the ergodicity of the chain.

**Theorem 1.5** *The embedded Markov chain $\{N_i\}$ is ergodic iff*

$$\rho < 1.$$

*Proof.* Because of the recursive structure of equation (1.49) it is very convenient to use criteria based on mean drift or, in other words, the theory of Lyapunov functions. The main result of this theory is the following Foster's criterion (Pakes, A. G.(1969) Some conditions for ergodicity and recurrence of Markov chains. *Operations Research*, **17**, 1058–1061).

**Statement 1** *For an irreducible and aperiodic Markov chain $\xi_i$ with state space $S$, a sufficient condition for ergodicity is the existence of a nonnegative function $f(s), s \in S$, (this function is said to be a Lyapunov function or test function) and $\varepsilon > 0$ such that the mean drift*

$$x_s \equiv \mathrm{E}(f(\xi_{i+1}) - f(\xi_i) \mid \xi_i = s)$$

*is finite for all $s \in S$ and $x_s \le -\varepsilon$ for all $s \in S$ except perhaps a finite number.*

In the simple case when the state space $S$ is the set $Z_+$ of nonnegative integers it is usually sufficient to consider the function $f(n) = n$. It means that a chain is ergodic if its mean drift

$$x_n \equiv \mathrm{E}(\xi_{i+1} - \xi_i \mid \xi_i = n)$$

is less than some negative number $-\varepsilon$ for all $n \ge N$, where $N$ is a sufficiently large integer.

Of course, if there exists the limit

$$x = \lim_{n \to \infty} x_n$$

this condition holds iff $x < 0$.

For the Markov chain under consideration the mean drift introduced above is (see (1.49), (1.50) and (1.52)):

$$
\begin{aligned}
x_n &= \mathrm{E}(-B_{i+1} + \nu_{i+1} \mid N_i = n) \\
&= -\mathrm{E}(B_{i+1} \mid N_i = n) + \mathrm{E}(\nu_{i+1} \mid N_i = n) \\
&= -\mathrm{P}(B_{i+1} = 1 \mid N_i = n) + \mathrm{E}(\nu_{i+1}) \\
&= -\frac{n\mu}{\lambda + n\mu} + \rho,
\end{aligned}
$$

so that $\lim_{n \to \infty} x_n = -1 + \rho$. This limit is negative iff $\rho < 1$. Applying the above Foster's criterion we can guarantee that for $\rho < 1$ the embedded Markov chain is ergodic.

To prove that $\rho < 1$ is necessary for ergodicity we will use the following criterion (Sennott, L. I., Humblet, P. A. and Tweedie, R. L. (1983) Mean drift and the non-ergodicity of Markov chains. *Operations Research*, **31**, No. 4, 783–788).

**Statement 2** *An irreducible and aperiodic Markov chain $\xi_i$ with state space $Z_+$ is not ergodic if the mean drift*

$$x_n \equiv \mathrm{E}(\xi_{i+1} - \xi_i \mid \xi_i = n),$$

*is finite for all $n \in Z_+$, mean down drift*

$$\delta_n \equiv \sum_{m < n} (m - n) r_{nm} \geq -\mathrm{Const},$$

*i.e. is bounded below, and there exists $N$ such that $x_n \geq 0$ for $n \geq N$.*

For the $M/G/1$ retrial queue under consideration the down drift is bounded below since $N_{i+1} - N_i \geq -1$, and if $\rho \geq 1$ then

$$x_n = -\frac{n\mu}{\lambda + n\mu} + \rho \geq -\frac{n\mu}{\lambda + n\mu} + 1 = \frac{\lambda}{\lambda + n\mu} > 0.$$

$\square$

It would be interesting to investigate the nature of the chain in the case $\rho \geq 1$ in more detail.

In the case $\rho > 1$ the chain is obviously transient. A formal proof is the following. Using the main equation (1.49) we have:

$$N_i = N_0 - (B_1 + \ldots + B_i) + \nu_1 + \ldots + \nu_i$$

$$\geq \quad N_0 - i + \nu_1 + \ldots + \nu_i$$
$$= \quad N_0 + i \cdot \left( \frac{\nu_1 + \ldots + \nu_i}{i} - 1 \right).$$

By the law of large numbers, with probability 1 we have

$$\frac{\nu_1 + \ldots + \nu_i}{i} - 1 \to \rho - 1 > 0.$$

Thus with probability 1

$$N_0 + i \cdot \left( \frac{\nu_1 + \ldots + \nu_i}{i} - 1 \right) \to +\infty,$$

which implies that $N_i \to +\infty$.

For most well known queueing systems, the equality $\rho = 1$, where $\rho$ is 'traffic intensity' (which must be introduced in the proper way), usually implies null recurrence. But in the case under consideration the behaviour of the system in the case $\rho = 1$ depends on the retrial rate $\mu$.

To simplify the analysis, we restrict ourselves to the case of exponentially distributed service times $(B(x) = 1 - e^{-\nu x})$. Under this assumption the process $(C(t), N(t))$ is Markov and so we can consider a chain $\{\zeta_i\}$ embedded at jump times rather than a chain $\{N_i\}$ embedded at service completion epochs.

This chain has as state space the set $\{0, 1\} \times Z_+$ and the following one-step transition probabilities:

$$r_{(0,n)(1,n)} = \frac{\lambda}{\lambda + n\mu}, \quad r_{(0,n)(1,n-1)} = \frac{n\mu}{\lambda + n\mu},$$
$$r_{(1,n)(0,n)} = \frac{\nu}{\nu + \lambda}, \quad r_{(1,n)(1,n+1)} = \frac{\lambda}{\nu + \lambda}.$$

To classify the states of the chain $\{\zeta_i\}$ we will use the following criterion (Feller, W. (1968) *An Introduction to Probability Theory and Its Applications.* Vol. 1, Third Edition, John Wiley & Sons, Inc., section XV.8, page 402):

**Statement 3** *An irreducible Markov chain with state space $S$ and one-step transition probabilities $r_{sp}$ is transient iff the set of equations*

$$y_s = \sum_{p \neq s_0} r_{sp} y_p, \quad s \neq s_0, \tag{1.54}$$

*has a nonzero bounded solution $y_s$. Here as a special state $s_0$ any state $s \in S$ can be taken.*

For the embedded chain $\{\zeta_i\}$ choose as the state $s_0$ the state $(0,0)$. Then the set of equations (1.54) becomes:

$$a_n = \frac{\lambda}{\lambda + n\mu} b_n + \frac{n\mu}{\lambda + n\mu} b_{n-1}, \ n \geq 1, \tag{1.55}$$

$$b_n = \frac{\lambda}{\nu + \lambda} b_{n+1} + \frac{\nu}{\nu + \lambda} a_n \cdot (1 - \delta_{n,0}), \ n \geq 0, \tag{1.56}$$

where $a_n \equiv y_{(0,n)}$, $n \geq 1$, $b_n \equiv y_{(1,n)}$, $n \geq 0$. The first equation corresponds to $s = (0,n)$ and the second one corresponds to $s = (1,n)$. Eliminating from equations (1.55) and (1.56) variable $a_n$ we get after some algebra:

$$b_0 = \frac{\lambda}{\nu + \lambda} b_1, \tag{1.57}$$

$$b_n = p_n \cdot b_{n+1} + q_n \cdot b_{n-1}, \ n \geq 1, \tag{1.58}$$

where

$$p_n = \frac{\lambda(\lambda + n\mu)}{\lambda(\lambda + n\mu) + n\mu\nu}; \ q_n = 1 - p_n = \frac{n\mu\nu}{\lambda(\lambda + n\mu) + n\mu\nu}.$$

From (1.58) we have:

$$(b_{n+1} - b_n) p_n = (b_n - b_{n-1}) q_n,$$

i.e.

$$b_{n+1} - b_n = \frac{q_n \cdots q_1}{p_n \cdots p_1} (b_1 - b_0), \ n \geq 1.$$

Using (1.57) we get

$$b_{n+1} = \sum_{k=1}^{n} \frac{q_1 \cdots q_k}{p_1 \cdots p_k} \cdot \frac{\nu b_0}{\lambda} + \frac{\nu + \lambda}{\lambda} b_0. \tag{1.59}$$

Equations (1.55) and (1.59) imply that a nonzero bounded solution of the equations (1.55), (1.56) exists iff the series

$$S = \sum_{n=1}^{\infty} \frac{q_1 \cdots q_n}{p_1 \cdots p_n} = \sum_{n=1}^{\infty} \prod_{k=1}^{n} \frac{k\mu\nu}{\lambda(\lambda + k\mu)}$$

converges, which it does iff $\mu < \nu$ . Indeed, let $\mu \geq \nu$. Since we consider the case $\rho = 1$, we have:

$$\frac{k\mu\nu}{\lambda(\lambda + k\mu)} = \frac{k\mu}{\lambda + k\mu} \geq \frac{k\mu}{\mu + k\mu} = \frac{k}{k+1},$$

so that

$$S \geq \sum_{n=1}^{\infty} \prod_{k=1}^{n} \frac{k}{k+1} = \sum_{n=1}^{\infty} \frac{1}{n+1} = \infty,$$

i.e. the series $S$ diverges. Now let $\mu < \nu$. Then

$$\left(\frac{p_n}{q_n} - 1\right) n = \lambda/\mu = \nu/\mu > 1,$$

so that the series $S$ converges.

Thus, in the case $\rho = 1$ the embedded Markov chain is transient iff $\mu < \nu$ and it is null recurrent iff $\mu \geq \nu$.

Appearance of the transition recurrence/transience in the case $\rho = 1$ is related to the fact that in this case mean drift $x_n$ is $O\left(\frac{1}{n}\right)$ as $n \to \infty$ and therefore second moments of jumps start to play a role.

### 1.3.2 The stationary distribution of $\{N_i\}$

Our second goal is to find the stationary distribution $\pi_n$ of the embedded Markov chain $\{N_i\}$.

Kolmogorov equations for the distribution $\pi_n$ are

$$
\begin{aligned}
\pi_n &= \sum_{m=0}^{n} \pi_m \frac{\lambda}{\lambda + m\mu} k_{n-m} \\
&+ \sum_{m=1}^{n+1} \pi_m \frac{m\mu}{\lambda + m\mu} k_{n-m+1}, \quad n = 0, 1, \dots \quad (1.60)
\end{aligned}
$$

Because of the presence of convolutions, equation (1.60) can be transformed with the help of the generating functions

$$\varphi(z) = \sum_{n=0}^{\infty} z^n \pi_n, \quad \psi(z) = \sum_{n=0}^{\infty} z^n \frac{\pi_n}{\lambda + n\mu}$$

to

$$\varphi(z) = k(z) \cdot (\lambda\psi(z) + \mu\psi'(z)). \quad (1.61)$$

Since

$$\varphi(z) = \lambda\psi(z) + \mu z\psi'(z),$$

we get the following equation for the generating function $\psi(z)$:

$$\mu[k(z) - z]\psi'(z) = \lambda[1 - k(z)]\psi(z). \quad (1.62)$$

Equation (1.61) can also be obtained directly from the recurrence relation (1.49):

$$\begin{aligned}
\varphi(z) &\equiv Ez^{N_i} = Ez^{N_{i-1}-B_i+\nu_i} \\
&= \sum_{n=0}^{\infty} E(z^{N_{i-1}-B_i+\nu_i}|N_{i-1}=n) \cdot P(N_{i-1}=n) \\
&= \sum_{n=0}^{\infty} z^n E(z^{-B_i}|N_{i-1}=n) \cdot Ez^{\nu_i} \cdot \pi_n \\
&= \sum_{n=0}^{\infty} z^n \left( \frac{\lambda}{\lambda+n\mu} \cdot 1 + \frac{n\mu}{\lambda+n\mu} \cdot z^{-1} \right) \cdot k(z) \cdot \pi_n \\
&= k(z) \cdot \left( \lambda \sum_{n=0}^{\infty} \frac{\pi_n}{\lambda+n\mu} z^n + \mu \sum_{n=0}^{\infty} \frac{\pi_n}{\lambda+n\mu} n z^{n-1} \right) \\
&= k(z) \cdot (\lambda\psi(z) + \mu\psi'(z)) .
\end{aligned}$$

Equation (1.61) can also be obtained with the help of the method of collective marks (the method of supplementary events), which will be extensively used in subsequent sections. Let us paint customers 'red' with probability $z$ and 'green' with probability $1 - z$; colours for different customers are chosen independently and do not depend on the functioning of the system. Then

$$\begin{aligned}
\varphi(z) =\ & P\,\{\text{all sources of repeated calls are 'red'} \\
& \text{at time of some departure}\}; \\[2mm]
\lambda\psi(z) =\ & \sum_{n=0}^{\infty} z^n \frac{\lambda}{\lambda+n\mu} \pi_n \\
=\ & P\,\{\text{all sources of repeated calls are 'red' at time} \\
& \text{of some departure and the next service will} \\
& \text{be started by a primary call}\}; \\[2mm]
\mu\psi'(z) =\ & \sum_{n=0}^{\infty} z^{n-1} \frac{n\mu}{\lambda+n\mu} \pi_n \\
=\ & P\,\{\text{all sources of repeated calls, except for one,} \\
& \text{are 'red' at time of some departure and the next} \\
& \text{service will be started by a repeated call}\}.
\end{aligned}$$

Event 'all sources of repeated calls are 'red' at time of some, say $(l + 1)$th, departure' can occur only if

(1) all sources of repeated calls are 'red' at time of the previous departure and

(2) 'green' customers did not arrive during a service period.

Besides we must note that if the $(l+1)$th customer came from the orbit, then its colour does not matter. It means that the first event must be divided into two mutually exclusive events:

(1a) $(l+1)$th customer is primary and all customers in the orbit at time $\eta_l$ are 'red';

(1b) $(l+1)$th call is repeated and all customers in orbit at time $\eta_l$ are 'red', except for the customer who enters service; its colour does not matter.

Probability of the first event is $\lambda\psi(z)$, the second is $\mu\psi'(z)$.

To find the probability that 'green' customers did not arrive during a service period, introduce a flow of 'catastrophes', assuming that each arrival of a 'green' customer means a 'catastrophe'. It is easy to see that we are dealing with a thinning of the initial flow of primary calls, and thus, the flow of 'catastrophes' has rate $s = \lambda(1 - z)$. Thus the probability that 'green' customers did not arrive during a service period is the probability that 'catastrophes' did not occur during a service period, and therefore equals $\beta(s) \equiv \beta(\lambda - \lambda z)$. Since events (1a)∪(1b) and (2) are independent,

$$\varphi(z) = (\mathrm{P}\{(1a)\} + \mathrm{P}\{(1b)\}) \cdot \mathrm{P}\{(2)\},$$

which yields equation (1.61).

Returning now to equation (1.62) we note that it is identical to equation (1.31) and thus in the case $\rho < 1$ we have

$$\psi(z) = \psi(1) \cdot \exp\left\{\frac{\lambda}{\mu} \int_1^z \frac{1 - k(u)}{k(u) - u} du\right\}. \qquad (1.63)$$

From this,

$$\varphi(z) = \lambda\psi(z) + \mu z\psi'(z) = \lambda k(z) \cdot \frac{1 - z}{k(z) - z}\psi(z).$$

Since $\varphi(1) = 1$, we have:

$$\psi(1) = \sum_{n=0}^{\infty} \frac{\pi_n}{\lambda + n\mu} = \frac{1 - \rho}{\lambda}. \qquad (1.64)$$

Finally we get the following formula for the generating function of the stationary distribution of the embedded Markov chain:

$$\varphi(z) = (1 - \rho)\frac{1 - z}{k(z) - z}k(z) \exp\left\{\frac{\lambda}{\mu} \int_1^z \frac{1 - k(u)}{k(u) - u} du\right\}. \qquad (1.65)$$

Comparing this equation and equation (1.36) we find that the stationary distribution $\pi_n$ of the embedded Markov chain is identical to the stationary distribution $Q_n$ of the number of customers in the system at an arbitrary time. In particular, in the steady state

$$\mathrm{E}N_i = \rho + \frac{\lambda^2}{1-\rho}\left(\frac{\beta_1}{\mu} + \frac{\beta_2}{2}\right). \qquad (1.66)$$

This formula can also be obtained directly from the main recursive equation (1.49).

First, take mean values of both sides of (1.49):

$$\mathrm{E}N_i = \mathrm{E}N_{i-1} - \mathrm{E}B_i + \mathrm{E}\nu_i.$$

Since in the steady state $\mathrm{E}N_i$ does not depend on $i$ and $\mathrm{E}\nu_i = \rho$, we get:

$$\mathrm{E}B_i = \rho.$$

Note that this relation is equivalent to (1.64).

Now take mean values of the squares of both sides of (1.49):

$$\begin{aligned}
\mathrm{E}N_i^2 &= \mathrm{E}N_{i-1}^2 + \mathrm{E}B_i^2 + \mathrm{E}\nu_i^2 \\
&\quad - 2\mathrm{E}(N_{i-1}B_i) + 2\mathrm{E}(N_{i-1}\nu_i) - 2\mathrm{E}(B_i\nu_i). \qquad (1.67)
\end{aligned}$$

In the steady state $\mathrm{E}N_i^2 = \mathrm{E}N_{i-1}^2$. Besides,

- since $B_i$ is a Bernoulli random variable, $\mathrm{E}B_i^2 = \mathrm{E}B_i = \rho$;
- $\mathrm{E}\nu_i^2 = k''(1) + k'(1) = \lambda^2\beta_2 + \rho$;
- since $\nu_i$ does not depend on $N_{i-1}$ and $B_i$, we have:

$$\begin{aligned}
\mathrm{E}(N_{i-1}\nu_i) &= \mathrm{E}N_{i-1}\cdot\mathrm{E}\nu_i = \rho\cdot\mathrm{E}N_i; \\
\mathrm{E}(B_i\nu_i) &= \mathrm{E}B_i\cdot\mathrm{E}\nu_i = \rho^2;
\end{aligned}$$

- and finally:

$$\begin{aligned}
\mathrm{E}(N_{i-1}B_i) &= \sum_{n=0}^{\infty}\mathrm{E}(N_{i-1}B_i|N_{i-1}=n)\cdot P(N_{i-1}=n) \\
&= \sum_{n=0}^{\infty}n\mathrm{E}(B_i|N_{i-1}=n)\pi_n \\
&= \sum_{n=0}^{\infty}n\frac{n\mu}{\lambda+n\mu}\pi_n = \sum_{n=0}^{\infty}n\left(1-\frac{\lambda}{\lambda+n\mu}\right)\pi_n \\
&= \sum_{n=0}^{\infty}n\pi_n - \frac{\lambda}{\mu}\sum_{n=0}^{\infty}\frac{n\mu}{\lambda+n\mu}\pi_n
\end{aligned}$$

$$= \quad \text{E}N_i - \frac{\lambda}{\mu}\text{E}B_i = \text{E}N_i - \frac{\lambda\rho}{\mu}.$$

Now (1.67) becomes:

$$2(1-\rho)\text{E}N_i = 2\rho(1-\rho) + \lambda^2\beta_2 + \frac{2\lambda\rho}{\mu},$$

which is equivalent to (1.66).

It should be noted that the above considerations can be used as a basis of another proof of Theorem 1.5.

Feller's criterion (Feller, W. (1968) *An Introduction to Probability Theory and Its Applications.* Vol. 1, Third Edition, John Wiley & Sons, Inc., section XV.7, page 393) states that an irreducible and aperiodic Markov chain is ergodic iff it has a stationary distribution.

Now we have already shown that if $\rho < 1$ then the stationary distribution exists. Thus, applying Feller's criterion we can guarantee that the chain $\{N_i\}$ is ergodic for $\rho < 1$.

Now let $\rho > 1$. Consider the coefficient $f(z) = k(z) - z \equiv \beta(\lambda - \lambda z) - z$ in the left-hand side of equation (1.62). It is easy to see that:

1. $f(0) = \beta(\lambda) > 0$;

2. $f(1) = 0$;

3. $f'(1) = -\lambda\beta'(0) - 1 = \rho - 1 > 0$;

4. $f''(z) = \lambda^2\beta''(\lambda - \lambda z) \geq 0$ and hence the function $f(z)$ is convex.

Thus there exists $z_0 \in (0,1)$ such that $k(z_0) = z_0$. From (1.62) we have that $\psi(z_0) = 0$, i.e. all $\pi_n = 0$ and therefore the set of equations (1.60) does not have a nontrivial solution.

If $\rho = 1$ then from (1.62) we have:

$$\psi(1) = \frac{\mu}{\lambda}\psi'(1) \cdot \lim_{z \to 1-0} \frac{k(z) - z}{1 - k(z)} = \frac{\mu}{\lambda}\psi'(1) \cdot \frac{1-\rho}{\rho} = 0$$

and again all $\pi_n = 0$. Therefore in the case $\rho = 1$, as in the case $\rho > 1$, the set of equations (1.60) does not have a nontrivial solution.

Using Feller's criterion we can guarantee that in the case $\rho \geq 1$ the embedded Markov chain is not ergodic.

## 1.4 Limit theorems for the stationary distribution of the queue length

Although the performance characteristics of the system under consideration are available in explicit form, they are cumbersome (the above formulas include integrals of Laplace transforms, in more complex models solutions of functional equations appear, etc.). This yields, in particular, that explicit formula (1.32) for the generating function $p(z) = \sum_{n=0}^{\infty} z^n q_n$ of the stationary distribution $q_n$ of the queue length does not reveal the nature of the distribution $q_n$.

However in some domains of the system parameters the distribution $q_n$ can be approximated by standard distributions, such as Gaussian distribution or the gamma distribution. With this goal, in this section we investigate the asymptotic behaviour of the queue length under limit values of various parameters.

### 1.4.1 Heavy traffic

First consider the case of heavy traffic when arrival rate $\lambda$ increases in such a way that $\rho \to 1 - 0$.

**Theorem 1.6** *If the $M/G/1$ type retrial queue is in the steady state and $\beta_2 < \infty$ then*

$$\lim_{\lambda \to 1/\beta_1 - 0} \mathbf{E} e^{-s(1-\rho)N(t)} = \left( 1 + \frac{\beta_2}{2\beta_1^2} s \right)^{-1 - \frac{2\beta_1}{\mu \beta_2}}, \qquad (1.68)$$

*i.e. under heavy traffic queue length $N(t)$ asymptotically has a gamma distribution.*

*Proof.* Let $\varepsilon = 1 - \rho$. The Laplace transform of the scaled random variable $\varepsilon N(t) \equiv (1 - \rho)N(t)$ can be obtained from equation (1.32) putting $z = e^{-\varepsilon s}$:

$$
\begin{aligned}
\mathbf{E} e^{-s\varepsilon N(t)} &= \varepsilon \frac{1 - e^{-\varepsilon s}}{k\left(e^{-\varepsilon s}\right) - e^{-\varepsilon s}} \\
&\quad \times \exp\left\{ \frac{\lambda}{\mu} \int_1^{e^{-\varepsilon s}} \frac{1 - k(u)}{k(u) - u} du \right\}. \qquad (1.69)
\end{aligned}
$$

As $\lambda \to 1/\beta_1 - 0$, the variable $t = \lambda\left(1 - e^{-\varepsilon s}\right) \to 0$. On the other hand, if $\beta_2 < \infty$ then as $t \to 0$ the following asymptotic expansion holds:

$$\beta(t) = 1 - \beta_1 t + \frac{\beta_2}{2}t^2 + o(t^2).$$

Thus

$$
\begin{aligned}
k\left(e^{-\varepsilon s}\right) &\equiv \beta(\lambda(1 - e^{-\varepsilon s})) \\
&= 1 - \varepsilon s + \varepsilon^2\left[s^2\left(\frac{1}{2} + \frac{\beta_2}{2\beta_1^2}\right) + s\right] + o(\varepsilon^2).
\end{aligned}
$$

This implies that

$$\lim_{\lambda \to 1/\beta_1 - 0} \varepsilon\frac{1 - e^{-\varepsilon s}}{k(e^{-\varepsilon s}) - e^{-\varepsilon s}} = \frac{1}{1 + \frac{\beta_2}{2\beta_1^2}s}. \qquad (1.70)$$

To find the limit of the exponential term in the right-hand side of equation (1.69) we must find

$$\lim_{\lambda \to 1/\beta_1 - 0} \int_1^{e^{-\varepsilon s}} \frac{1 - k(u)}{k(u) - u}\,du. \qquad (1.71)$$

Introducing a new variable $v = \frac{1-u}{1-e^{-\varepsilon s}}$ we can transform the integral in (1.71) to the form:

$$\int_0^1 \frac{1 - \beta(\lambda(1 - e^{-\varepsilon s})v)}{\beta(\lambda(1 - e^{-\varepsilon s})v) - 1 + (1 - e^{-\varepsilon s})v} \cdot \left(e^{-\varepsilon s} - 1\right)\,dv. \qquad (1.72)$$

If $\beta_2 < \infty$, then uniformly with respect to $s$ in any finite interval $[0; S]$ we have

$$\beta(\varepsilon s) = 1 - \beta_1 \varepsilon s + \frac{\beta_2}{2}\varepsilon^2 s^2 + \varepsilon^2 \cdot o(1).$$

Indeed, introduce a function $f(t) = \beta(t) - 1 + \beta_1 t - \frac{\beta_2}{2}t^2$. Then:
   (a) $f(0) = 0, f'(0) = 0, f''(0) = 0$,
   (b) $f'''(t) = \beta'''(t)$ and thus $f'''(t)$ is finite and negative for $t > 0$.
   Therefore, $f''(t)$ is decreasing as $t \geq 0$. This in turn implies that $f''(t) < 0$ for $t > 0$, i.e. $f'(t)$ is decreasing for $t \geq 0$ From this we have that $f'(t) < 0$, i.e. $f(t)$ is decreasing, so that $f(t) < 0$. Hence, for $t \leq T$ we have $f(T) \leq f(t) \leq 0$. Besides, existence of $\beta_2$ implies that $\lim_{t\to 0} \frac{f(t)}{t^2} = 0$.
   Now let $t = \varepsilon s, T = \varepsilon S, 0 \leq s \leq S$. Then

$$S^2 \frac{f(\varepsilon S)}{\varepsilon^2 S^2} \leq \frac{f(\varepsilon s)}{\varepsilon^2} \leq 0.$$

As $\varepsilon \to 0$ the variable $\frac{f(\varepsilon S)}{\varepsilon^2 S^2} \to 0$ and thus $\frac{f(\varepsilon s)}{\varepsilon^2} \to 0$ uniformly with respect to $s \in [0; S]$.

Put $s = v$, $\varepsilon = \lambda(1 - e^{-\varepsilon s})$. Then, as $\lambda \to 1/\beta_1 - 0$, uniformly with respect to $v \in [0; 1]$ we have:

$$\beta(\lambda(1 - e^{-\varepsilon s})v) = 1 - \varepsilon sv + \varepsilon^2 \left( sv + \frac{s^2 v}{2} + \frac{\beta_2}{2\beta_1^2} s^2 v^2 \right) + \varepsilon^2 \cdot o(1).$$
$$(1.73)$$

Inserting (1.73) into (1.72) we get that the function in the integral in (1.72) uniformly converges to the function

$$-\frac{s}{1 + \frac{\beta_2}{2\beta_1^2} sv}.$$

Thus the integral (1.72) has a limit equal to

$$-\int_0^1 \frac{s}{1 + \frac{\beta_2}{2\beta_1^2} sv} dv = -\frac{2\beta_1^2}{\beta_2} \ln \left( 1 + \frac{\beta_2}{2\beta_1^2} s \right). \qquad (1.74)$$

Now equation (1.68) follows from (1.69), (1.70) and (1.74). $\quad\square$

### 1.4.2 Low rate of retrials

The following theorem describes the queue length distribution in the case of low rate of retrials, that is, as $\mu \to 0$.

**Theorem 1.7** *If $\beta_2 < \infty$ then as $\mu \to 0$ the queue length is asymptotically Gaussian with mean $\frac{\lambda\rho}{(1-\rho)\mu}$ and variance $\frac{\lambda^3\beta_2 + 2\lambda\rho - 2\lambda\rho^2}{2(1-\rho)^2\mu}$.*

*Proof.* Let

$$N^*(u) = \frac{N(u) - \frac{\lambda\rho}{(1-\rho)\mu}}{\frac{1}{\sqrt{\mu}}} = \sqrt{\mu} N(u) - \frac{\lambda\rho}{(1-\rho)\sqrt{\mu}}.$$

The characteristic function $\mathrm{E}\exp\{itN^*(u)\}$ can be expressed in terms of the generating function $p(z)$ as follows:

$$\mathrm{E}\exp\{itN^*(u)\} = p\left(e^{it\sqrt{\mu}}\right) \cdot e^{-it\frac{\lambda\rho}{(1-\rho)\sqrt{\mu}}}$$

$$= \exp\left\{ \frac{\lambda}{\mu} \int_1^{e^{it\sqrt{\mu}}} \frac{1 - k(u)}{k(u) - u} du - it\frac{\lambda\rho}{(1-\rho)\sqrt{\mu}} \right\}$$

$$\times (1 - \rho) \frac{1 - e^{it\sqrt{\mu}}}{k\left(e^{it\sqrt{\mu}}\right) - e^{it\sqrt{\mu}}}. \qquad (1.75)$$

If $\mu \to 0$ then $z = e^{it\sqrt{\mu}} \to 1$ and thus

$$(1 - \rho) \frac{1 - e^{it\sqrt{\mu}}}{k\left(e^{it\sqrt{\mu}}\right) - e^{it\sqrt{\mu}}} \to 1.$$

To calculate the limit of the exponent in the right-hand side of (1.75) we transform the argument of the exponential function as follows:

$$\frac{\lambda}{\mu} \int_1^{e^{it\sqrt{\mu}}} \frac{1-k(u)}{k(u)-u} du - it\frac{\lambda\rho}{(1-\rho)\sqrt{\mu}}$$

$$= \frac{\lambda}{\mu} \int_1^{e^{it\sqrt{\mu}}} \left(\frac{1-k(u)}{k(u)-u} - \frac{\rho}{1-\rho}\right) du$$

$$+ \frac{\lambda\rho}{1-\rho} \cdot \frac{e^{it\sqrt{\mu}} - 1 - it\sqrt{\mu}}{\mu}. \tag{1.76}$$

The second term in the right-hand side of (1.76) has the limit (as $\mu \to 0$) equal to $-\frac{\lambda\rho t^2}{2(1-\rho)}$.

To calculate the limit of the first term introduce a function

$$f(\mu) = \int_1^{e^{it\sqrt{\mu}}} \left(\frac{1-k(u)}{k(u)-u} - \frac{\rho}{1-\rho}\right) du.$$

It is easy to see that:

- $f(0) = 0$;

- $f'(0) = -\frac{\lambda^2\beta_2}{4(1-\rho)^2}t^2$.

Thus as $\mu \to 0$ we have:

$$f(\mu) = f(0) + \mu f'(0) + o(\mu) = -\mu\frac{\lambda^2\beta_2}{4(1-\rho)^2}t^2 + o(\mu).$$

This implies that the first term in the right-hand side of (1.76) is

$$\frac{\lambda}{\mu}f(\mu) = -\frac{\lambda^3\beta_2}{4(1-\rho)^2}t^2 + o(1).$$

Finally we have

$$\lim_{\mu \to 0} E \exp\{itN^*(u)\} = \exp\left\{-\frac{t^2}{2} \cdot \frac{\lambda^3\beta_2 + 2\lambda\rho - 2\lambda\rho^2}{2(1-\rho)^2}\right\}.$$

The right-hand side of this equation is the characteristic function of a Gaussian random variable with mean equal to 0 and variance $\frac{\lambda^3\beta_2 + 2\lambda\rho - 2\lambda\rho^2}{2(1-\rho)^2}$, which implies that the theorem holds.    □

### 1.4.3 High rate of retrials

In real situations subscribers repeat their calls practically immediately. So an investigation of asymptotic behavior of retrial queues under high intensity of repetition is of special practical interest.

Usually, as $\mu \to \infty$ a stationary characteristic $a(\mu)$ of a retrial queue converges to a limit $a(\infty)$, which is the corresponding stationary characteristic of a certain 'limit' system. Intuitively this system can be easily identified. For example, as $\mu \to \infty$ the main $M/G/1$ retrial queue can be thought of as the standard $M/G/1/\infty$ queueing system. Thus, if we denote by $q_n$ the queue length distribution in an $M/G/1$ retrial queue and $q_n(\infty)$ the queue length distribution in the corresponding stationary $M/G/1/\infty$ queue then we can expect that

$$\lim_{\mu \to \infty} q_n = q_n(\infty). \qquad (1.77)$$

This formula is equivalent to the corresponding limit for the generating functions:

$$\lim_{\mu \to \infty} p(z) = p^{(\infty)}(z).$$

Since $p^{(\infty)}(z) = (1 - \rho)\frac{1-z}{\beta(\lambda - \lambda z) - z}$ (by the Pollaczek–Khinchin formula), this relation trivially follows from formula (1.32) for $p(z)$.

The distribution $q_n(\infty)$ is studied in full detail in queueing theory and so can be considered as standard (like Gaussian, gamma, etc.).

The following result for the rate of convergence of the distribution $q_n$ to the distribution $q_n(\infty)$ is more interesting than the obvious relation (1.77).

**Theorem 1.8** *As $\mu \to \infty$ the distance*

$$\sum_{n=0}^{\infty} |q_n - q_n(\infty)|$$

*between distributions $q_n$ and $q_n(\infty)$ is $O\left(\frac{1}{\mu}\right)$. To be more exact, the following inequalities hold:*

$$2\frac{1-\rho}{\beta(\lambda)}\left[1 - \exp\left\{-\frac{\lambda}{\mu}\int_0^1 \frac{1 - k(u)}{k(u) - u}du\right\}\right]$$
$$\leq \sum_{n=0}^{\infty} |q_n - q_n(\infty)|$$

$$\leq 2 \left[ 1 - \exp\left\{ -\frac{\lambda}{\mu} \int_0^1 \frac{1 - k(u)}{k(u) - u} du \right\} \right]. \tag{1.78}$$

*Proof.* The proof will be based on the property of stochastic decomposition of the number of sources of repeated calls in the steady state

$$N_\mu = N_\infty + R_\mu$$

which was established in section 1.2. In terms of distributions

$$q_n = \mathrm{P}(N_\mu = n), q_n(\infty) = \mathrm{P}(N_\infty = n), \widehat{q}_n = \mathrm{P}(R_\mu = n)$$

this property means that $q_n$ is a convolution of the distributions $q_n(\infty)$ and $\widehat{q}_n$:

$$q_n = \sum_{k=0}^{n} q_k(\infty)\widehat{q}_{n-k}. \tag{1.79}$$

Recall that the distribution $\widehat{q}_n$ in fact coincides with the conditional distribution of the number of sources of repeated calls given that the server is free:

$$\widehat{q}_n = \mathrm{P}(N(t) = n \mid C(t) = 0) = p_{0n}/(1 - \rho),$$

which in particular implies that its generating function $\mathrm{E}z^{R_\mu}$ is given by the formula:

$$\sum_{n=0}^{\infty} z^n \widehat{q}_n = \exp\left\{ \frac{\lambda}{\mu} \int_1^z \frac{1 - k(u)}{k(u) - u} du \right\}. \tag{1.80}$$

Rewriting formula (1.79) as

$$q_n - q_n(\infty) = q_n(\infty)\widehat{q}_0 - q_n(\infty) + \sum_{k=0}^{n-1} q_k(\infty)\widehat{q}_{n-k},$$

we see that

$$
\begin{aligned}
|q_n - q_n(\infty)| &\leq |q_n(\infty)\widehat{q}_0 - q_n(\infty)| + \sum_{k=0}^{n-1} q_k(\infty)\widehat{q}_{n-k} \\
&= q_n(\infty)(1 - \widehat{q}_0) + q_n - q_n(\infty)\widehat{q}_0 \\
&= q_n(\infty) \cdot (1 - 2 \cdot \widehat{q}_0) + q_n.
\end{aligned}
$$

Therefore,

$$\sum_{n=0}^{\infty} |q_n - q_n(\infty)| \leq (1 - 2\widehat{q}_0) \sum_{n=0}^{\infty} q_n(\infty) + \sum_{n=0}^{\infty} q_n.$$

But both $q_n$ and $q_n(\infty)$ are probability distributions. Thus both sums in the right-hand side of this inequality equal 1, i.e.

$$\sum_{n=0}^{\infty} |q_n - q_n(\infty)| \leq 2 \cdot (1 - \widehat{q}_0). \tag{1.81}$$

Putting $z = 0$ in equation (1.80) we get an explicit formula for $\widehat{q}_0$:

$$\widehat{q}_0 = \exp\left\{ -\frac{\lambda}{\mu} \int_0^1 \frac{1 - k(u)}{k(u) - u} du \right\}. \tag{1.82}$$

Now from (1.81) and (1.82) we get the first part of inequality (1.78).

To get an estimate from below we use the obvious inequality $|a - b| \geq a - b$:

$$\begin{aligned}
\sum_{n=0}^{\infty} |q_n - q_n(\infty)| &= |q_0 - q_0(\infty)| + \sum_{n=1}^{\infty} |q_n - q_n(\infty)| \\
&\geq |q_0 - q_0(\infty)| + \sum_{n=1}^{\infty} (q_n - q_n(\infty)) \\
&= |q_0 - q_0(\infty)| + q_0(\infty) - q_0.
\end{aligned}$$

But,

$$\begin{aligned}
q_0(\infty) &= \frac{1 - \rho}{\beta(\lambda)}, \\
q_0 &= q_0(\infty) \cdot \widehat{q}_0 < q_0(\infty).
\end{aligned}$$

Thus,

$$|q_0(\mu) - q_0(\infty)| = q_0(\infty) - q_0 = q_0(\infty) \cdot (1 - \widehat{q}_0),$$

so that

$$\sum_{n=0}^{\infty} |q_n - q_n(\infty)| \geq 2 \cdot (q_0(\infty) - q_0) = 2 \cdot \frac{1 - \rho}{\beta(\lambda)} \cdot (1 - \widehat{q}_0).$$

$\square$

The probability $\widehat{q}_0$ can be estimated with the help of an approach based on stochastic orderings (see next section 1.5) in simpler terms. In particular, always

$$1 - \widehat{q}_0 \leq 1 - \exp\left\{ -\frac{\lambda}{\mu} \int_0^1 \frac{1 - e^{-\rho(1-u)}}{e^{-\rho(1-u)} - u} du \right\}$$

and if $B(x)$ is a NBUE distribution then

$$1 - \widehat{q}_0 \geq 1 - (1 - \rho)^{\lambda/\mu}.$$

## 1.5  Stochastic inequalities

### 1.5.1  Monotonicity properties of the embedded Markov chain

Let $T$ be the transition operator of the embedded Markov chain $\{N_i\}$, which to every distribution $p = (p_n)_{n \geq 0}$ associates a distribution $Tp = q = (q_m)_{m \geq 0}$ such that $q_m = \sum_n p_n r_{nm}$.

Here the $r_{nm}$ are one-step transition probabilities of the chain which as we have established in section 1.3, equation (1.53), are given by

$$r_{nm} = \frac{\lambda}{\lambda + n\mu} k_{m-n} + \frac{n\mu}{\lambda + n\mu} k_{m-n+1}.$$

**Theorem 1.9**  *The operator $T$ is monotone with respect to the strong stochastic ordering $\leq_{\mathrm{st}}$ and the convex ordering $\leq_{\mathrm{v}}$, i.e. for any two distributions $p^{(1)}$, $p^{(2)}$ the inequality $p^{(1)} \leq_{\mathrm{st}} p^{(2)}$ (or $p^{(1)} \leq_{\mathrm{v}} p^{(2)}$) implies that $Tp^{(1)} \leq_{\mathrm{st}} Tp^{(2)}$ (respectively, $Tp^{(1)} \leq_{\mathrm{v}} Tp^{(2)}$).*

*Proof.* We will use the following general theorem from section 4.2 of the monograph by Stoyan, D. (1983), *Comparison methods for queues and other stochastic models,* Wiley, New York.

**Statement 4**  *An operator $T$ is monotone with respect to $\leq_{\mathrm{st}}$ iff $\bar{r}_{n-1m} \leq \bar{r}_{nm}$ for all $n, m$, and is monotone with respect to $\leq_{\mathrm{v}}$ iff $2\bar{\bar{r}}_{nm} \leq \bar{\bar{r}}_{n-1m} + \bar{\bar{r}}_{n+1m}$ for all $n, m$, where $\bar{r}_{nm} = \sum_{l=m}^{\infty} r_{nl}$, $\bar{\bar{r}}_{nm} = \sum_{l=m}^{\infty} \bar{r}_{nl}$.*

In our case

$$
\begin{aligned}
\bar{r}_{nm} &= \frac{\lambda}{\lambda + n\mu} \bar{k}_{m-n} + \frac{n\mu}{\lambda + n\mu} \bar{k}_{m-n+1} \\
&= \bar{k}_{m-n} - \frac{n\mu}{\lambda + n\mu} k_{m-n} \\
&= \bar{k}_{m-n+1} + \frac{\lambda}{\lambda + n\mu} k_{m-n}.
\end{aligned}
$$

Thus,

$$
\begin{aligned}
\bar{r}_{nm} - \bar{r}_{n-1m} &= \bar{k}_{m-n+1} + \frac{\lambda}{\lambda + n\mu} k_{m-n} - \bar{k}_{m-n+1} \\
&\quad + \frac{(n-1)\mu}{\lambda + (n-1)\mu} k_{m-n+1} \geq 0,
\end{aligned}
$$

and hence $T$ is monotone with respect to $\leq_{\mathrm{st}}$ .

Furthermore,

$$\bar{\bar{r}}_{nm} = \frac{\lambda}{\lambda + n\mu}\bar{\bar{k}}_{m-n} + \frac{n\mu}{\lambda + n\mu}\bar{\bar{k}}_{m-n+1}$$

$$= \bar{\bar{k}}_{m-n} - \frac{n\mu}{\lambda + n\mu}\bar{k}_{m-n}$$

$$= \bar{\bar{k}}_{m-n+1} + \frac{\lambda}{\lambda + n\mu}\bar{k}_{m-n}.$$

Thus

$$\bar{\bar{r}}_{n-1m} + \bar{\bar{r}}_{n+1m} - 2\bar{\bar{r}}_{nm}$$

$$= \bar{\bar{k}}_{m-n+1} - \frac{(n-1)\mu}{\lambda + (n-1)\mu}\bar{k}_{m-n+1}$$

$$+\bar{\bar{k}}_{m-n} + \frac{\lambda}{\lambda + (n+1)\mu}\bar{k}_{m-n-1}$$

$$-\bar{\bar{k}}_{m-n} + \frac{n\mu}{\lambda + n\mu}\bar{k}_{m-n}$$

$$-\bar{\bar{k}}_{m-n+1} - \frac{\lambda}{\lambda + n\mu}\bar{k}_{m-n}$$

$$= \bar{k}_{m-n} \cdot \left\{ -\frac{(n-1)\mu}{\lambda + (n-1)\mu} + \frac{n\mu}{\lambda + n\mu} \right.$$

$$\left. +\frac{\lambda}{\lambda + (n+1)\mu} - \frac{\lambda}{\lambda + n\mu} \right\}$$

$$+\frac{(n-1)\mu}{\lambda + (n-1)\mu}k_{m-n} + \frac{\lambda}{\lambda + (n+1)\mu}k_{m-n-1}$$

$$= \bar{k}_{m-n} \cdot \frac{2\lambda\mu^2}{(\lambda + (n-1)\mu)(\lambda + n\mu)(\lambda + (n+1)\mu)}$$

$$+\frac{(n-1)\mu}{\lambda + (n-1)\mu}k_{m-n} + \frac{\lambda}{\lambda + (n+1)\mu}k_{m-n-1} \geq 0,$$

and thus $T$ is monotone with respect to $\leq_{\mathrm{v}}$. $\qquad\square$

This theorem implies in particular that if at time $t = 0$ the system was empty, then the number of customers in the system at departure times forms a monotonically increasing sequence with respect to the strong stochastic ordering:

$$N_0 = 0 \leq_{\mathrm{st}} N_1 \leq_{\mathrm{st}} N_2 \leq_{\mathrm{st}} N_3 \leq_{\mathrm{st}} \cdots.$$

Now suppose we have two $M/G/1$ retrial queues with parameters $\lambda^{(1)}, \mu^{(1)}, B^{(1)}(x)$ and $\lambda^{(2)}, \mu^{(2)}, B^{(2)}(x)$ respectively. Denote by $T^{(1)}$, $T^{(2)}$ the transition operators of the corresponding embedded Markov chains.

**Theorem 1.10** *If $\lambda^{(1)} \leq \lambda^{(2)}$, $\mu^{(1)} \geq \mu^{(2)}$, $B^{(1)}(x) \leq_s B^{(2)}(x)$, where 's' is either 'st' or 'v', then $T^{(1)} \leq_s T^{(2)}$, i.e. for any distribution p we have: $T^{(1)}p \leq_s T^{(2)}p$.*

*Proof.* By the general theorem 4.2.3 of the monograph by Stoyan we have to establish that for the corresponding one-step transition probabilities $r_{nm}^{(1)}$, $r_{nm}^{(2)}$ the following numerical inequalities hold:

$$\bar{r}_{nm}^{(1)} \leq \bar{r}_{nm}^{(2)} \text{ (for the ordering } \leq_{st} ),$$

$$\bar{\bar{r}}_{nm}^{(1)} \leq \bar{\bar{r}}_{nm}^{(2)} \text{ (for the ordering } \leq_v ),$$

or equivalently

$$\frac{\lambda^{(1)}}{\lambda^{(1)} + n\mu^{(1)}} \bar{k}_{m-n}^{(1)} + \frac{n\mu^{(1)}}{\lambda^{(1)} + n\mu^{(1)}} \bar{k}_{m-n+1}^{(1)}$$
$$\leq \frac{\lambda^{(2)}}{\lambda^{(2)} + n\mu^{(2)}} \bar{k}_{m-n}^{(2)} + \frac{n\mu^{(2)}}{\lambda^{(2)} + n\mu^{(2)}} \bar{k}_{m-n+1}^{(2)} \quad (1.83)$$

(for the ordering $\leq_{st}$),

$$\frac{\lambda^{(1)}}{\lambda^{(1)} + n\mu^{(1)}} \bar{\bar{k}}_{m-n}^{(1)} + \frac{n\mu^{(1)}}{\lambda^{(1)} + n\mu^{(1)}} \bar{\bar{k}}_{m-n+1}^{(1)}$$
$$\leq \frac{\lambda^{(2)}}{\lambda^{(2)} + n\mu^{(2)}} \bar{\bar{k}}_{m-n}^{(2)} + \frac{n\mu^{(2)}}{\lambda^{(2)} + n\mu^{(2)}} \bar{\bar{k}}_{m-n+1}^{(2)} \quad (1.84)$$

(for the ordering $\leq_v$). In the subsequent analysis we will use the following result which can be easily established with the help of Theorems 1.2.2 and 1.3.1 of the above-cited monograph by Stoyan.

**Statement 5** *If $\lambda^{(1)} \leq \lambda^{(2)}$, $B^{(1)}(x) \leq_s B^{(2)}(x)$ , where 's' is one of the symbols 'st', 'v' then the same stochastic inequality holds for the corresponding distributions $k_n^{(1)}$, $k_n^{(2)}$ of the number of new arrivals during a service period, i.e. $\{k_n^{(1)}\} \leq_s \{k_n^{(2)}\}$.*

By Statement 5, $\bar{k}_n^{(1)} \leq \bar{k}_n^{(2)}$ for all $n \geq 0$ (in the case when s=st) and $\bar{\bar{k}}_n^{(1)} \leq \bar{\bar{k}}_n^{(2)}$ for all $n \geq 0$ (in the case when s=v). Besides, $\lambda^{(1)} \leq \lambda^{(2)}$, $\mu^{(1)} \geq \mu^{(2)}$ imply that $\frac{\lambda^{(1)}}{\mu^{(1)}} \leq \frac{\lambda^{(2)}}{\mu^{(2)}}$. Since the function $\frac{x}{x+n}$ is increasing, we have: $\frac{\lambda^{(1)}}{\lambda^{(1)} + n\mu^{(1)}} \leq \frac{\lambda^{(2)}}{\lambda^{(2)} + n\mu^{(2)}}$. Using these inequalities we get:

$$\frac{\lambda^{(1)}}{\lambda^{(1)} + n\mu^{(1)}} \bar{k}_{m-n}^{(1)} + \frac{n\mu^{(1)}}{\lambda^{(1)} + n\mu^{(1)}} \bar{k}_{m-n+1}^{(1)}$$

$$= \overline{k}^{(1)}_{m-n+1} + \frac{\lambda^{(1)}}{\lambda^{(1)} + n\mu^{(1)}} k^{(1)}_{m-n}$$

$$\leq \overline{k}^{(1)}_{m-n+1} + \frac{\lambda^{(2)}}{\lambda^{(2)} + n\mu^{(2)}} k^{(1)}_{m-n}$$

$$= \frac{\lambda^{(2)}}{\lambda^{(2)} + n\mu^{(2)}} \overline{k}^{(1)}_{m-n} + \frac{n\mu^{(2)}}{\lambda^{(2)} + n\mu^{(2)}} \overline{k}^{(1)}_{m-n+1}$$

$$\leq \frac{\lambda^{(2)}}{\lambda^{(2)} + n\mu^{(2)}} \overline{k}^{(2)}_{m-n} + \frac{n\mu^{(2)}}{\lambda^{(2)} + n\mu^{(2)}} \overline{k}^{(2)}_{m-n+1},$$

and hence (1.83) holds.

In a similar way inequality (1.84) also holds.   $\square$

### 1.5.2 Stochastic inequalities for the stationary number of customers in the system

**Theorem 1.11** *Suppose that we have two $M/G/1$ retrial queues having parameters $\lambda^{(1)}, \mu^{(1)}, B^{(1)}(x)$ and $\lambda^{(2)}, \mu^{(2)}, B^{(2)}(x)$ respectively and let $Q_n^{(1)}$, $Q_n^{(2)}$ be the corresponding stationary distributions of the number of customers in the system. Then $\lambda^{(1)} \leq \lambda^{(2)}$, $\mu^{(1)} \geq \mu^{(2)}$, $B^{(1)}(x) \leq_s B^{(2)}(x)$, where 's' is either 'st' or 'v', imply that $\left\{ Q_n^{(1)} \right\} \leq_s \left\{ Q_n^{(2)} \right\}$.*

*Proof.* As we have shown, the stationary distribution of the number of customers in each system is identical to the stationary distribution of the corresponding embedded Markov chain. Thus we can apply the general theory of section 4.2 of the monograph by Stoyan. After this Theorem 1.11 follows from Theorems 1.9 and 1.10.

Based on Theorem 1.11 we can establish insensitive stochastic bounds for the stationary distribution of the number of customers in the system.

**Theorem 1.12** *For any $M/G/1$ retrial queue the distribution $Q_n$ is greater relative to the ordering $\leq_V$ than the distribution with the generating function*

$$Q^*(z) = (1 - \rho)e^{\rho(z-1)} \frac{1 - z}{e^{\rho(z-1)} - z}$$

$$\times \exp\left\{ \frac{\lambda}{\mu} \int_1^z \frac{1 - e^{\rho(u-1)}}{e^{\rho(u-1)} - u} du \right\}.$$

*Proof.* Consider an auxiliary $M/D/1$ retrial queue with the same arrival rate $\lambda$, retrial rate $\mu$, mean service time $\beta_1$, but with deterministic service time distribution $B^*(x)$ :

$$B^*(x) = \begin{cases} 0, & \text{if } x \le \beta_1, \\ 1, & \text{if } x > \beta_1. \end{cases}$$

It is well known (see section 1.9 of the monograph by Stoyan) that $B^*(x) \le_V B(x)$ and the result follows from Theorem 1.11.   □

**Theorem 1.13** *If in the $M/G/1$ retrial queue, the service time distribution $B(x)$ is NBUE (new better than used in expectation) (or NWUE – new worse than used in expectation) then the distribution $Q_n$ is less (respectively, greater) relative to the ordering $\le_V$ than the negative binomial distribution*

$$Q_n^* = \frac{(\rho/\mu)^n}{n!} \prod_{i=1}^{n} (\lambda + i\mu)(1 - \rho)^{\frac{\lambda}{\mu}+1}.$$

*Proof.* Consider an auxiliary $M/M/1$ retrial queue with the same arrival rate $\lambda$, retrial rate $\mu$, mean service time $\beta_1$, but with exponentially distributed service times:

$$B^*(x) = \begin{cases} 1 - e^{-x/\beta_1}, & \text{if } x \ge 0, \\ 0, & \text{if } x < 0. \end{cases}$$

If $B(x)$ is NBUE then $B(x) \le_V B^*(x)$ (the inequality is reversed if $B(x)$ is NWUE). Thus $Q_n$ is less (respectively, greater if $B(x)$ is NWUE) than the corresponding distribution in the $M/M/1$ retrial queue. The former distribution is negative binomial and is given by the above formula.   □

### 1.5.3 Stochastic inequalities for other distributions

With the help of the above approach we can get stochastic inequalities for other distributions associated with the main $M/G/1$ retrial queue.

Consider, for example, the conditional distribution $\widehat{q}_n$ of the stationary queue given that the server is free. This distribution has also appeared in the stochastic decomposition formula for the stationary queue length. As we saw its generating function $\widehat{p}(z) = \sum_{n=0}^{\infty} z^n \widehat{q}_n$ is given by

$$\widehat{p}(z) = \exp\left\{ -\frac{\lambda}{\mu} \int_z^1 \frac{1 - k(u)}{k(u) - u} du \right\}.$$

**Theorem 1.14** *Suppose we have two $M/G/1$ retrial queues with parameters $\lambda^{(1)}, \mu^{(1)}, B^{(1)}(x)$ and $\lambda^{(2)}, \mu^{(2)}, B^{(2)}(x)$. If $\lambda^{(1)} \leq \lambda^{(2)}$, $\mu^{(1)} \geq \mu^{(2)}$, $B^{(1)}(x) \leq_L B^{(2)}(x)$, where $\leq_L$ is the Laplace ordering, then $\widehat{q}_n^{(1)} \leq_L \widehat{q}_n^{(2)}$.*

*Proof.* The stochastic inequality $B^{(1)}(x) \leq_L B^{(2)}(x)$ means that for the corresponding Laplace–Stieltjes transforms the ordinary inequality $\beta^{(1)}(s) \geq \beta^{(2)}(s)$ holds. From this for $z \in [0,1]$ we have:

$$\beta^{(1)}(\lambda^{(1)} - \lambda^{(1)}z) \geq \beta^{(2)}(\lambda^{(1)} - \lambda^{(1)}z) \geq \beta^{(2)}(\lambda^{(2)} - \lambda^{(2)}z),$$

or equivalently,

$$k^{(1)}(z) \geq k^{(2)}(z) \text{ for all } z \in [0,1],$$

where $k^{(1)}(z), k^{(2)}(z)$ are the corresponding distributions of the number of new arrivals during a service time. This implies that

$$\frac{1 - k^{(1)}(u)}{k^{(1)}(u) - u} \leq \frac{1 - k^{(2)}(u)}{k^{(2)}(u) - u}$$

Besides, $\lambda^{(1)}/\mu^{(1)} \leq \lambda^{(2)}/\mu^{(2)}$ and thus

$$\frac{\lambda^{(1)}}{\mu^{(1)}} \int_z^1 \frac{1 - k^{(1)}(u)}{k^{(1)}(u) - u} du \leq \frac{\lambda^{(2)}}{\mu^{(2)}} \int_z^1 \frac{1 - k^{(2)}(u)}{k^{(2)}(u) - u} du.$$

Taking into account the above formula for $\widehat{p}(z)$ we get:

$$\widehat{p}^{(1)}(z) \geq \widehat{p}^{(2)}(z) \text{ for all } z \in [0;1],$$

which means the stochastic inequality $\left\{\widehat{q}_n^{(1)}\right\} \leq_L \left\{\widehat{q}_n^{(2)}\right\}$. $\qquad\square$

**Theorem 1.15** *For any $M/G/1$ retrial queue the distribution $\widehat{q}_n$ is less relative to the ordering $\leq_L$ than the distribution with generating function*

$$\exp\left\{-\frac{\lambda}{\mu} \int_z^1 \frac{1 - e^{\rho(u-1)}}{e^{\rho(u-1)} - u} du\right\},$$

*and if $B(x)$ is NBUE then the distribution $\widehat{q}_n$ is greater relative to the ordering $\leq_L$ than the distribution*

$$\widehat{q}_n^* = \frac{(\rho/\mu)^n}{n!} \prod_{i=0}^{n-1} (\lambda + i\mu)(1-\rho)^{\lambda/\mu}.$$

*Proof.* Consider auxiliary $M/D/1$ and $M/M/1$ retrial queues with the same arrival rates $\lambda$, retrial rates $\mu$, mean service times $\beta_1$.

Since $B(x)$ is always less, relative to the ordering $\leq_L$, than a deterministic distribution with the same mean value, based on Theorem 1.14 we can guarantee the first inequality.

If $B(x)$ is NBUE then $B(x)$ is greater relative to the ordering $\leq_L$ than the exponential distribution with the same mean. For $M/M/1$ the distribution $\widehat{q}_n^*$ can be obtained from equation (1.1), and so based on Theorem 1.14 we can guarantee the second inequality as well.                                                                      □

Theorem 1.15 can also be proved using the ordering $\leq_V$.

With this goal consider an auxiliary $M/G/1$ retrial queue with the same arrival rate $\lambda$ and retrial rate $\mu$. Service time distribution $B^*(x)$ in this system can be arbitrary, but must have the same mean value: $\beta_1^* = \beta_1$.

If $B(x) \leq_V B^*(x)$ (or $B^*(x) \leq_V B(x)$) then $\{k_n\} \leq_V \{k_n^*\}$ (correspondingly, $\{k_n^*\} \leq_V \{k_n\}$).

Introduce distributions $l_n$, $l_n^*$ by the formulas:

$$l_n = \frac{1}{\rho}\overline{k}_{n+1}, \quad l_n^* = \frac{1}{\rho}\overline{k}_{n+1}^*,$$

where

$$\overline{k}_n = \sum_{i=n}^{\infty} k_i.$$

It is easy to see that $\{k_n\} \leq_V \{k_n^*\}$ (or $\{k_n^*\} \leq_V \{k_n\}$) implies that $\{l_n\} \leq_{st} \{l_n^*\}$ (respectively, $\{l_n^*\} \leq_{st} \{l_n\}$). From this we have: $\{l_n\} \leq_L \{l_n^*\}$ (respectively, $\{l_n^*\} \leq_L \{l_n\}$). But

$$l(z) \equiv \sum_{n=0}^{\infty} l_n z^n = \frac{1}{\rho} \cdot \frac{1 - k(z)}{1 - z}.$$

So we can rewrite the formula for $\widehat{p}(z)$ as:

$$\widehat{p}(z) = \exp\left\{ -\frac{\lambda}{\mu} \int_z^1 \frac{\rho l(u)}{1 - \rho l(u)} du \right\}.$$

If $\{l_n\} \leq_L \{l_n^*\}$ then $l(u) \geq l^*(u)$ and so $\frac{\rho l(u)}{1-\rho l(u)} \geq \frac{\rho l^*(u)}{1-\rho l^*(u)}$. This implies that $\widehat{p}(z) \leq \widehat{p}^*(z)$, i.e. $\{\widehat{q}_n^*\} \leq_L \{\widehat{q}_n\}$. Similarly, if $\{l_n^*\} \leq_L \{l_n\}$ then $\{\widehat{q}_n\} \leq_L \{\widehat{q}_n^*\}$. Thus,

$$B(x) \leq_V B^*(x) \quad \Rightarrow \quad \{\widehat{q}_n^*\} \leq_L \{\widehat{q}_n\},$$
$$B^*(x) \leq_V B(x) \quad \Rightarrow \quad \{\widehat{q}_n\} \leq_L \{\widehat{q}_n^*\}.$$

It is well known that if $B^*(x)$ is a deterministic distribution, then

$B^*(x) \leq_V B(x)$, and if $B^*(x)$ is an exponential distribution and $B(x)$ is NBUE then $B(x) \leq_V B^*(x)$. Thus

$$\widehat{q}_n \leq_L \widehat{q}_n^{(\text{deterministic})},$$

and if $B(x)$ is NBUE then

$$\widehat{q}_n^{(\text{exponential})} \leq_L \widehat{q}_n.$$

$\square$

It should be noted that the distribution $\widehat{q}_n$ is closely connected with the busy period of the $M/G/1$ retrial queue, which will be investigated in the next section, and some inequalities for the busy period can be obtained based on the above theorems for the distribution $\widehat{q}_n$.

## 1.6 The busy period

A busy period is defined as the period that starts at an epoch when a call enters an empty system (when the server is free and there is no source of repeated calls) and ends at the next departure epoch at which the system is empty. The busy period defined in this way consists of alternating service periods and periods during which the server is free and there are sources of repeated calls in the system.

More general is the concept of $k$-busy period. This is a busy period which starts when a call enters the system with idle server and $(k-1)$ sources of repeated calls. Thus an ordinary busy period can be defined as a 1-busy period.

The main characteristics of a $k$-busy period are:

- $L^{(k)}$, the length of the $k$-busy period;

- $I^{(k)}$ , the number of demands which were served during this $k$-busy period.

For the ordinary busy period (which corresponds to the case $k = 1$) we will omit the index '(1)' in variables $L^{(1)}$ and $I^{(1)}$.

The following theorem describes the joint distribution of these random variables in terms of the Laplace transform

$$\pi^{(k)}(s, y) \equiv \mathrm{E}\left\{ e^{-sL^{(k)}} \cdot y^{I^{(k)}} \right\}.$$

**Theorem 1.16**

$$\pi^{(k)}(s,y) = \frac{\displaystyle\int_0^{\pi_\infty(s,y)} \frac{yu^{k-1}\beta(s+\lambda-\lambda u)}{[y\beta(s+\lambda-\lambda u)-u]e(s,u,y)}\,du}{\displaystyle\int_0^{\pi_\infty(s,y)} \frac{1}{[y\beta(s+\lambda-\lambda u)-u]e(s,u,y)}\,du}, \tag{1.85}$$

where $\pi_\infty(s,y)$ is the solution of the equation $y\beta(s+\lambda-\lambda\pi_\infty) = \pi_\infty$ in the interval $0 \le \pi_\infty \le 1$, which is continuous at the point $s = 0$, $y = 1$; and $e(s,z,y)$ is given by the formula

$$e(s,z,y) = \exp\left\{ \frac{1}{\mu}\int_0^z \frac{s+\lambda-\lambda y\beta(s+\lambda-\lambda u)}{y\beta(s+\lambda-\lambda u)-u}\,du \right\}. \tag{1.86}$$

*In particular,*

- *if either $\rho > 1$ or $\rho = 1, \mu < 2\beta_1/\beta_2$ then*

$$P(L^{(k)} = \infty) = P(I^{(k)} = \infty) > 0;$$

- *if $\rho = 1$ and $\mu \ge 2\beta_1/\beta_2$ then*

$$L^{(k)} < \infty, \quad I^{(k)} < \infty$$

*with probability 1, but*

$$EL^{(k)} = EI^{(k)} = \infty;$$

- *if $\rho < 1$ then the length of the k-busy period and the number of customers served during the k-busy period are finite and have finite mean values*

$$\begin{aligned}
EL^{(k)} &= \frac{k\beta_1}{1-\rho} + \frac{1}{1-\rho}\int_0^1 \frac{1-u^{k-1}\beta(\lambda-\lambda u)}{\mu(\beta(\lambda-\lambda u)-u)} \\
&\quad \times \exp\left\{ -\frac{\lambda}{\mu}\int_1^u \frac{1-\beta(\lambda-\lambda v)}{\beta(\lambda-\lambda v)-v}\,dv \right\} du, \tag{1.87}
\end{aligned}$$

$$\begin{aligned}
EI^{(k)} &= \frac{k}{1-\rho} + \frac{\lambda}{1-\rho}\int_0^1 \frac{1-u^{k-1}\beta(\lambda-\lambda u)}{\mu(\beta(\lambda-\lambda u)-u)} \\
&\quad \times \exp\left\{ -\frac{\lambda}{\mu}\int_1^u \frac{1-\beta(\lambda-\lambda v)}{\beta(\lambda-\lambda v)-v}\,dv \right\} du
\end{aligned}$$

$$= k + \lambda E L^{(k)}. \qquad (1.88)$$

*correspondingly.*

*Proof.* Assume that a $k$-busy period starts at time $t = 0$, and let $s > 0$, $y \in [0;1]$ be some parameters. Consider a Poisson flow of 'catastrophes' with rate $s$ which is independent of the system functioning. Besides, we will colour all arriving customers 'red' with probability $y$ and 'green' with probability $1 - y$.

**Lemma 1.1** *The Laplace transforms $\pi^{(k)}(s,y)$ satisfy the following set of equations:*

$$\pi^{(k)}(s,y) = y \sum_{n=0}^{\infty} k_n(s) \left[ \frac{\lambda}{s + \lambda + (k-1+n)\mu} \pi^{(k+n)}(s,y) \right.$$
$$+ \left. \frac{(k-1+n)\mu}{s + \lambda + (k-1+n)\mu} \pi^{(k-1+n)}(s,y) \right],$$
$$\text{if } k \geq 2, \qquad (1.89)$$

$$\pi^{(1)}(s,y) = y k_0(s) + y \sum_{n=1}^{\infty} k_n(s) \left[ \frac{\lambda}{s + \lambda + n\mu} \pi^{(n+1)}(s,y) \right.$$
$$+ \left. \frac{n\mu}{s + \lambda + n\mu} \pi^{(n)}(s,y) \right], \qquad (1.90)$$

*where*

$$k_n(s) = \int_0^{\infty} \frac{(\lambda x)^n}{n!} e^{-(s+\lambda)x} dB(x). \qquad (1.91)$$

*Proof.* First note that the variable $k_n(s)$ can be thought of as the probability that during a service period exactly $n$ new primary calls arrive in the system and no 'catastrophes' occur.

The variable $\pi^{(k)}(s,y)$ can be thought of as the probability that during a $k$-busy period no 'catastrophes' occur and all served calls are 'red'. This event (denote it by $E$) can be realized only if:

(1) the first served customer is 'red' (probability of this event is $y$);

(2) during the service period of this customer no 'catastrophes' occur and exactly $n$ new primary calls arrive (probability of this event is $k_n(s)$), and thus at time of the first departure there are $k - 1 + n$ sources of repeated calls;

(3a) until the beginning of the second service period no 'catastrophes' occur, the second call is a primary (probability of this event is $\dfrac{\lambda}{s + \lambda + (k - 1 + n)\mu})$ and during the $(k + n)$-busy period opened by the second call, no 'catastrophes' occur and only 'red' customers are served; or

(3b) until the beginning of the second service period no 'catastrophes' occur, the second call is repeated (probability of this event is $\dfrac{(k - 1 + n)\mu}{s + \lambda + (k - 1 + n)\mu})$ and during the $(k - 1 + n)$-busy period opened by the second call, no 'catastrophes' occur and only 'red' customers are served.

Since the event $E$ can be realized only in accordance with the above scheme (for some $n \geq 0$), this immediately implies (1.89).

In the case $k = 1$ realization of the second event for $n = 0$ means the end of the busy period. This special case is described by the formula (1.90).                                                                    □

Equations (1.89), (1.90) can be rewritten in matrix form as

$$
\begin{aligned}
(\pi^{(1)}(s, y), \pi^{(2)}(s, y), \ldots)^T &= \mathbf{A}(s, y)(\pi^{(1)}(s, y), \pi^{(2)}(s, y), \ldots)^T \\
&+ (yk_0(s), 0, 0, \ldots)^T,
\end{aligned}
$$

where the matrix $\mathbf{A}(s, y)$ is constructed from (1.89), (1.90) in the usual way. The matrix $\mathbf{A}(s, y)$ can be thought of as an operator which acts on the space of all bounded sequences. Its norm is $y\beta(s)$ and thus, if $(y, s) \neq (1, 0)$ (i.e. $y$ and $s$ are not equal to 1 and 0 respectively simultaneously) the operator $\mathbf{I} - \mathbf{A}(s, y)$ is invertible and

$$
\begin{aligned}
(\pi^{(1)}(s, y), \pi^{(2)}(s, y), \ldots)^T &= [\mathbf{I} - \mathbf{A}(s, y)]^{-1} \\
&\times (yk_0(s), 0, 0, \ldots)^T.
\end{aligned}
$$

Because only the first coordinate of the vector $(yk_0(s), 0, \ldots)$ is nonzero it is sufficient to find only the first column of the matrix $(\mathbf{I} - \mathbf{A})^{-1}$. Then $\pi^{(k)}(s, y) = yk_0(s) \cdot \left[(\mathbf{I} - \mathbf{A})^{-1}\right]_{k,1}$, where $\left[(\mathbf{I} - \mathbf{A})^{-1}\right]_{k,1}$ denotes the $k$th element of the first column of the matrix $(\mathbf{I} - \mathbf{A})^{-1}$.

To find the matrix $[\mathbf{I} - \mathbf{A}(s, y)]^{-1}$ we solve the adjoint equation

$$
\mathbf{x} \cdot [\mathbf{I} - \mathbf{A}(s, y)] = \mathbf{v},
$$

where $\mathbf{v} = (v_1, v_2, \ldots)$ is some vector.

Then the first column of the matrix $(\mathbf{I} - \mathbf{A})^{-1}$ can be found from the equation

$$
x_1 = \sum_k v_k [(\mathbf{I} - \mathbf{A})^{-1}]_{k,1},
$$

i.e. $\pi^{(k)}(s, y)$ is equal to $yk_0(s)$ multiplying by the coefficient of $v_k$ in the above expansion of $x_1$.

In coordinate-wise form the equation $\mathbf{x} \cdot [\mathbf{I} - \mathbf{A}(s, y)] = \mathbf{v}$ is

$$x_1 - x_1 \frac{\mu y}{s + \lambda + \mu} k_1(s) - x_2 \frac{\mu y k_0(s)}{s + \lambda + \mu} = v_1,$$

$$x_i - \sum_{n=1}^{i+1} x_n \left\{ \frac{\lambda y k_{i-n}(s)}{s + \lambda + (i-1)\mu} + \frac{i \mu y k_{i-n+1}(s)}{s + \lambda + i\mu} \right\} = v_i, \; i \geq 2.$$

For the generating function $x(z) = \sum_{i=1}^{\infty} z^i x_i$ these equations become:

$$\begin{aligned} x(z) &= \frac{\lambda}{\mu} yz X(z) + yz X'(z) \\ &+ v(z) - x_1 z \frac{\lambda y k_0(s)}{s + \lambda}, \end{aligned} \quad (1.92)$$

where

$$X(z) = \sum_{i=0}^{\infty} \frac{\mu z^i}{s + \lambda + i\mu} \sum_{n=1}^{i+1} x_n k_{i-n+1}(s),$$

$$v(z) = \sum_{i=1}^{\infty} v_i z^i.$$

The function $X(z)$ can be expressed in terms of function $x(z)$ as follows:

$$X(z) = z^{-\frac{s+\lambda}{\mu}} \int_0^z x(u) \beta(s + \lambda - \lambda u) u^{\frac{s+\lambda}{\mu} - 2} du. \quad (1.93)$$

To prove this relation we first note that

$$\begin{aligned} \sum_{i=1}^{\infty} u^i \sum_{n=1}^{i} x_n k_{i-n}(s) &= \sum_{n=1}^{\infty} u^n x_n \sum_{i=n}^{\infty} u^{i-n} k_{i-n}(s) \\ &= x(u) \cdot \beta(s + \lambda - \lambda u). \end{aligned}$$

Multiplying this equality by $u^{\frac{s+\lambda}{\mu} - 2}$ and integrating with respect to $u$ from 0 to $z$, after some algebra we get (1.93).

Since

$$x(z) = \frac{\left[ X(z) \cdot z^{\frac{s+\lambda}{\mu}} \right]'}{\beta(s + \lambda - \lambda z) z^{\frac{s+\lambda}{\mu} - 2}} = \frac{\mu z X'(z) + (s + \lambda) X(z)}{\mu \beta(s + \lambda - \lambda z)} z,$$

equation (1.92) can be rewritten as

$$
\begin{aligned}
\mu[y\beta(s+\lambda-\lambda z)-z]X'(z) &= [s+\lambda-\lambda y\beta(s+\lambda-\lambda z)]X(z) \\
&+ \left[x_1\frac{\lambda y\beta(s+\lambda)}{s+\lambda}-\frac{v(z)}{z}\right] \\
&\times \ \mu\beta(s+\lambda-\lambda z). \quad (1.94)
\end{aligned}
$$

Consider the coefficient $f(z) = y\beta(s+\lambda-\lambda z)-z$. It is easy to see that

- $f(0) = y\beta(s+\lambda) \geq 0$,

- $f(1) = y\beta(s)-1 < 0$ (if $(s,y) \neq (0,1)$),

- $f''(z) = \lambda^2 y\beta''(s+\lambda-\lambda z) \geq 0$, i.e. function $f(z)$ is convex.

Thus in the case $(s,y) \neq (0,1)$ function $f(z)$ has exactly one root in the interval $[0,1]$. It is well known that this root $z = \pi_\infty(s,y)$ is $\mathrm{E}e^{-sL_\infty}y^{I_\infty}$, where $L_\infty$ and $I_\infty$ are the length of the busy period and the number of customers served during this busy period respectively, in the standard $M/G/1$ queue.

Since $X(0) = \frac{\mu\beta(s+\lambda)}{s+\lambda}x_1$, for $0 \leq z < \pi_\infty(s,y)$ the solution of equation (1.94) is

$$
\begin{aligned}
X(z) &= \frac{\mu\beta(s+\lambda)}{s+\lambda}x_1 + e(s,z,y) \\
&\times \int_0^z \frac{x_1\beta(s+\lambda)-\frac{v(u)}{u}\beta(s+\lambda-\lambda u)}{[y\beta(s+\lambda-\lambda u)-u]\,e(s,u,y)}du, \quad (1.95)
\end{aligned}
$$

where the function $e(s,z,y)$ is given by (1.86).

As $u \to \pi_\infty(s,y)-0$ the integrand in the right-hand side of (1.86), is equivalent to

$$
\frac{1}{\pi_\infty(s,y)-u}.
$$

Thus the integral

$$
\int_0^{\pi_\infty(s,y)} \frac{s+\lambda-\lambda y\beta(s+\lambda-\lambda u)}{y\beta(s+\lambda-\lambda u)-u}du
$$

diverges, i.e. as $z \to \pi_\infty(s,y)-0$ the function $e(s,z,y) \to +\infty$. On the other hand, $X(\pi_\infty(s,y)) < \infty$. Therefore, the integral in the right-hand side of (1.95) must tend to 0 as $z \to \pi_\infty(s,y)-0$.

Because integrals

$$\int\limits_{0}^{\pi_\infty(s,y)} \frac{1}{[y\beta(s+\lambda-\lambda u)-u]\,e(s,u,y)}\,du$$

and

$$\int\limits_{0}^{\pi_\infty(s,y)} \frac{\frac{v(u)}{u}\beta(s+\lambda-\lambda u)}{[y\beta(s+\lambda-\lambda u)-u]\,e(s,u,y)}\,du$$

are finite, we get:

$$x_1 = \frac{\displaystyle\int\limits_{0}^{\pi_\infty(s,y)} \frac{\frac{v(u)}{u}\beta(s+\lambda-\lambda u)}{[y\beta(s+\lambda-\lambda u)-u]e(s,u,y)}\,du}{\displaystyle\int\limits_{0}^{\pi_\infty(s,y)} \frac{\beta(s+\lambda)}{[y\beta(s+\lambda-\lambda u)-u]e(s,u,y)}\,du}$$

$$= \frac{\displaystyle\sum_{k=1}^{\infty} v_k \cdot \int\limits_{0}^{\pi_\infty(s,y)} \frac{u^{k-1}\beta(s+\lambda-\lambda u)}{[y\beta(s+\lambda-\lambda u)-u]e(s,u,y)}\,du}{\displaystyle\int\limits_{0}^{\pi_\infty(s,y)} \frac{\beta(s+\lambda)}{[y\beta(s+\lambda-\lambda u)-u]e(s,u,y)}\,du}.$$

Thus the $k$th element of the first column of matrix $[\mathbf{I} - \mathbf{A}(s,y)]^{-1}$ is

$$\frac{\displaystyle\int\limits_{0}^{\pi_\infty(s,y)} \frac{u^{k-1}\beta(s+\lambda-\lambda u)}{[y\beta(s+\lambda-\lambda u)-u]e(s,u,y)}\,du}{\displaystyle\int\limits_{0}^{\pi_\infty(s,y)} \frac{\beta(s+\lambda)}{[y\beta(s+\lambda-\lambda u)-u]e(s,u,y)}\,du}.$$

which immediately implies equation (1.85).                        $\square$

For an ordinary busy period (which corresponds to the case $k = 1$) formulas (1.85), (1.87) and (1.88) can be simplified to get

$$\frac{\mu}{s+\lambda-\lambda \mathrm{E}e^{-sL}y^I} = \int\limits_{0}^{\pi_\infty(s,y)} \frac{1}{y\beta(s+\lambda-\lambda u)-u}$$

$$\times \exp\left\{-\frac{1}{\mu}\int_0^u \frac{s+\lambda-\lambda y\beta(s+\lambda-\lambda v)}{y\beta(s+\lambda-\lambda v)-v}\,dv\right\}\,du. \qquad (1.96)$$

$$\mathrm{E}L = \frac{1}{\lambda(1-\rho)}\exp\left\{\frac{\lambda}{\mu}\int\limits_{0}^{1}\frac{1-k(u)}{k(u)-u}\,du\right\} - \frac{1}{\lambda}, \qquad (1.97)$$

$$\mathrm{E}I = \frac{1}{1-\rho} \exp\left\{\frac{\lambda}{\mu} \int_0^1 \frac{1-k(u)}{k(u)-u} du\right\}. \qquad (1.98)$$

It should be noted that, as implied by the general theory of Markov chains, finiteness of $L$ and $I$ is connected with recurrence of the embedded Markov chain and finiteness of $\mathrm{E}L$ and $\mathrm{E}I$ is equivalent to positive recurrence of the chain. Besides, $\mathrm{E}I = 1/\pi_0$, where $\{\pi_n\}$ is the stationary distribution of the embedded Markov chain. Thus the above conclusions and equation (1.98) are consistent with the results obtained in Section 1.3. Besides, the analysis of the embedded Markov chain implies that if $\rho = 1$ then $I$ and $L$ are finite with probability 1 iff $\mu \geq 2\beta_1/\beta_2$ (although in section 3 we investigated this problem only in the case of exponential service, the result can be generalized easily).

The above expressions (even those for $\mathrm{E}L$ and $\mathrm{E}I$) are not convenient for computational purposes because of the presence of the Laplace transform $\beta(s)$. For this reason we will obtain the following insensitive estimates (which require only the mean value $\beta_1$ of the service time distribution).

**Theorem 1.17** *For every service time distribution*

$$\mathrm{E}I \leq \frac{1}{1-\rho} \exp\left\{\frac{\lambda}{\mu} \int_0^1 \frac{1-e^{-\rho(1-u)}}{e^{-\rho(1-u)}-u} du\right\}$$

*and if $B(x)$ is NBUE, then*

$$\mathrm{E}I \geq (1-\rho)^{-\frac{\lambda}{\mu}-1}.$$

*Proof.* Note that

$$\mathrm{E}I = \frac{1}{(1-\rho)\widehat{q}_0} = \frac{1}{(1-\rho)\widehat{p}(0)},$$

where $\widehat{p}(z) = \sum_{n=0}^{\infty} z^n \widehat{q}_n$ is the generating function of the conditional distribution of the number of customers in orbit given that the server is free, and the theorem follows from Theorem 1.15 of Section 1.5. $\qquad \square$

Using Theorem 1.14 of Section 1.5 we can get a more general result.

**Theorem 1.18** *Suppose we have two M/G/1 retrial queues with parameters $\lambda^{(1)}$, $\mu^{(1)}$, $B^{(1)}(x)$ and $\lambda^{(2)}$, $\mu^{(2)}$, $B^{(2)}(x)$. If $\lambda^{(1)} \leq \lambda^{(2)}$, $\mu^{(1)} \leq \mu^{(2)}$, $B^{(1)}(x) \leq_L B^{(2)}(x)$, then for the mean number* *of customers served during a busy period in these systems,* $EI'$ *and* $EI''$, *we have:*

$$EI' \leq EI''.$$

*Proof.* Under the assumptions of the theorem, $\beta_1^{(1)} \leq \beta_1^{(2)}$ and thus $\rho^{(1)} \leq \rho^{(2)}$, which implies that

$$\frac{1}{1 - \rho^{(1)}} \leq \frac{1}{1 - \rho^{(2)}}.$$

Besides, direct application of Theorem 1.14 of Section 1.5 gives

$$\widehat{q}_0^{(1)} \geq \widehat{q}_0^{(2)},$$

or equivalently

$$\frac{1}{\widehat{q}_0^{(1)}} \leq \frac{1}{\widehat{q}_0^{(2)}}.$$

Because

$$EI' = \frac{1}{1 - \rho^{(1)}} \cdot \frac{1}{\widehat{q}_0^{(1)}}, \ EI'' = \frac{1}{1 - \rho^{(2)}} \cdot \frac{1}{\widehat{q}_0^{(2)}},$$

the theorem follows immediately. □

The above approach allows investigation of the busy period in more detail. Introduce, for example, the following random variables:

(1) $A^{(k)}$ as the total length of all idle periods during a $k$-busy period;

(2) $B^{(k)}$ as the total length of all service times during this $k$-busy period.

Let $\pi^{(k)}(a, b, y) = Ee^{-aA^{(k)}}e^{-bB^{(k)}}y^{I^{(k)}}$. Since

$$L^{(k)} = A^{(k)} + B^{(k)},$$

we can express the functions $\pi^{(k)}(s, y) = Ee^{-sL^{(k)}}y^{I^{(k)}}$ as follows:

$$\pi^{(k)}(s, y) = \pi^{(k)}(s, s, y).$$

The functions $\pi^{(k)}(a, b, y)$ satisfy the following set of equations which generalize equations (1.89) and (1.90):

$$\begin{aligned}
\pi^{(k)}(a, b, y) &= y \sum_{n=0}^{\infty} k_n(b) \left[ \frac{\lambda}{a + \lambda + (k - 1 + n)\mu} \pi^{(k+n)}(a, b, y) \right. \\
&+ \left. \frac{(k - 1 + n)\mu}{a + \lambda + (k - 1 + n)\mu} \pi^{(k+n-1)}(a, b, y) \right], \\
& \quad \text{if } k \geq 2,
\end{aligned}$$

$$\pi^{(1)}(a,b,y) = yk_0(b) + y\sum_{n=1}^{\infty} k_n(b)\left[\frac{\lambda}{a+\lambda+n\mu}\pi^{(n+1)}(a,b,y)\right.$$

$$+ \left.\frac{n\mu}{a+\lambda+n\mu}\pi^{(n+1)}(a,b,y)\right].$$

As before, rewrite these equations in matrix form:

$$(\pi^1(a,b,y),\ldots)^T = \mathbf{A}(a,b,y)\cdot(\pi^1(a,b,y),\ldots)^T$$
$$+ (yk_0(b),0,\ldots)^T.$$

Consider the adjoint equation

$$\mathbf{x}\cdot[\mathbf{I}-\mathbf{A}(a,b,y)] = \mathbf{v},$$

or in coordinatewise form

$$x_1 - x_1\frac{\mu y}{a+\lambda+\mu}k_1(b) - x_2\frac{\mu yk_0(b)}{a+\lambda+\mu} = v_1$$

$$x_i - \sum_{n=1}^{i+1}x_n\left\{\frac{\lambda yk_{i-n}(b)}{a+\lambda+(i-1)\mu} + \frac{i\mu yk_{i-n+1}(b)}{a+\lambda+i\mu}\right\} = v_i, \ i\geq 2.$$

For generating functions $x(z) = \sum_{i=1}^{\infty}x_iz^i$, $v(z) = \sum_{i=1}^{\infty}v_iz^i$ these equations give:

$$x(z) = \frac{\lambda yz}{\mu}X(z) + yzX'(z)$$

$$+ v(z) - x_1z\frac{\lambda yk_0(b)}{a+\lambda},$$

where

$$X(z) = \sum_{i=0}^{\infty}\frac{\mu z^i}{a+\lambda+i\mu}\sum_{n=1}^{i+1}x_nk_{i-n+1}(b).$$

As before, one can easily show that

$$X(z) = z^{-\frac{a+\lambda}{\mu}}\int_0^z x(u)\beta(b+\lambda-\lambda u)u^{\frac{a+\lambda}{\mu}-2}du,$$

which yields the following equation:

$$\mu[y\beta(b+\lambda-\lambda z)-z]X'(z) = [a+\lambda-\lambda y\beta(b+\lambda-\lambda z)]X(z)$$

$$+ \left[x_1\frac{\lambda y\beta(b+\lambda)}{a+\lambda} - \frac{v(z)}{z}\right]\mu\beta(b+\lambda-\lambda z).$$

Since $X(0) = \frac{\mu\beta(b+\lambda)}{a+\lambda}x_1$, for $0 \le z < \pi_\infty(b,y)$ we have:

$$X(z) = \frac{\mu\beta(b+\lambda)}{a+\lambda}x_1 + e(a,b,z,y)$$
$$\times \int_0^z \frac{x_1\beta(b+\lambda) - \frac{v(u)}{u}\beta(b+\lambda-\lambda u)}{[y\beta(b+\lambda-\lambda u) - u]\,e(a,b,u,y)}\,du,$$

where

$$e(a,b,z,y) = \exp\left\{ \frac{1}{\mu}\int_0^z \frac{a+\lambda-\lambda y\beta(b+\lambda-\lambda u)}{y\beta(b+\lambda-\lambda u) - u}\,du \right\}.$$

The function $e(a,b,z,y) \to +\infty$ as $z \to \pi_\infty(b,y) - 0$. Thus

$$x_1 = \frac{\displaystyle\int_0^{\pi_\infty(b,y)} \frac{\frac{v(u)}{u}\beta(b+\lambda-\lambda u)}{[y\beta(b+\lambda-\lambda u)-u]e(a,b,u,y)}\,du}{\displaystyle\int_0^{\pi_\infty(b,y)} \frac{\beta(b+\lambda)}{[y\beta(b+\lambda-\lambda u)-u]e(a,b,u,y)}\,du}$$

$$= \frac{\displaystyle\sum_{k=1}^\infty v_k \int_0^{\pi_\infty(b,y)} \frac{u^{k-1}\beta(b+\lambda-\lambda u)}{[y\beta(b+\lambda-\lambda u)-u]e(a,b,u,y)}\,du}{\displaystyle\int_0^{\pi_\infty(b,y)} \frac{\beta(b+\lambda)}{[y\beta(b+\lambda-\lambda u)-u]e(a,b,u,y)}\,du},$$

which implies that

$$\pi^{(k)}(a,b,y) = \frac{\displaystyle\int_0^{\pi_\infty(b,y)} \frac{yu^{k-1}\beta(b+\lambda-\lambda u)}{[y\beta(b+\lambda-\lambda u)-u]e(a,b,u,y)}\,du}{\displaystyle\int_0^{\pi_\infty(b,y)} \frac{1}{[y\beta(b+\lambda-\lambda u)-u]e(a,b,u,y)}\,du}.$$

## 1.7 The number of customers in the system in the nonstationary regime

In this section we consider the transient distribution

$$p_{0ni}(t) = \mathrm{P}\{C(t)=0, N(t)=n, I(t)=i\},$$
$$p_{1ni}(t,x) = \frac{d}{dx}\mathrm{P}\{C(t)=1, N(t)=n, I(t)=i, \xi(t) < x\}$$

of the process $(C(t),\xi(t),N(t),I(t))$ in nonstationary regime. To avoid unnecessary complication with minor details we will assume that the system was empty at the initial epoch $t_0 = 0$, i.e. $C(0) = 0$, $N(0) = 0$, $I(0) = 0$.

**Theorem 1.19** *Laplace transforms*

$$p_0^*(s, z, y) \equiv \int_0^\infty e^{-st} p_0(t, z, y) dt,$$

$$p_1^*(s, z, y, x) \equiv \int_0^\infty e^{-st} p_1(t, z, y, x) dt$$

*of generating functions*

$$p_0(t, z, y) \equiv \sum_{n=0}^\infty z^n \sum_{i=0}^\infty y^i p_{0ni}(t),$$

$$p_1(t, z, y, x) \equiv \sum_{n=0}^\infty z^n \sum_{i=0}^\infty y^i p_{1ni}(t, x)$$

*are given by the following formulas:*

$$p_0^*(s, z, y) = \begin{cases} \int_z^{\pi_\infty(s,y)} \dfrac{1}{\mu[y\beta(s+\lambda-\lambda u)-u]} \\ \qquad \times \exp\left\{\int_u^z \dfrac{s+\lambda-\lambda y\beta(s+\lambda-\lambda v)}{\mu[y\beta(s+\lambda-\lambda v)-v]} dv\right\} du, \\ \quad \text{if } z \neq \pi_\infty(s, y) \\[2ex] \dfrac{1}{s+\lambda-\lambda\pi_\infty(s,y)}, \quad \text{if } z = \pi_\infty(s, y). \end{cases} \qquad (1.99)$$

$$p_1^*(s, z, y, x) = \frac{[(s+\lambda-\lambda z)p_0^*(s, z, y) - 1]}{y\beta(s+\lambda-\lambda z) - z}$$
$$\times [1 - B(x)] e^{-(s+\lambda-\lambda z)x}. \qquad (1.100)$$

*Proof.* **A.** First we investigate the process only at departure times during the busy period. Let $s > 0$. Consider an auxiliary Poisson flow of 'catastrophes' with rate $s$. Denote by $\pi_{ni}^{(k)}(s)$ the probability that at the time $\eta_i$ of the $i$th departure from the beginning of a $k$-busy period there are $n \geq 0$ customers in orbit, no 'catastrophes' have occurred until this time and the $k$-busy period did not end before time $\eta_i$.

The probabilities $\pi_{ni}^{(k)}(s)$ can be found with the help of the following recursive formulas:

$$\pi_{n1}^{(k)}(s) = k_{n-k+1}(s), \ n \geq 0 \qquad (1.101)$$

$$\pi_{ni}^{(k)}(s) = \sum_{m=1}^{n} \pi_{m,i-1}^{(k)}(s) \frac{\lambda}{s+\lambda+m\mu} k_{n-m}(s)$$

$$+ \sum_{m=1}^{n+1} \pi_{m,i-1}^{(k)}(s) \frac{m\mu}{s+\lambda+m\mu} k_{n-m+1}(s)$$

$$n \geq 0, \ i \geq 2. \tag{1.102}$$

As an example, we give a proof of equation (1.101).

The probability $\pi_{n1}^{(k)}(s)$ can be viewed as the probability that at time of the first departure there are $n$ customers in orbit and 'catastrophes' did not occur. This event can occur only if during the first service period:

(1) 'catastrophes' did not occur;

(2) exactly $n - k + 1$ new primary calls arrived.

The probability of this event is $k_{n-k+1}(s)$.

To solve equations (1.101), (1.102) we introduce the generating function

$$f^{(k)}(s,z,y) = \sum_{n=1}^{\infty} z^n \sum_{i=1}^{\infty} y^i \frac{\pi_{ni}^{(k)}(s)}{s+\lambda+n\mu}. \tag{1.103}$$

This series converges if $s$ and $y$ are not simultaneously equal to 0 and 1 respectively (below, this condition will be assumed to be valid).

Then equations (1.101), (1.102) give:

$$\mu \left[ y\beta(s+\lambda-\lambda z) - z \right] \frac{\partial f^{(k)}(s,z,y)}{\partial z}$$

$$= \left[ s+\lambda - \lambda y\beta(s+\lambda-\lambda z) \right] f^{(k)}(s,z,y)$$

$$+\pi^{(k)}(s,y) - yz^{k-1}\beta(s+\lambda-\lambda z). \tag{1.104}$$

Since $f^{(k)}(s,0,y) = 0$, for $z \in [0; \pi_\infty(s,y))$ the solution of equation (1.104) is given by the formula:

$$f^{(k)}(s,z,y) = e(s,z,y) \int_0^z \frac{\pi^{(k)}(s,y) - yu^{k-1}\beta(s+\lambda-\lambda u)}{\mu \left[ y\beta(s+\lambda-\lambda u) - u \right] e(s,u,y)} du, \tag{1.105}$$

where the function $e(s,z,y)$ is given by (1.86).

If $s > 0$ then $\lim_{z \to \pi_\infty(s,y)-0} e(s,z,y) = +\infty$. On the other hand $f^{(k)}(s,\pi_\infty(s,y),y) < \infty$. Thus the integral in the right-hand side of equation (1.105) must tend to zero as $z \to \pi_\infty(s,y) - 0$, i.e.

$$\int_0^{\pi_\infty(s,y)} \frac{\pi^{(k)}(s,y) - yu^{k-1}\beta(s+\lambda-\lambda u)}{\mu\left[y\beta(s+\lambda-\lambda u) - u\right]e(s,u,y)} du = 0. \qquad (1.106)$$

Using (1.106) we can rewrite (1.105) as follows:

$$\begin{aligned}
f^{(k)}(s,y,u) &= \int_{\pi_\infty(s,y)}^z \frac{\pi^{(k)}(s,y) - yu^{k-1}\beta(s+\lambda-\lambda u)}{\mu\left[y\beta(s+\lambda-\lambda u) - u\right]} \\
&\times \exp\left\{\int_u^z \frac{s+\lambda-\lambda y\beta(s+\lambda-\lambda v)}{\mu[y\beta(s+\lambda-\lambda v) - v]} dv\right\} du, \\
& \qquad 0 \le z < \pi_\infty(s,y).
\end{aligned} \qquad (1.107)$$

Consider now the interval $\pi_\infty(s,y) < z \le 1$. For this $z$ the coefficient $y\beta(s+\lambda-\lambda z) - z \ne 0$ (in fact it is negative) and thus,

$$\begin{aligned}
f^{(k)}(s,z,y) &= e_1(s,z,y) \cdot \Big\{ C \\
&+ \int_1^z \frac{\pi^{(k)}(s,y) - yu^{k-1}\beta(s+\lambda-\lambda u)}{\mu\left[y\beta(s+\lambda-\lambda u) - u\right]e_1(s,u,y)} du \Big\}, \\
& \qquad \pi_\infty(s,y) < z \le 1,
\end{aligned} \qquad (1.108)$$

where

$$\begin{aligned}
e_1(s,z,y) &= \exp\left\{\int_1^z \frac{s+\lambda-\lambda y\beta(s+\lambda-\lambda u)}{\mu[y\beta(s+\lambda-\lambda u) - u]} du\right\}, \\
& \qquad \pi_\infty(s,y) < z \le 1.
\end{aligned}$$

As $z \to \pi_\infty(s,y) + 0$ the function $e_1(s,z,y) \to +\infty$. On the other hand $f^{(k)}(s,\pi_\infty(s,y),y) < \infty$. Thus,

$$C = \int_{\pi_\infty(s,y)}^1 \frac{\pi^{(k)}(s,y) - yu^{k-1}\beta(s+\lambda-\lambda u)}{\mu\left[y\beta(s+\lambda-\lambda u) - u\right]e_1(s,u,y)} du,$$

which allows us to transform (1.108) to the same form as (1.107). Thus we can guarantee that (1.107) holds for all $z \ne \pi_\infty(s,y)$.

For $z = \pi_\infty(s, y)$ we have directly from (1.104) :

$$f^{(k)}(s, z, y) = \frac{\pi_\infty^k(s, y) - \pi^{(k)}(s, y)}{s + \lambda - \lambda \pi_\infty(s, y)}.$$

**B.** We next investigate the process at an arbitrary time $t$ during a busy period.

Assume that a $k$-busy period starts at time $t = 0$. Define:

$$P_{0ni}^{(k)}(t) \equiv \mathrm{P}\left\{ L^{(k)} > t, C(t) = 0, N(t) = n, I(t) = i \right\},$$

$$P_{1ni}^{(k)}(t, x)$$
$$\equiv \frac{\partial}{\partial x}\mathrm{P}\left\{ L^{(k)} > t, C(t) = 1, \xi(t) < x, N(t) = n, I(t) = i \right\},$$

and introduce the corresponding Laplace transforms:

$$\varphi_{0ni}^{(k)}(s) = \int\limits_0^\infty e^{-st} P_{0ni}(t)dt, \quad \varphi_{1ni}^{(k)}(s, x) = \int\limits_0^\infty e^{-st} P_{1ni}(t, x)dt.$$

Using the language of supplementary events we can think of $s\varphi_{0ni}^{(k)}(s)$ as the probability that at time $\tau_s$, when the first 'catastrophe' occurred, the server was free, there were $n$ customers in orbit, exactly $i$ customers had completed service by this time, and the $k$-busy period had not expired. Similarly we can think of $s\varphi_{1ni}^{(k)}(s, x)dx$ as the probability that at time $\tau_s$, when the first 'catastrophe' occurred, the server was busy, the elapsed service time $\xi \in (x, x + dx)$, there were $n$ customers in orbit, exactly $i$ customers had completed service, and the $k$-busy period had not expired.

Taking into account the probabilistic meaning of Laplace transforms $\pi_{ni}^{(k)}(s)$ we have the following equations:

$$\varphi_{0ni}^{(k)}(s) = \frac{\pi_{ni}^{(k)}(s)}{s + \lambda + n\mu}, \quad n \geq 1, i \geq 1,$$

$$\varphi_{1ni}^{(k)}(s, x) = \left\{ \sum_{m=1}^n \pi_{mi}^{(k)}(s)\frac{\lambda}{s + \lambda + m\mu} \cdot \frac{(\lambda x)^{n-m}}{(n-m)!}e^{-\lambda x} \right.$$

$$\left. + \sum_{m=1}^{n+1} \pi_{mi}^{(k)}(s)\frac{m\mu}{s + \lambda + m\mu}\frac{(\lambda x)^{n-m+1}}{(n-m+1)!}e^{-\lambda x} \right\}$$

$$\times \; [1 - B(x)]e^{-sx}, \quad n \geq 0, i \geq 1,$$

$$\varphi_{1n0}^{(k)}(s,x) = \frac{(\lambda x)^{n-k+1}}{(n-k+1)!} e^{-\lambda x} \cdot [1 - B(x)] e^{-sx}, \; n \geq 0, i = 0.$$

To solve these equations introduce generating functions

$$\varphi_0^{(k)}(s,z,y) = \sum_{i=1}^{\infty} y^i \sum_{n=1}^{\infty} z^n \varphi_{0ni}^{(k)}(s),$$

$$\varphi_1^{(k)}(s,z,y,x) = \sum_{i=0}^{\infty} y^i \sum_{n=0}^{\infty} z^n \varphi_{1ni}^{(k)}(s,x).$$

Then the above equations for $\varphi_{0ni}^{(k)}(s)$, $\varphi_{1ni}^{(k)}(s)$ become

$$\varphi_0^{(k)}(s,z,y) = f^{(k)}(s,z,y), \tag{1.109}$$

where the function $f^{(k)}(s,z,y)$ was introduced earlier by (1.103), and

$$\varphi_1^{(k)}(s,z,y,x) = \left[ z^{k-1} + \lambda \varphi_0^{(k)}(s,z,y) + \mu \frac{\partial \varphi_0^{(k)}(s,z,y)}{\partial z} \right]$$
$$\times \; (1 - B(x)) e^{-(s+\lambda-\lambda z)x}.$$

With the help of (1.104), this equation for $\varphi_1^{(k)}(s,z,y,x)$ can be reduced to the following form:

$$\varphi_1^{(k)}(s,z,y,x) = \frac{\pi^{(k)}(s,y) - z^k + (s+\lambda-\lambda z)\varphi_0^{(k)}(s,z,y)}{y\beta(s+\lambda-\lambda z) - z}$$
$$\times \; (1 - B(x)) e^{-(s+\lambda-\lambda z)x}. \tag{1.110}$$

Equations (1.107), (1.109), (1.110) fully describe the process during a $k$-busy period.

**C.** Now we can get final formulas for $p_0^*(s,z,y)$ and $p_1^*(s,z,y)$. With this goal we:

(1) introduce a flow of 'catastrophes' with rate $s > 0$,

(2) paint customers in the system 'red' with probability $z$ (and 'white' with probability $1 - z$),

(3) paint served customers 'green' with probability $y$ (and 'yellow' with probability $1 - y$).

Denote by $A$ the event {at moment $\tau_s$ when the first 'catastrophe' has occurred the server was free, all customers in the system are 'red', all customers served by this moment are 'green'} and define a busy cycle as the interval between two successive busy period completions. Let $\eta(t)$ be the number of busy periods (or equivalently, busy cycles) until time $t$.

Then

$$sp_0^*(s, z, y) = P\{A\} = \sum_{n=0}^{\infty} P\{A; \eta(\tau_s) = n\}.$$

But $P\{A; \eta(\tau_s) = n\}$ equals the probability that during the first $n$ busy cycles catastrophes did not occur, only 'green' customers were served; at time $\tau_s$ when the first catastrophe occurred, only red customers were in the system, the busy cycle had not expired, only 'green' customers had been served. This probability can be written as

$$\left[\frac{\lambda}{s+\lambda}\pi^{(1)}(s, y)\right]^n \cdot \left[\frac{s}{s+\lambda} + \frac{\lambda}{s+\lambda}s\varphi_0^{(1)}(s, z, y)\right].$$

Thus

$$p_0^*(s, z, y) = \frac{1 + \lambda\varphi_0^{(1)}(s, z, y)}{s + \lambda - \lambda\pi^{(1)}(s, y)}. \tag{1.111}$$

Similarly

$$\begin{aligned}
p_1^*(s, z, y, x) &= \frac{\lambda\varphi_1^{(1)}(s, z, y, x)}{s + \lambda - \lambda\pi^{(1)}(s, y)} \\
&= \frac{\lambda\pi^{(1)}(s, y) - \lambda z + (s + \lambda - \lambda z)\lambda\varphi_0^{(1)}(s, z, y)}{[y\beta(s + \lambda - \lambda z) - z][s + \lambda - \lambda\pi^{(1)}(s, y)]} \\
&\quad \times (1 - B(x))e^{-(s+\lambda-\lambda z)x} \\
&= \frac{(s + \lambda - \lambda z) \cdot p_0^*(s, z, y) - 1}{y\beta(s + \lambda - \lambda z) - z} \cdot (1 - B(x))e^{-(s+\lambda-\lambda z)x}.
\end{aligned}$$

**D.** To complete the proof we must transform formula (1.111) for $p_0^*(s, z, y)$ to the form (1.99).

In the case $z = \pi_\infty(s, y)$ we have:

$$\begin{aligned}
p_0^*(s, \pi_\infty(s, y), y) &= \frac{1 + \lambda f^{(1)}(s, \pi_\infty(s, y), y)}{s + \lambda - \lambda\pi^{(1)}(s, y)} \\
&= \frac{1 + \lambda\dfrac{\pi_\infty(s, y) - \pi^{(1)}(s, y)}{s + \lambda - \lambda\pi_\infty(s, y)}}{s + \lambda - \lambda\pi^{(1)}(s, y)} \\
&= \frac{1}{s + \lambda - \lambda\pi^{(1)}(s, y)}.
\end{aligned}$$

In the case $z \neq \pi_\infty(s,y)$ we first note that

$$\int\limits_z^{\pi_\infty(s,y)} d_u \exp\left\{\int\limits_u^z \frac{s+\lambda-\lambda y\beta(s+\lambda-\lambda v)}{\mu[y\beta(s+\lambda-\lambda v)-v]}dv\right\} = -1,$$

(where $d_u$ means that $u$ is considered as the variable of integration) or, equivalently,

$$\int\limits_z^{\pi_\infty(s,y)} \frac{s+\lambda-\lambda y\beta(s+\lambda-\lambda u)}{y\beta(s+\lambda-\lambda u)-u}$$

$$\times \exp\left\{\int\limits_u^z \frac{s+\lambda-\lambda y\beta(s+\lambda-\lambda v)}{\mu[y\beta(s+\lambda-\lambda v)-v]}dv\right\} du = \mu.$$

Taking into account (1.111), (1.109) and (1.107) we see that

$$\left[s+\lambda-\lambda\pi^{(1)}(s,y)\right]p_0^*(s,z,y)$$

equals

$$1+\lambda\pi^{(1)}(s,y)\cdot \int\limits_{\pi_\infty(s,y)}^z \frac{1}{\mu\left[y\beta(s+\lambda-\lambda u)-u\right]}\exp\{...\}\,du$$

$$-\int\limits_{\pi_\infty(s,y)}^z \frac{\lambda y\beta(s+\lambda-\lambda u)}{\mu\left[y\beta(s+\lambda-\lambda u)-u\right]}\exp\{...\}\,du$$

$$=1+\lambda\pi^{(1)}(s,y)\cdot \int\limits_{\pi_\infty(s,y)}^z \frac{1}{\mu\left[y\beta(s+\lambda-\lambda u)-u\right]}\exp\{...\}\,du$$

$$-1-(s+\lambda)\int\limits_{\pi_\infty(s,y)}^z \frac{1}{\mu\left[y\beta(s+\lambda-\lambda u)-u\right]}\exp\{...\}\,du$$

$$=\int\limits_z^{\pi_\infty(s,y)} \frac{s+\lambda-\lambda\pi^{(1)}(s,y)}{\mu\left[y\beta(s+\lambda-\lambda u)-u\right]}\exp\{...\}\,du,$$

which yields the desired result.                                    $\square$

Analysis of the queueing process at an arbitrary time during a busy period can be performed with the help of Kolmogorov equations for probabilities $P_{0ni}^{(k)}(t)$ and $P_{1ni}^{(k)}(t,x)$. These equations are as follows:

$$\frac{dP_{0ni}^{(k)}(t)}{dt} = -(\lambda + n\mu)P_{0ni}^{(k)}(t) + \int_0^\infty P_{1,n,i-1}^{(k)}(t,x)b(x)dx,$$

$$n \geq 1, i \geq 1,$$

$$\frac{d\Pi_i^{(k)}(t)}{dt} = \int_0^\infty P_{1,0,i-1}^{(k)}(t,x)b(x)dx, \; i \geq 1,$$

$$\frac{\partial P_{1ni}^{(k)}(t,x)}{\partial t} = -\left(\frac{\partial}{\partial x} + \lambda + b(x)\right)P_{1ni}^{(k)}(t,x) + \lambda P_{1,n-1,i}^{(k)}(t,x),$$

$$n \geq 0, i \geq 0,$$

$$P_{1ni}^{(k)}(t,0) = \lambda P_{0ni}^{(k)}(t) + (n+1)\mu P_{0,n+1,i}^{(k)}(t), \; n \geq 0, i \geq 1,$$

$$P_{1n0}^{(k)}(t,0) = 0,$$

where

$$\Pi_i^{(k)}(t)dt = P\left(L^{(k)} \in (t,t+dt), I^{(k)} = i\right),$$

$$\pi_i^{(k)}(s) = E\left(e^{-sL^{(k)}}; I^{(k)} = i\right).$$

The initial conditions are

$$P_{0ni}^{(k)}(0) = 0,$$

$$P_{1ni}^{(k)}(0,x) = \delta(x)\delta_{n,k-1}\delta_{i,0},$$

where $\delta(x)$ is the Dirac delta function and $\delta_{i,j}$ is Kronecker's delta.

For the Laplace transforms $\varphi_{0ni}^{(k)}(s)$ and $\varphi_{1ni}^{(k)}(s,x)$ these equations become:

$$(s + \lambda + n\mu)\varphi_{0ni}^{(k)}(s) = \int_0^\infty \varphi_{1,n,i-1}^{(k)}(s,x)b(x)dx, \; n \geq 1, i \geq 1,$$

$$\pi_i^{(k)}(s) = \int_0^\infty \varphi_{1,0,i-1}^{(k)}(s,x)b(x)dx, \; i \geq 1,$$

$$\frac{\partial \varphi_{1ni}^{(k)}(s,x)}{\partial x} = -(s + \lambda + b(x))\varphi_{1ni}^{(k)}(s,x)$$

$$+ \lambda \varphi_{1,n-1,i}^{(k)}(s,x) + \delta(x)\delta_{n,k-1}\delta_{i,0},$$

$$n \geq 0, i \geq 0,$$

$$\varphi_{1ni}^{(k)}(s,0) = \lambda \varphi_{0ni}^{(k)}(s) + (n+1)\mu \varphi_{0,n+1,i}^{(k)}(s),$$

$$n \geq 0, i \geq 1,$$

$$\varphi_{1n0}^{(k)}(s,0) = 0,$$

and for generating functions $\varphi_0^{(k)}(s,z,y)$, $\varphi_1^{(k)}(s,z,y,x)$ we have:

$$\mu z \frac{\partial \varphi_0^{(k)}(s,z,y)}{\partial z} = y \int_0^\infty \varphi_1^{(k)}(s,z,y,x)b(x)dx$$

$$- (s+\lambda)\varphi_0^{(k)}(s,z,y) - \pi^{(k)}(s,y), \quad (1.112)$$

$$\frac{\partial \varphi_1^{(k)}(s,z,y,x)}{\partial x} = -(s+\lambda-\lambda z+b(x))\varphi_1^{(k)}(s,z,y,x)$$

$$+ \delta(x)z^{k-1}, \quad (1.113)$$

$$\varphi_1^{(k)}(s,z,y,0) = \lambda\varphi_0^{(k)}(s,z,y) + \mu\frac{\partial \varphi_0^{(k)}(s,z,y)}{\partial z}. \quad (1.114)$$

From equation (1.113) we find the form of dependence of the function $\varphi_1^{(k)}(s,z,y,x)$ upon variable $x$:

$$\varphi_1^{(k)}(s,z,y,x) = (1-B(x))e^{-(s+\lambda-\lambda z)x}\left\{\varphi_1^{(k)}(s,z,y,0) + z^{k-1}\right\}. \quad (1.115)$$

This allows us to rewrite equation (1.112) as

$$\mu z \frac{\partial \varphi_0^{(k)}(s,z,y)}{\partial z} = y\beta(s+\lambda-\lambda z)\left\{\varphi_1^{(k)}(s,z,y,0) + z^{k-1}\right\}$$

$$- (s+\lambda)\varphi_0^{(k)}(s,z,y) - \pi^{(k)}(s,y). \quad (1.116)$$

Eliminating $\varphi_1^{(k)}(s,z,y,0)$ with the help of equation (1.114), we transform equations (1.115) and (1.116) to basic equations (1.110) and (1.104), which yield formulas (1.99) and (1.100) for functions $p_0^*(s,z,y)$ and $p_1^*(s,z,y,x)$.

It is worth noting that equation (1.85) for the joint distribution of the length of busy period and the number of customers served during the busy period can be obtained directly from analysis of the queueing process during a busy period as follows.

Since integrals

$$\int_0^{\pi_\infty(s,y)} \frac{1}{\mu\left[y\beta(s+\lambda-\lambda u) - u\right]e(s,u,y)}du$$

and

$$\int\limits_0^{\pi_\infty(s,y)} \frac{yu^{k-1}\beta(s+\lambda-\lambda u)}{\mu\left[y\beta(s+\lambda-\lambda u)-u\right]e(s,u,y)}\,du$$

are finite, we have from (1.106) that

$$\pi^{(k)}(s,y) = \frac{\displaystyle\int\limits_0^{\pi_\infty(s,y)} \frac{yu^{k-1}\beta(s+\lambda-\lambda u)}{\mu[y\beta(s+\lambda-\lambda u)-u]e(s,u,y)}\,du}{\displaystyle\int\limits_0^{\pi_\infty(s,y)} \frac{1}{\mu[y\beta(s+\lambda-\lambda u)-u]e(s,u,y)}\,du}.$$

Analysis of the queueing process at an arbitrary time during a busy period provides a technically convenient method for getting moments of random variables $L^{(k)}$ and $I^{(k)}$ in the case $\rho < 1$. Namely, putting $z = \pi_\infty(s,y)$ in equation (1.104) we have:

$$\pi^{(k)}(s,y) = [\pi_\infty(s,y)]^k - [s+\lambda-\lambda\pi_\infty(s,y)]\,f^{(k)}(s,\pi_\infty(s,y),y),$$

so that

$$\begin{aligned}
\frac{\pi^{(k)}(s,1)-1}{s} &= \frac{[\pi_\infty(s,1)]^k - 1}{s} \\
&\quad - \frac{s+\lambda-\lambda\pi_\infty(s,1)}{s} f^{(k)}(s,\pi_\infty(s,1),1).
\end{aligned}$$

Now let $s \to 0$. Taking into account the relations

$$\pi_\infty(0,1) = 1, \quad \frac{\partial\pi_\infty(0,1)}{\partial s} = -\mathrm{E}L_\infty = -\frac{\beta_1}{1-\rho},$$

we have:

$$\mathrm{E}L^{(k)} = \frac{k\beta_1}{1-\rho} + \frac{f^{(k)}(0,1,1)}{1-\rho}. \tag{1.117}$$

Similarly,

$$\begin{aligned}
\mathrm{E}I^{(k)} &= \lim_{y\to 1} \frac{\pi^{(k)}(0,y)-1}{y-1} \\
&= \frac{k}{1-\rho} + \frac{\lambda f^{(k)}(0,1,1)}{1-\rho}. \tag{1.118}
\end{aligned}$$

On the other hand putting $s=0, y=1$ in equation (1.104) we get the following differential equation for $\varphi_0^{(k)}(0,z,1)$:

$$\mu\left[k(z)-z\right]\frac{\partial\varphi_0^{(k)}(0,z,1)}{\partial z} = \lambda\left[1-k(z)\right]\varphi_0^{(k)}(0,z,1)+1-z^{k-1}k(z),$$

which implies that

$$\varphi_0^{(k)}(0, z, 1) = \int\limits_0^z \frac{1 - u^{k-1}k(u)}{\mu(k(u) - u)} \exp\left\{\frac{\lambda}{\mu} \int\limits_u^z \frac{1 - k(v)}{k(v) - v} dv\right\} du.$$

$$(1.119)$$

Equations (1.117), (1.118) and (1.119) immediately imply formulas (1.87) and (1.88) for the mean length of the $k$-busy period and the mean number of customers served during the $k$-busy period.

Now consider some results which follow from the above main theorem.

If we put $y = 1$, $z = 1$ in the main formula (1.99) we get the following formula for the Laplace transform $p_1^*(s) = \int_0^\infty e^{-st}p_1(t)dt$ of the nonstationary blocking probability $p_1(t) = 1 - p_0(t)$:

$$p_1^*(s) = \frac{1}{s} - \int\limits_1^{\pi_\infty(s)} \frac{1}{\mu\left[\beta\left(s + \lambda - \lambda u\right) - u\right]}$$

$$\times \exp\left\{\frac{1}{\mu}\int\limits_u^1 \frac{s + \lambda - \lambda\beta(s + \lambda - \lambda v)}{\beta(s + \lambda - \lambda v) - v} dv\right\} du.$$

Also, for the Laplace transform of the generating function of the number of customers in orbit we have

$$p^*(s, z) \equiv \int\limits_0^\infty e^{-st} \sum_{n=0}^\infty z^n P(N(t) = n)dt$$

$$= p_0^*(s, z, 1) + \int\limits_0^\infty p_1^*(s, z, 1, x)dx$$

$$= (1 - z)\frac{p_0^*(s, z, 1)}{\beta(s + \lambda - \lambda z) - z}$$

$$- \frac{1 - \beta(s + \lambda - \lambda z)}{\left[\beta(s + \lambda - \lambda z) - z\right]\left[s + \lambda - \lambda z\right]}.$$

Differentiating this relation at the point $z = 1$, we get the following formula for the Laplace transform of the mean queue length $EN(t)$ in nonstationary regime:

$$\int\limits_0^\infty e^{-st}EN(t)dt = \frac{\lambda}{s^2} - \frac{p_1^*(s)}{1 - \beta(s)}.$$

If $\rho < 1$ then there exists the limit

$$\lim_{t \to \infty} EN(t) = \frac{\lambda^2}{1 - \rho} \left( \frac{\beta_1}{\mu} + \frac{\beta_2}{2} \right).$$

On the other hand,

$$
\begin{aligned}
\lim_{t \to \infty} EN(t) &= \lim_{s \to 0} s \int_0^\infty e^{-st} EN(t) dt \\
&= \lim_{s \to 0} \left[ \frac{\lambda}{s} - p_1^*(s) \frac{s}{1 - \beta(s)} \right] \\
&= \lim_{s \to 0} \frac{s}{1 - \beta(s)} \cdot \lim_{s \to 0} \left[ \frac{\lambda(1 - \beta(s))}{s^2} - p_1^*(s) \right] \\
&= \frac{1}{\beta_1} \lim_{s \to 0} \left( \frac{\lambda(1 - \beta(s))}{s^2} - \frac{\rho}{s} \right) + \frac{1}{\beta_1} \lim_{s \to 0} \left[ \frac{\rho}{s} - p_1^*(s) \right] \\
&= \frac{1}{\beta_1} \cdot \frac{-\lambda \beta_2}{2} + \frac{1}{\beta_1} \lim_{s \to 0} \int_0^\infty e^{-st} \left[ \rho - p_1(t) \right] dt \\
&= -\frac{\lambda \beta_2}{2 \beta_1} + \frac{1}{\beta_1} \cdot \int_0^\infty \left[ \rho - p_1(t) \right] dt.
\end{aligned}
$$

This yields the following integral estimate of the proximity of transient and stationary blocking probabilities:

$$\int_0^\infty \left[ \rho - p_1(t) \right] dt = \frac{\rho^2/\mu + \lambda \beta_2/2}{1 - \rho}.$$

This results can also be rewritten in the following form which is useful for traffic measurement. Let

$$\xi_T = \frac{1}{T} \int_0^T C(t) dt.$$

Then,

$$
\begin{aligned}
E\xi_T &= \frac{1}{T} \int_0^T P(C(t) = 1) dt = \frac{1}{T} \int_0^T p_1(t) dt \\
&= \rho + \frac{1}{T} \int_0^T \left[ p_1(t) - \rho \right] dt
\end{aligned}
$$

$$= \rho + \frac{1}{T} \left\{ \int_0^\infty [p_1(t) - \rho]\, dt + o(1) \right\}$$

$$= \rho - \frac{1}{T} \cdot \frac{\rho^2/\mu + \lambda\beta_2/2}{1 - \rho} + o\left(\frac{1}{T}\right).$$

## 1.8 The waiting time process

Waiting time in retrial queues is a much more difficult to analyse than the number of customers in the system. Whereas the mean waiting time $W$ in the steady state can be easily obtained with the help of Little's formula:

$$W = \frac{EN(t)}{\lambda} = \frac{\lambda}{1 - \rho}\left(\frac{\beta_1}{\mu} + \frac{\beta_2}{2}\right),$$

even calculation of the variance of the waiting time is not trivial. Mainly this is connected with the fact that retrial queues are queues with random overtaking.

### 1.8.1 Waiting time of a tagged customer

Suppose that at the moment of departure of some customer (considered as the 0th customer) there are $n \geq 1$ other customers in the system. Tag one of them and denote by $T_n$ its waiting time.

**Theorem 1.20** *The Laplace transform of the waiting time of the tagged customer is given by the formula:*

$$Ee^{-sT_n} = \int_1^{\pi_\infty(s)} \frac{u^{n-1}}{\beta(s + \lambda - \lambda u) - u}$$

$$\times \quad \exp\left\{ \int_u^1 \frac{\mu + s + \lambda - \lambda\beta(s + \lambda - \lambda v)}{\mu[\beta(s + \lambda - \lambda v) - v]}\, dv \right\} du,$$

*where $\pi_\infty(s)$ is the Laplace transform of the length of a busy period in the corresponding standard $M/G/1$ queue.*

*Proof.* To find the Laplace transform of the random variable $T_n$ we use the method of collective marks. Fix some $s > 0$ and introduce an additional Poisson process with intensity $s$, which is independent of the functioning of the system. The events of this Poisson process will be called 'catastrophes'. Denote by $p_m^{(i)}(s)$ the probability of the following event: 'at the time of the $i$th departure there are $m \geq 1$ demands in the system including the tagged one and until this time no 'catastrophe' had occurred'. This event (denote it by $E$) can be realized only if:

(1) no 'catastrophe' occurred till the time of the $(i-1)$th departure, the tagged customer was in the system at that time jointly with $l - 1$ $(1 \leq l \leq m + 1)$ other customers (probability of this event is $p_l^{(i-1)}(s)$);

(2a) from the time of the $(i-1)$th departure until the beginning of the service of the $i$th customer no 'catastrophe' has occurred, the $i$th customer is a primary call (probability of this event is $\frac{\lambda}{s + \lambda + l\mu}$), during its service time no 'catastrophe' has occurred and exactly $m - l$ new primary calls arrived (probability of this event is $k_{m-l}(s)$); or

(2b) from the time of the $(i - 1)$th departure until the beginning of the service of the $i$th customer no 'catastrophe' occurred, the $i$th call is a repeated call, but not from the tagged customer (probability of this event is $\frac{(l-1)\mu}{s+\lambda+l\mu}$), during its service time no 'catastrophe' occurred and exactly $m - l + 1$ new primary calls arrived (probability of this event is $k_{m-l+1}(s)$).

Since event $E$ can be realized only in accordance with the above scheme (for some $l = 1, \ldots, m + 1$), this implies that the following main equation holds:

$$p_m^{(i)}(s) = \sum_{l=1}^{m+1} p_l^{(i-1)}(s) \frac{\lambda k_{m-l}(s) + (l - 1)\mu k_{m-l+1}(s)}{s + \lambda + l\mu}, \quad i \geq 1.$$

For the initial case $i = 0$ we obviously have

$$p_m^{(0)}(s) = \delta_{m,n}.$$

Introducing the generating function

$$p(s, z, y) = \sum_{i=0}^{+\infty} y^i \sum_{m=1}^{+\infty} z^m \frac{1}{s + \lambda + m\mu} p_m^{(i)}(s)$$

we can transform this equation to

$$\mu \left[ y\beta(s + \lambda - \lambda z) - z \right] p_z'(s, z, y)$$
$$= \left[ s + \lambda - \left( \lambda - \frac{\mu}{z} \right) y\beta(s + \lambda - \lambda z) \right] p(s, z, y)$$
$$- z^n. \tag{1.120}$$

The solution of equation (1.120), which we are interested in, has to be bounded on the set $|z| \leq 1$, $(s,y) \neq (0,1)$.

Equation (1.120) is very similar to the equation (1.94). Thus we can apply the method used in section 1.6 to solve equation (1.94).

First of all note that the coefficient of the derivative in the left-hand side of (1.120) is the same as the coefficient of the derivative in the left-hand side of (1.94). Thus this coefficient equals zero on the interval $0 \leq z \leq 1$ only at the point $z = \pi_\infty(s,y)$ (provided that $(s,y) \neq (0,1)$), where, as we defined in section 1.6, $\pi_\infty(s,y) = Ee^{-sL_\infty}y^{I_\infty}$ is the joint Laplace transform/generating function of the length of a busy period, $L_\infty$, and the number of calls served during this busy period, $I_\infty$, in the standard $M/G/1/\infty$ queue with arrival rate $\lambda$ and service time distribution function $B(x)$. If $0 \leq z < \pi_\infty(s,y)$, then the coefficient $y\beta(s+\lambda-\lambda z) - z$ is strictly positive. Hence the general solution of (1.120) on the interval $0 \leq z < \pi_\infty(s,y)$ is

$$p(s,z,y) = \widehat{e}(s,z,y)\left\{C - \int\limits_0^z \frac{u^n}{\mu\left[y\beta(s+\lambda-\lambda u) - u\right]\widehat{e}(s,u,y)}du\right\},$$
(1.121)

where

$$\widehat{e}(s,z,y) = \exp\left\{\int\limits_0^z \frac{s+\lambda - \left(\lambda - \frac{\mu}{u}\right)y\beta(s+\lambda-\lambda u)}{\mu[y\beta(s+\lambda-\lambda u) - u]}du\right\},$$
$$0 \leq z < \pi_\infty(s,y).$$
(1.122)

As $u \to \pi_\infty(s,y)$ the integrand in the right-hand side of (1.122) is infinite of order $1/(\pi_\infty(s,y) - u)$. Thus the integral

$$\int\limits_0^{\pi_\infty(s,y)} \frac{s+\lambda - \left(\lambda - \frac{\mu}{u}\right)y\beta(s+\lambda-\lambda u)}{y\beta(s+\lambda-\lambda u) - u}du$$

diverges, i.e. as $z \to \pi_\infty(s,y)$ function $\widehat{e}(s,z,y) \to +\infty$. On the other hand, $p(s,\pi_\infty(s,y),y) < \infty$. Therefore the expression in brackets in the right-hand side of (1.121) must tend to 0 as as $z \to \pi_\infty(s,y)$. This allows us to determine the constant $C$:

$$C = \int\limits_0^{\pi_\infty(s,y)} \frac{u^n}{\mu\left[y\beta(s+\lambda-\lambda u) - u\right]\widehat{e}(s,u,y)}du,$$

which in turn implies the following final formula for the generating function $p(s, z, y)$ on the interval $0 \leq z < \pi_\infty(s, y)$:

$$p(s, z, y) = \widehat{e}(s, z, y) \int_z^{\pi_\infty(s,y)} \frac{u^n}{\mu\left[y\beta(s + \lambda - \lambda u) - u\right]\widehat{e}(s, u, y)} du,$$

$$0 \leq z < \pi_\infty(s, y). \tag{1.123}$$

Consider now the case $\pi_\infty(s, y) < z \leq 1$. When $\pi_\infty(s, y) < z \leq 1$ the coefficient $y\beta(s + \lambda - \lambda z) - z$ is strictly negative and so

$$p(s, z, y) = \widehat{e}_1(s, z, y)\left\{ C_1 - \int_1^z \frac{u^n}{\mu[y\beta(s + \lambda - \lambda u) - u]\widehat{e}_1(s, u, y)} du \right\},$$

where

$$\widehat{e}_1(s, z, y) = \exp\left\{ \frac{1}{\mu} \int_1^z \frac{s + \lambda - \left(\lambda - \frac{\mu}{u}\right)y\beta(s + \lambda - \lambda u)}{y\beta(s + \lambda - \lambda u) - u} du \right\},$$

$$\pi_\infty(s, y) < z \leq 1.$$

In the same manner as above, it follows from these equations that for $\pi_\infty(s, y) < z \leq 1$

$$p(s, z, y) = \widehat{e}_1(s, z, y) \int_z^{\pi_\infty(s,y)} \frac{u^n}{\mu[y\beta(s + \lambda - \lambda u) - u]\widehat{e}_1(s, u, y)} du. \tag{1.124}$$

After some algebra, (1.123) and (1.124) can be reduced to the joint formula

$$p(s, z, y) = \int_z^{\pi_\infty(s,y)} \frac{zu^{n-1}}{\mu\left[y\beta(s + \lambda - \lambda u) - u\right]}$$

$$\times \exp\left\{ \int_u^z \frac{\mu + s + \lambda - \lambda y\beta(s + \lambda - \lambda v)}{\mu\left[y\beta(s + \lambda - \lambda v) - v\right]} dv \right\} du,$$

if $z \neq \pi_\infty(s, y)$.

For $z = \pi_\infty(s,y)$, we have directly from (1.120) that

$$p(s, \pi_\infty(s,y), y) = \frac{\pi_\infty^n(s,y)}{\mu + s + \lambda - \lambda\pi_\infty(s,y)}.$$

Now the Laplace transform of the waiting time $T_n$ of the tagged customer can be calculated as follows:

$\mathrm{E}\exp(-sT_n)$

$= \mathrm{P}(\text{during period } T_n \text{ 'catastrophes' did not occur})$

$= \displaystyle\sum_{i=0}^{\infty} \mathrm{P}(\text{during period } T_n \text{ 'catastrophes' did not occur}$

and exactly $i$ customers were served before the tagged one)

$= \displaystyle\sum_{i=0}^{\infty}\sum_{m=1}^{\infty} \mathrm{P}(\text{during period } T_n \text{ 'catastrophes' did not}$

occur, exactly $i$ customers were served before the tagged

one, there are $m - 1$ other customers in the queue

at the moment $T_n$)

$= \displaystyle\sum_{i=0}^{\infty}\sum_{m=1}^{\infty} p_m^{(i)}(s)\frac{\mu}{s + \lambda + m\mu} = \mu p(s,1,1)$

$$= \int\limits_{1}^{\pi_\infty(s)} \frac{u^{n-1}}{\beta(s + \lambda - \lambda u)}$$

$$\times \exp\left\{\int\limits_{u}^{1} \frac{\mu + s + \lambda - \lambda\beta(s + \lambda - \lambda v)}{\mu\left[\beta(s + \lambda - \lambda v) - v\right]}dv\right\}du,$$

where $\pi_\infty(s) = \pi_\infty(s,1) = \mathrm{E}\exp(-sL_\infty)$ is the Laplace transform of the length of a busy period in the corresponding standard $M/G/1/\infty$ queue. $\qquad\square$

It should be noted that the dependence of $\mathrm{E}\exp(-sT_n)$ on $n$ has the form

$$\mathrm{E}\exp(-sT_n) = \int\limits_{1}^{\pi_\infty(s)} u^{n-1}f(s,u)du. \qquad (1.125)$$

Another important fact is that Theorem 1.20 is valid without any restrictions on system parameters; in particular, it holds for $\rho \geq 1$.

*1.8.2 The virtual waiting time*

The virtual waiting time $W(t)$ at time $t$ is defined as the waiting time for a primary call which enters the system at this moment.

**Theorem 1.21** *In the steady state, the Laplace transform of the virtual waiting time,* $Ee^{-sW(t)}$, *equals*

$$1 - \rho + \frac{1-\rho}{s} \int\limits_{1}^{\pi_\infty(s)} \frac{\lambda(1-u)\left[\beta(\lambda - \lambda u) - \beta(s + \lambda - \lambda u)\right]}{[\beta(\lambda - \lambda u) - u]\,[\beta(s + \lambda - \lambda u) - u]}$$

$$\times \exp\left\{ \int\limits_{u}^{1} \frac{\mu + s + \lambda - \lambda v}{\mu[\beta(s + \lambda - \lambda v) - v]} dv \right\}$$

$$\times \exp\left\{ \int\limits_{1}^{u} \frac{\lambda(1-v)}{\mu[\beta(\lambda - \lambda v) - v]} dv \right\} du.$$

*Proof.* From the formula of total probability, $Ee^{-sW(t)}$ can be expressed as

$$E\left\{ e^{-sW(t)} | C(t) = 0 \right\} \cdot P(C(t) = 0)$$

$$+ \int\limits_{0}^{\infty} \sum_{n=0}^{\infty} E\left\{ e^{-sW(t)} | C(t) = 1, N(t) = n, \xi(t) = x \right\}$$

$$\times d_x P(C(t) = 1, N(t) = n, \xi(t) < x).$$

Obviously,

$$E\left\{ e^{-sW(t)} | C(t) = 0 \right\} = 1.$$

The term

$$E\left\{ e^{-sW(t)} | C(t) = 1, N(t) = n, \xi(t) = x \right\}$$

equals

$$\int\limits_{x}^{\infty} \frac{dB(y)}{1 - B(x)} e^{-s(y-x)} \sum_{m=0}^{\infty} \frac{(\lambda(y-x))^m}{m!} e^{-\lambda(y-x)} Ee^{-sT_{n+m+1}}.$$

Using formula (1.125) we can rewrite this expression as

$$\int_{1}^{\pi_\infty(s)} f(s, u) u^n \int_{x}^{\infty} \frac{dB(y)}{1 - B(x)} e^{-(s + \lambda - \lambda u)(y-x)} du.$$

Since in the steady state $P(C(t) = 0) = 1 - \rho$ and generating function $p_1(z, x) \equiv \sum_{n=0}^{\infty} z^n p_{1n}(x)$ is known from Theorem 1.2 (section 1.2), we get:

$$Ee^{-sW(t)} = 1 - \rho + \int_0^{\infty} \sum_{n=0}^{\infty} p_{1n}(x)dx$$

$$\times \int_1^{\pi_{\infty}(s)} du f(s, u) u^n \int_x^{\infty} \frac{dB(y)}{1 - B(x)} e^{-(s+\lambda-\lambda u)(y-x)}$$

$$= 1 - \rho$$

$$+ \int_0^{\infty} dx \int_1^{\pi_{\infty}(s)} du \int_x^{\infty} dB(y) \frac{f(s, u)p_1(u, x)}{1 - B(x)} e^{-(s+\lambda-\lambda u)(y-x)}$$

$$= 1 - \rho$$

$$+ \int_0^{\infty} dx \int_1^{\pi_{\infty}(s)} du \int_x^{\infty} dB(y) f(s, u) p_1(u, 0) e^{-(s+\lambda-\lambda u)y+sx}$$

$$= 1 - \rho$$

$$+ \int_1^{\pi_{\infty}(s)} du \int_0^{\infty} dB(y) \int_0^y dx f(s, u) p_1(u, 0) e^{-(s+\lambda-\lambda u)y+sx}$$

$$= 1 - \rho + \int_1^{\pi_{\infty}(s)} f(s, u) p_1(u, 0) \frac{\beta(\lambda - \lambda u) - \beta(s + \lambda - \lambda u)}{s} du.$$

Taking into account the explicit formulas for $f(s, u)$ and $p_1(u, x)$ the theorem follows.                                                                                                   □

It is easy to see that the above proof without any essential changes can be applied to obtain $Ee^{-sW(t)}$ in the nonstationary regime.

First, similarly to the stationary case, $Ee^{-sW(t)}$ can be written as

$$\int_0^{\infty} \sum_{n=0}^{\infty} E\left\{e^{-sW(t)}|C(t) = 1, N(t) = n, \xi(t) = x\right\} p_{1n}(t, x)dx$$

$$+ E\left\{e^{-sW(t)}|C(t) = 0\right\} p_0(t),$$

where

$$p_0(t) = P(C(t) = 0),$$

$$p_{1n}(t, x) = \frac{d}{dx}P(C(t) = 1, N(t) = n, \xi(t) < x).$$

Thus

$$Ee^{-sW(t)} = p_0(t) + \int\limits_0^\infty dx \int\limits_x^\infty dB(y) \int\limits_1^{\pi_\infty(s)} du$$

$$\times e^{-(s+\lambda-\lambda u)(y-x)} \frac{p_1(t, u, x)f(s, u)}{1 - B(x)},$$

where $p_1(t, z, x) = \sum_{n=0}^\infty z^n p_{1n}(t, x)$.

The nonstationary probabilities $p_0(t)$ and $p_{1n}(t, x)$ are known in terms of Laplace transforms

$$p_0^*(s) = \int_0^\infty e^{-st}p_0(t)dt$$

$$p_1^*(s, z, x) = \int_0^\infty e^{-st}p_1(t, z, x)dt$$

(see section 1.7, Theorem 1.19). So let us introduce the Laplace transform $\int_0^\infty e^{-\sigma t}Ee^{-sW(t)}dt$ in order to describe the dependence on time. For this function we have:

$$\int\limits_0^{+\infty} e^{-\sigma t}E\exp(-sW(t))dt = p_0^*(\sigma)$$

$$+ \int\limits_0^\infty dx \int\limits_x^\infty dB(y) \int\limits_1^{\pi_\infty(s)} du \cdot e^{-(s+\lambda-\lambda u)(y-x)} \frac{f(s, u)p_1^*(\sigma, u, x)}{1 - B(x)}.$$

Substituting into this formula the explicit formulas for $p_0^*(\sigma)$ and $p_1^*(\sigma, u, x)$ obtained in Theorem 1.19 we get the explicit formula for the function $\int_0^\infty e^{-\sigma t}Ee^{-sW(t)}dt$, which describes the virtual waiting time in a nonstationary regime. This formula is not given here because of its complexity.

## 1.8.3 Phase transitions associated with waiting time

Phase transitions in physical systems are studied in relatively full detail. Similar phenomena take place in queueing systems as well.

The best known and most studied among them is the transition from ergodicity to nonergodicity when offered traffic increases. Below we will describe a new type of phase transition in queueing systems. This transition is connected with finiteness of waiting time $W(t)$ (and its mean value) for fixed finite time $t$. It is easy to see that, for the standard $M/G/1$ queue with FIFO discipline, $W(t)$ and $EW(t)$ are finite for any offered traffic $\rho$. For the standard $M/G/1$ queue with LIFO discipline $W(t) < \infty$ iff $\rho \leq 1$ and $EW(t) < \infty$ iff $\rho < 1$. Retrial queues are one of the simplest classes of queues for which the answers to these questions are not so trivial.

**Theorem 1.22** *If the time $t$ is fixed then independently of the system load $\rho$, the virtual waiting time $W(t)$ is finite almost surely. But $EW(t) < \infty$ if and only if $\rho < 2$.*

*Proof.* It is clear that the finiteness of the variables $W(t)$ (and their mean values) at fixed times $t$ is equivalent to the finiteness of variables $T_n$ (and their mean values) under fixed $n$. To study the variables $T_n$, let us introduce the Markov chain $\{\xi_k\}$ with state space $Z_+$, which has an absorbing state $i = 0$ and the following one-step transition probabilities:

$$p_{i,0} = \frac{\mu}{\lambda + i\mu}, \ i \geq 1,$$

$$p_{i,i+n} = \frac{\lambda}{\lambda + i\mu}k_n + \frac{(i-1)}{\lambda + i\mu}k_{n+1}, \ n \geq -1, i \geq 1.$$

The state $i \geq 1$ can be thought of as the presence in the system at some departure moment of exactly $i$ sources of repeated calls including the tagged one, and the transition into the absorbing state $i = 0$ occurs when the tagged customer enters the server. Now the problem reduces to the study of conditions for finiteness of the number of steps before the transition of the chain $\{\xi_k\}$ into state 0 from state $n$, i.e. to the recurrence of the chain. Besides, it is clear that the finiteness of $ET_n$ is equivalent to the ergodicity of the chain $\{\xi_k\}$. To study these problems, as in section 1.3.1, it is convenient to use criteria based on mean drifts.

The following criterion (Pakes, A. G.(1969) Some conditions for ergodicity and recurrence of Markov chains. *Operations Research*, **17**, 1058–1061) gives a sufficient condition for recurrence.

**Statement 6** *Let $\{\xi_k\}$ be an irreducible and aperiodic Markov chain with state space $Z_+$. Assume that there exists a Lyapunov* *function* $f(i), i \in Z_+$, *such that* $\lim_{i \to \infty} f(i) = +\infty$, *the mean drift*

$$x_i \equiv \mathrm{E}(f(\xi_{k+1}) - f(\xi_k) \mid \xi_k = i),$$

*is finite for all* $i \in Z_+$ *and* $x_i \leq 0$ *for all* $i \in Z_+$ *except perhaps a finite number. Then the chain is recurrent.*

For the chain under investigation consider the Lyapunov function $\varphi(i) = \log(i + 1)$. Then the mean drift from the point $i$ is

$$
\begin{aligned}
x_i &= -\frac{\mu}{\lambda + i\mu} \log(i + 1) \\
&+ \sum_{n=-1}^{\infty} \left( \frac{\lambda}{\lambda + i\mu} k_n + \frac{(i-1)\mu}{\lambda + i\mu} k_{n+1} \right) \log \frac{i + n + 1}{i + 1} \\
&\leq \frac{1}{\lambda + i\mu} \left[ -\mu \log(i + 1) + \frac{\lambda}{i+1}\rho + \frac{(i-1)\mu}{i+1}(\rho - 1) \right].
\end{aligned}
$$

For sufficiently large values of $i$, the variable $x_i$ is negative, so that chain $\{\xi_k\}$ is recurrent.

Consider next the Lyapunov function $\varphi(i) = i$. The mean drift from the point $i$ is now equal to

$$
\begin{aligned}
x_i &= -\frac{\mu}{\lambda + i\mu} i + \sum_{n=-1}^{+\infty} \left( \frac{\lambda}{\lambda + i\mu} k_n + \frac{(i-1)\mu}{\lambda + i\mu} k_{n+1} \right) n \\
&= -\frac{\mu i}{\lambda + i\mu} + \frac{\lambda}{\lambda + i\mu}\rho + \frac{(i-1)\mu}{\lambda + i\mu}(\rho - 1) \\
&\to \rho - 2 \quad \text{when } i \to \infty.
\end{aligned}
$$

Thus if $\rho < 2$, for large values of $i$, $x_i \leq -\varepsilon < 0$ and therefore, by Foster's criterion, the chain $\{\xi_k\}$ is ergodic. If $\rho > 2$ then for large values of $i$, $x_i \geq \varepsilon > 0$. Since mean down drifts $\delta_i = \sum_{j \leq i} p_{ij}(j - i)$ are bounded from below, we can guarantee the nonergodicity of the chain $\{\xi_k\}$.

The nonergodicity of the chain $\{\xi_k\}$ in the boundary case $\rho = 2$ could also have been deduced by means of a Lyapunov function, but much more delicate reasoning would have been required. Therefore we shall use another approach which is of interest on its own right because it leads to another interesting problem.

Differentiating formula for $\mathrm{E} \exp(-sT_n)$ given in Theorem 1.20 we have:

$$\mathrm{E}T_n = \frac{(2 + \rho)/\mu + (n - 1)\beta_1}{2 - \rho}, \tag{1.126}$$

so that $\mathrm{E}T_n = \infty$ if $\rho = 2$. $\qquad \square$

It should be noted that the point $\rho = 2$ of phase transition between finiteness and infiniteness of $EW(t)$ on increasing the load $\rho$ is not identical to the point $\rho = 1$ of the transition between ergodicity and nonergodicity.

Besides, it can be proved that $E\left[W(t)\right]^k$ is finite if and only if $\rho < (k+1)/k$ (of course, if $\beta_k < \infty$). Thus in fact we have a spectrum of phase transitions. As $k \to \infty$ the points of these phase transitions converge to the point $\rho = 1$ of the phase transition between ergodicity and nonergodicity.

### 1.8.4 Limit theorems for the waiting time

Formula (1.126) for $ET_n$ shows that $ET_n$ is a linear function of $n$ and $ET_n/n \to \beta_1/(2-\rho)$ when $n \to \infty$. A natural question to be answered is what is the limit distribution of random variable $T_n/n$ when $n \to \infty$.

**Theorem 1.23** *When $n \to \infty$ the distribution of the random variable $T_n/(n\beta_1)$ converges weakly to the distribution with density*

$$
f(t) = \begin{cases}
[1 - (1-\rho)t]^{\rho/(1-\rho)}, & 0 \le t \le \frac{1}{1-\rho}, & \text{if } \rho < 1; \\[2mm]
e^{-t}, & & \text{if } \rho = 1; \\[2mm]
(1 + (\rho-1)t)^{-\rho/(\rho-1)}, & & \text{if } \rho > 1.
\end{cases}
$$

*Proof.* The function

$$
p(s, z, y) = \sum_{i=0}^{+\infty} y^i \sum_{m=1}^{+\infty} z^m \frac{1}{s + \lambda + m\mu} p_m^{(i)}(s)
$$

which was introduced in the proof of Theorem 1.20 depends also on $n$, and has a probabilistic interpretation as

$$
\frac{1}{\mu} E e^{-sT_n} z^{N(T_n - 0)} y^{I(T_n)}.
$$

Thus for the joint Laplace transform

$$
\varphi_n(s, t, u) = E \exp\left\{-sT_n/(n\beta_1) - tN(T_n - 0)/n - uI(T_n)/n\right\}
$$

of scaled random variables $T_n/(n\beta_1)$, $N(T_n - 0)/n$, $I(T_n)/n$, from (1.120), the following equation can be deduced:

$$
-\left[e^{-u/n}\beta\left(\frac{s}{n\beta_1} + \lambda - \lambda e^{-t/n}\right) - e^{-t/n}\right] n e^{t/n} \frac{\partial \varphi_n(s, t, u)}{\partial t}
$$

$$= \left[ \frac{s}{n\beta_1} + \lambda - (\lambda - \mu e^{t/n}) e^{-u/n} \beta \left( \frac{s}{n\beta_1} + \lambda - \lambda e^{-t/n} \right) \right]$$

$$\times \frac{1}{\mu} \varphi_n(s,t,u) - e^{-t}.$$

For

$$\psi(s,t,u) = \lim_{n \to \infty} \varphi_n(s,t,u)$$

from this equation we get

$$[u + s + (\rho - 1)t] \, \psi_t'(s,t,u) = \psi(s,t,u) - e^{-t}.$$

The solution of this equation must be bounded for $s, t, u \geq 0$. Taking this into account, $\psi(s,t,u)$ (depending on the sign of $\rho - 1$) is given by the following formulas:

1. if $\rho > 1$, then

$$\psi(s,t,u) = (u + s + (\rho - 1)t)^{1/(\rho-1)}$$

$$\times \int_t^\infty e^{-x} [u + s + (\rho - 1)x]^{-\rho/(\rho-1)} \, dx;$$

2. if $\rho = 1$, then

$$\psi(s,t,u) = \frac{1}{1 + u + s} e^{-t};$$

3. if $\rho < 1$ and $0 \leq t < \frac{u+s}{1-\rho}$, then

$$\psi(s,t,u) = (u + s - (1 - \rho)t)^{-1/(1-\rho)}$$

$$\times \int_t^{(u+s)/(1-\rho)} e^{-x} [u + s(1 - \rho)x]^{\rho/(1-\rho)} \, dx;$$

4. if $\rho < 1$ and $\frac{u+s}{1-\rho} < t < \infty$, then

$$\psi(s,t,u) = ((1 - \rho)t - (u + s))^{-1/(1-\rho)}$$

$$\times \int_{(u+s)/(1-\rho)}^t e^{-x} [(1 - \rho)x - (u + s)]^{\rho/(1-\rho)} \, dx;$$

5. if $\rho < 1$ and $t = \frac{u+s}{1-\rho}$, then

$$\psi(s,t,u) = \exp(-(u + s)/(1 - \rho)).$$

Putting $t = u = 0$ we get the Laplace transform of the limit distribution of the random variable $T_n/(n\beta_1)$. It is easily inverted and gives the required result.                                                                  □

### 1.8.5 Characterization of the waiting time by the number of retrials.

For systems with repeated calls it is natural to measure the waiting time by the number of retrials $R(t)$ which have to be made by a primary call arriving into the system at time $t$, before it is served. $R(t)$ is an important quantity in itself because in real telephone systems it determines the additional load on control devices. The process $R(t)$ can be studied by the same methods as were applied to analyse the virtual waiting time $W(t)$.

First study the number of retrials made by a tagged customer. Assume that at the moment of departure of some customer (which is considered as the 0th customer) there are $n \geq 1$ other customers in the system. Tag one of them and denote by $R_n$ the number of retrials which this customer makes during the period $T_n$, i.e. until it starts to be served. To study $R_n$, let us introduce an additional event, but now, instead of 'catastrophes', it deals with 'colour' of demands. To be more exact, fix some $x \in [0, 1]$ and declare repeated calls arriving from the tagged source as 'red' with probability $x$ and as 'blue' with probability $1 - x$ (painting is made independently of the functioning of the system and the colour of the other calls). Next introduce probabilities $r_m^{(i)}(x)$ that, at the time of the $i$th departure, there are $m$ sources in the system including the tagged one, and the tagged source has produced 'red' repeated calls only. The following main recursive formula is valid:

$$r_m^{(i)}(x) = \sum_{l=1}^{m+1} r_l^{(i-1)}(x) \frac{\lambda k_{m-l}(x) + (l-1)\mu k_{m-l+1}(x)}{\lambda + l\mu}, \quad (1.127)$$

where

$$k_n(x) = \int_0^\infty \frac{(\lambda t)^n}{n!} e^{-\lambda t} e^{-\mu(1-x)t} dB(t)$$

is the probability that, during the service time of an arbitrary customer, exactly $n$ new primary calls arrived in the system and the tagged source generated 'red' repeated calls only. Obviously, $k_{-1}(x) = 0$.

The initial condition for equation (1.127) is

$$r_m^{(0)}(x) = \delta_{m,n}.$$

For the generating function

$$r(s,z,y) = \sum_{i=0}^{\infty} y^i \sum_{m=1}^{\infty} z^m \frac{r_m^{(i)}(x)}{\lambda + m\mu}$$

these equations yield

$$\mu\left[y\beta(\lambda - \lambda z + \mu - \mu x) - z\right] r_z'(x,z,y)$$
$$= \left[\lambda - (\lambda - \mu/z)y\beta(\lambda - \lambda z + \mu - \mu x)\right] r(x,z,y) - z^n.$$

The required solution must be bounded on the set $|z| \leq 1$. This yields that

$$r(x,z,y) = \int\limits_{z}^{\pi_\infty(\mu - \mu x, y)} \frac{z u^{n-1}}{\mu\left[y\beta(\lambda - \lambda u + \mu - \mu x) - u\right]}$$
$$\times \quad \exp\left\{\int\limits_{u}^{x} \frac{\mu + \lambda - \lambda y\beta(\lambda - \lambda v + \mu - \mu x)}{\mu\left[y\beta(\lambda - \lambda v + \mu - \mu x) - v\right]} dv\right\} du,$$

if $z \neq \pi_\infty(\mu - \mu x, y)$, and

$$r(x,z,y) = \frac{(\pi_\infty(\mu - \mu x, y))^n}{\mu + \lambda - \lambda\pi_\infty(\mu - \mu x, y)},$$

if $z = \pi_\infty(\mu - \mu x, y)$.

Since $\mathrm{E}x^{R_n}$ can be thought of as the probability that all the repeated calls which are produced by the tagged source during its waiting time have a 'red' colour, we have:

$$\mathrm{E}x^{R_n} = \sum_{i=0}^{\infty} \sum_{m=1}^{\infty} r_m^{(i)}(x)\frac{\mu x}{\lambda + m\mu} = \mu x r(x,1,1)$$

$$= \int\limits_{1}^{\pi_\infty(\mu - \mu x)} \frac{x u^{n-1}}{\beta(\lambda - \lambda u + \mu - \mu x) - u}$$
$$\times \quad \exp\left\{\int\limits_{u}^{1} \frac{\mu + \lambda - \lambda\beta(\lambda - \lambda v + \mu - \mu x)}{\mu\left[\beta(\lambda - \lambda v + \mu - \mu x) - v\right]} dv\right\} du.$$

Differentiating with respect to $x$, we get the mean value $\mathrm{E}R_n$; if

$\rho < 2$, then

$$\mathrm{ER}_n = \frac{2 + \rho + (n-1)\mu\beta_1}{2 - \rho}.$$

A word-for-word repetition of the proof of Theorem 1.21 gives the distribution of $R(t)$ in the steady state:

$$
\begin{aligned}
\mathrm{E}x^{R(t)} &= 1 - \rho + \frac{1-\rho}{\mu - \mu x}x \\
&\times \int_1^{\pi_\infty(\mu - \mu x)} \frac{\lambda(1-u)\,[\beta(\lambda - \lambda u) - \beta(\mu - \mu x + \lambda - \lambda u)]}{\beta\,[(\lambda - \lambda u) - u]\,[\beta(\mu - \mu x + \lambda - \lambda u) - u]} \\
&\times \exp\left\{\int_u^1 \frac{\mu + \lambda - \lambda v}{\mu\,[\beta(\mu - \mu x + \lambda - \lambda v) - v]}dv\right\} \\
&\times \exp\left\{\int_1^u \frac{\lambda(1-v)}{\mu\,[\beta(\lambda - \lambda v) - v]}dv\right\}\,du.
\end{aligned}
$$

## 1.9 The departure process

### 1.9.1 The structure of the departure process

The departure process is defined as the sequence of the times $\eta_1, \eta_2, \ldots$ at which customers leave the system after service, or, equivalently, as the sequence $T_i = \eta_i - \eta_{i-1}$ of the interdeparture times.

The interval $T_i$ consists of two parts:

- idle period $R_i$ until the start of service for the $i$th customer;

- service time $S_i$ of the $i$th customer.

The random variable $S_i$ does not depend on events which occurred in the system before time $\xi_i$, when the $i$th service starts. In particular, $S_i$ does not depend on $R_i$. The distribution of $S_i$ is the service time distribution $B(x)$ with Laplace–Stieltjes transform $\beta(s)$, mean $\beta_1$, variance $\sigma^2 = \beta_2 - \beta_1^2$.

The random variable $R_i$ depends on the history of the system until time $\eta_{i-1}$ only through the number of customers in orbit at this time, $N_{i-1}$, and has the conditional distribution

$$\mathrm{P}(R_i < x | N_{i-1} = n) = 1 - e^{-(\lambda + n\mu)x}$$

with the mean

$$E(R_i | N_{i-1} = n) = \frac{1}{\lambda + n\mu}.$$

These remarks about the structure of the interdeparture intervals are the basis of all further considerations.

### 1.9.2 The distribution of interdeparture intervals

First find the distribution of the random variable $T_i$. Let $\omega_i(s) = Ee^{-sT_i}$ be its Laplace transform. Then

$$\begin{aligned}
\omega_i(s) &= Ee^{-s(R_i + S_i)} = Ee^{-sS_i} \cdot Ee^{-sR_i} \\
&= \beta(s) \cdot \sum_{n=0}^{\infty} E\left(e^{-sR_i} | N_{i-1} = n\right) \cdot P(N_{i-1} = n) \\
&= \beta(s) \cdot \sum_{n=0}^{\infty} \frac{\lambda + n\mu}{s + \lambda + n\mu} P(N_{i-1} = n).
\end{aligned} \tag{1.128}$$

**Lemma 1.2** *Random variables $T_i$ are identically distributed iff random variables $N_i$ are identically distributed.*

*Proof.* If $P(N_{i-1} = n)$ does not depend on $i$ then (1.128) implies that $\omega_i(s)$ does not depend on $i$, i.e. all $T_i$ have the same distribution.

Now let all $T_i$ have the same distribution:

$$\omega_i(s) = \omega_{i+1}(s).$$

Then from (1.128) we get

$$\sum_{n=0}^{\infty} \frac{\lambda + n\mu}{s + \lambda + n\mu} P(N_{i-1} = n) = \sum_{n=0}^{\infty} \frac{\lambda + n\mu}{s + \lambda + n\mu} P(N_i = n). \tag{1.129}$$

The functions in the both sides of (1.129) are analytic in the whole plane except for $s = -(\lambda + n\mu)$, $n = 0, 1, 2, \ldots$ Thus, (1.129) implies that

$$P(N_{i-1} = n) = P(N_i = n),$$

i.e. distribution of the random variable $N_i$ does not depend on $i$.
$\square$

For the system in the steady state we have from (1.128) the following formula for the Laplace transform $\omega(s)$ of the interdeparture intervals:

$$\omega(s) = \beta(s) \sum_{n=0}^{\infty} \pi_n \frac{\lambda + n\mu}{s + \lambda + n\mu},$$

$$= \beta(s) \left[ 1 - s \sum_{n=0}^{\infty} \frac{\pi_n}{s + \lambda + n\mu} \right], \qquad (1.130)$$

where $\pi_n$ is the stationary distribution of the embedded Markov chain.

Using equation (1.65) for generating function $\varphi(z) \equiv \sum_{n=0}^{\infty} z^n \pi_n$ we can get an explicit formula for $\omega(s)$. With this goal put $z = x^{\frac{\mu}{s+\lambda}}$ in relation $\varphi(z) \equiv \sum_{n=0}^{\infty} z^n \pi_n$ and integrate it with respect to $x$ from $x = 0$ to $x = 1$. After some algebra we have

$$\sum_{n=0}^{\infty} \frac{\pi_n}{s + \lambda + n\mu} = \frac{1}{\mu} \int_0^1 \varphi(u) u^{\frac{s+\lambda}{\mu}-1} du.$$

Now from (1.130) and (1.65) we get the following explicit formula for the distribution of interdeparture intervals:

$$\omega(s) = \beta(s) \left[ 1 - \frac{s}{\mu}(1-\rho) \int_0^1 u^{\frac{s+\lambda}{\mu}-1} k(u) \frac{1-u}{k(u)-u} \right.$$

$$\times \left. \exp\left\{ \frac{\lambda}{\mu} \int_1^u \frac{1-k(v)}{k(v)-v} dv \right\} du \right]. \qquad (1.131)$$

Formulas for the moments of $T_i$ can be obtained from (1.131), but it is more convenient to use the representation

$$T_i = R_i + S_i.$$

Taking into account formula (1.64) we get for the mean interdeparture interval:

$$ET_i = ES_i + ER_i = \beta_1 + \sum_{n=0}^{\infty} E(R_i | N_{i-1} = n)\pi_n$$

$$= \beta_1 + \sum_{n=0}^{\infty} \frac{\pi_n}{\lambda + n\mu} = \beta_1 + \frac{1-\rho}{\lambda} = \frac{1}{\lambda}. \qquad (1.132)$$

Of course, this formula is absolutely obvious intuitively, since in the steady state the mean interval between departures is equal to the mean interval between arrivals (which in turn equals $1/\lambda$).

For the variance, by the independence of $R_i$ and $S_i$, we have

$$\text{Var}T_i = \text{Var}S_i + \text{Var}R_i = \beta_2 - \beta_1^2 + ER_i^2 - (ER_i)^2$$

$$= \beta_2 - \beta_1^2 - \frac{(1-\rho)^2}{\lambda^2} + \sum_{n=0}^{\infty} \frac{2}{(\lambda + n\mu)^2} \pi_n.$$

To simplify this formula we use the generating function $\psi(z) = \sum_{n=0}^{\infty} z^n \frac{\pi_n}{\lambda + n\mu}$ introduced in section 1.3. Putting in (1.63) $z = x^{\frac{\mu}{\lambda}}$ and integrating with respect to $x$, we have

$$\sum_{n=0}^{\infty} \frac{\pi_n}{(\lambda + n\mu)^2} = \frac{1-\rho}{\lambda\mu} \int_0^1 \exp\left\{ \frac{\lambda}{\mu} \int_1^u \frac{1-k(v)}{k(v)-v} dv \right\} u^{\frac{\lambda}{\mu}-1} du,$$

which yields the following result:

$$\mathrm{Var}T_i = \left[ \lambda^2 \beta_2 - \rho^2 - (1-\rho)^2 + \frac{2\lambda(1-\rho)}{\mu} \right.$$

$$\left. \times \int_0^1 u^{\frac{\lambda}{\mu}-1} \exp\left\{ \frac{\lambda}{\mu} \int_1^u \frac{1-k(v)}{k(v)-v} dv \right\} du \right]. \quad (1.133)$$

It is possible to get simple inequalities for $\mathrm{Var}T_i$. Since

$$\sum_{n=0}^{\infty} \frac{\pi_n}{(\lambda + n\mu)^2} > \frac{\pi_0}{\lambda^2} = \frac{1-\rho}{\lambda^2} \exp\left\{ -\frac{\lambda}{\mu} \int_0^1 \frac{1-k(u)}{k(u)-u} du \right\}$$

and

$$\sum_{n=0}^{\infty} \frac{\pi_n}{(\lambda + n\mu)^2} = \frac{1}{\lambda^2} \sum_{n=0}^{\infty} \frac{\pi_n}{\left(1 + \frac{n\mu}{\lambda}\right)^2} < \frac{1}{\lambda^2} \sum_{n=0}^{\infty} \frac{\pi_n}{1 + \frac{n\mu}{\lambda}}$$

$$= \frac{1}{\lambda} \sum_{n=0}^{\infty} \frac{\pi_n}{\lambda + n\mu} = \frac{1-\rho}{\lambda^2},$$

we have:

$$\frac{\lambda^2 \beta_2 - \rho^2 - (1-\rho)^2 + 2\pi_0}{\lambda^2} < \mathrm{Var}T_i < \frac{\lambda^2 \beta_2 - 2\rho^2 + 1}{\lambda^2}. \quad (1.134)$$

When $\mu \to \infty$, the probability $\pi_0 \to 1 - \rho$ and this double inequality becomes an exact equality.

Relation (1.134) implies also that $\mathrm{Var}T_i < \mathrm{Var}T_i^{(\infty)}$, where $T_i^{(\infty)}$ is a random variable representing an interdeparture interval in the standard $M/G/1/\infty$ queue.

Consider now the case of exponential distribution of service time (this means that $\beta(s) = \frac{\nu}{s+\nu}$, $B(x) = 1 - e^{-\nu x}$, where $\nu = 1/\beta_1$ is the rate of service). Then formulas (1.131) and (1.134) become:

$$\omega(s) = \frac{\nu}{s + \nu} \left[ 1 - \frac{s}{\mu} \frac{(1 - \rho)^{\frac{\lambda}{\mu} + 1}}{\rho^{\frac{s + \lambda}{\mu}}} B_\rho \left( \frac{s + \lambda}{\mu}, -\frac{\lambda}{\mu} \right) \right]$$

and

$$2(1 - \rho)^{\frac{\lambda}{\mu} + 1} + 2\rho - 1 < \lambda^2 \cdot \text{Var} T_i < 1.$$

respectively. Here

$$B_\rho(a, b) = \int_0^\rho x^{a-1} (1 - x)^{b-1} dx$$

is an incomplete Beta function.

### 1.9.3 Covariance properties of the departure process

The inequality $\lambda^2 \text{Var} T_i < 1$ implies, in particular, that for the $M/M/1$ retrial queue interdeparture intervals cannot be exponential, and thus the departure process cannot be a Poisson process. It should be noted that for the standard $M/M/1/\infty$ queue the departure process is Poisson.

We will show now that the departure process from the $M/G/1$ type retrial queue cannot be even a renewal process (except for the trivial case of instantaneous service).

**Lemma 1.3** *The random variables $T_i$ and $T_{i+1}$ are independent iff random variables $T_i$ and $N_i$ are independent.*

*Proof.* The joint distribution of $T_i$ and $T_{i+1}$ (in terms of Laplace transforms) can be expressed as follows:

$$\begin{aligned}
\text{E} \left( e^{-sT_i - rT_{i+1}} \right) &= \text{E} \left( e^{-sT_i} e^{-rR_{i+1}} e^{-rS_{i+1}} \right) \\
&= \text{E} e^{-rS_{i+1}} \cdot \text{E} \left( e^{-sT_i} e^{-rR_{i+1}} \right) \\
&= \beta(r) \sum_{n=0}^\infty \text{E} \left( e^{-sT_i} e^{-rR_{i+1}}; N_i = n \right) \\
&= \beta(r) \sum_{n=0}^\infty \frac{\lambda + n\mu}{r + \lambda + n\mu} \text{E} \left( e^{-sT_i}; N_i = n \right).
\end{aligned}$$

If $T_i$ and $T_{i+1}$ are independent, then

$$\text{E} \left( e^{-sT_i} e^{-rT_{i+1}} \right) = \text{E} e^{-sT_i} \cdot \text{E} e^{-rT_{i+1}},$$

which yields that

$$\sum_{n=0}^{\infty} \frac{\lambda+n\mu}{r+\lambda+n\mu} \mathrm{E}\left(e^{-sT_i}; N_i = n\right)$$

$$= \sum_{n=0}^{\infty} \frac{\lambda+n\mu}{r+\lambda+n\mu} \mathrm{P}(N_i = n) \cdot \mathrm{E}e^{-sT_i}.$$

Since functions in both sides are analytic for all $r \neq -(\lambda+n\mu)$, we have:

$$\mathrm{E}\left(e^{-sT_i}; N_i = n\right) = \mathrm{P}(N_i = n) \cdot \mathrm{E}e^{-sT_i},$$

which means independence of $T_i$ and $N_i$.                             $\square$

We will show later on that independence of $T_i$ and $N_i$ implies that the departure process is renewal. This result holds also for the standard $M/G/1/N$ queue. But in this system independence of $T_i$ and $T_{i+1}$ means independence of $T_i$ and the event $\{N_i = 0\}$ and does not imply independence between $T_i$ and $N_i$. To guarantee the independence of $T_i$ and $N_i$ in an $M/G/1/N$ queue it is sufficient to require, for example, independence of $T_i$ and $T_{i+2}$.

**Lemma 1.4** *If $T_i$ and $N_i$ are independent for all $i$ then service time is equal to 0 (a.s.).*

*Proof.* First note that

$$\mathrm{E}\left(e^{-sT_i}; N_i = n\right) = \sum_{m=0}^{n} \mathrm{P}(N_{i-1} = m) \frac{\lambda k_{n-m}(s)}{s+\lambda+m\mu}$$

$$+ \sum_{m=1}^{n+1} \mathrm{P}(N_{i-1} = m) \frac{m\mu k_{n-m+1}(s)}{s+\lambda+m\mu}.$$

Random variables $T_i$ and $N_i$ are independent iff

$$\mathrm{E}\left(e^{-sT_i}; N_i = n\right) = \mathrm{E}e^{-sT_i} \cdot \mathrm{P}(N_i = n). \qquad (1.135)$$

For $n = 0$ this equation can be rewritten as

$$\beta(s) \sum_{n=0}^{\infty} \mathrm{P}(N_{i-1} = n) \frac{\lambda+n\mu}{s+\lambda+n\mu} \cdot \mathrm{P}(N_i = 0)$$

$$= \mathrm{P}(N_{i-1} = 0) \cdot \frac{\lambda}{s+\lambda} \beta(s+\lambda)$$

$$+ \mathrm{P}(N_{i-1} = 1) \cdot \frac{\mu}{s+\lambda+\mu} \beta(s+\lambda). \qquad (1.136)$$

Let $i$ be large enough to guarantee that $\mathrm{P}(N_i = 0) > 0$. Then

(1.136) implies that $\beta(s)$ has finite derivatives at the point $s = 0$, i.e. service time distribution has moments of all orders.

From (1.135) we also have

$$P(N_i = 0) \cdot E\left(e^{-sT_i}; N_i = n\right) = P(N_i = n) \cdot E\left(e^{-sT_i}; N_i = 0\right),$$

i.e.

$$P(N_i = 0) \cdot \left[\sum_{m=0}^{n} \frac{\lambda P(N_i = m)}{s + \lambda + m\mu} \frac{(-\lambda)^{n-m}}{(n-m)!} \beta^{(n-m)}(s + \lambda)\right.$$

$$+ \left.\sum_{m=1}^{n+1} \frac{m\mu P(N_{i-1} = m)}{s + \lambda + m\mu} \frac{(-\lambda)^{n-m+1}}{(n-m+1)!} \beta^{(n-m+1)}(s + \lambda)\right]$$

$$= \left[\frac{\lambda P(N_{i-1} = 0)}{s + \lambda} + \frac{\mu P(N_{i-1} = 1)}{s + \lambda + \mu}\right]$$

$$\times P(N_i = n)\beta(s + \lambda). \tag{1.137}$$

All functions in (1.137) are analytic for $\operatorname{Re} s \geq -\lambda$. Thus we can guarantee that (1.137) holds for $\operatorname{Re} s \geq -\lambda$. Multiply both sides of (1.137) by $s + \lambda$ and put $s \to -\lambda$. Taking into account that $P(N_{i-1} = 0) \neq 0$ we get:

$$P(N_i = n) = \frac{\lambda^n}{n!} \beta_n P(N_i = 0). \tag{1.138}$$

Now consider (1.137) for $n = 1$ and replace $s + \lambda$ by $s$:

$$\left[\frac{1}{s} + \frac{\mu\beta_1}{s+\mu}\right] \beta'(s) = \left[\frac{\beta_1 - \mu\beta_1^2}{s+\mu} + \frac{\mu\beta_2}{s+2\mu} - \frac{\beta_1}{s}\right] \beta(s). \tag{1.139}$$

Since $\beta'(s) \leq 0$, the coefficient in the right-hand side of (1.139) must be nonpositive, which after some algebra gives that

$$(\beta_2 - \beta_1^2)s^2 + (\mu\beta_2 - 2\mu\beta_1^2 - \beta_1)s - 2\mu\beta_1 \leq 0 \text{ for all } s > 0.$$

Obviously this implies that $\beta_2 - \beta_1^2 \leq 0$, i.e. $\operatorname{Var} S_i = 0$. Thus the random variable $S_i$ is deterministic. In this case $\beta(s) = e^{-\beta_1 s}$ and (1.139) can be rewritten as

$$\frac{\beta_1}{s+\mu} + \frac{\mu\beta_2}{s+2\mu} = 0 \text{ for all } s > 0.$$

Since both terms are positive, $\beta_1 = 0$, i.e. service time is equal to 0 (a.s.). $\qquad\square$

From Lemmas 1.2, 1.3, 1.4 we get the following result.

**Theorem 1.24** *The departure process is a renewal process iff the* *service time is equal to 0 (a.s.) and the system is in the steady state.*

*Proof.* If the departure process is a renewal process then:

(1) all $T_i$ have the same distribution. Thus by Lemma 1.2 the system is in the steady state;

(2) $T_i$ and $T_{i+1}$ are independent. Thus by Lemma 1.3 $T_i$ and $N_i$ are independent, which by Lemma 1.4 implies that $S_i = 0$ (a.s.).

If $S_i = 0$ and the system is in the steady state then $N_i = 0$ (a.s.). It means that sources of repeated calls are absent and all primary calls are immediately admitted for service, i.e. the departure process is identical to the arrival process and thus it is Poisson (and therefore renewal). □

## 1.10 Estimation of retrial rate in the case of exponential service

An important feature of real queues with repeated calls is that they cannot be fully observed. Usually only a joint arrival flow of primary and repeated calls to the servers (where we cannot distinguish between primary and repeated calls) and holding times can be recorded. Obviously, this allows estimation of the rate of input flow (say, through observation of the departure flow) and the service time distribution. Estimation of the retrial rate is more difficult. Nevertheless for the stable model (when $\rho < 1$) this can be done with the help of the process $M(t)$ introduced in section 1.2. In the present section we consider this problem in the case of exponentially distributed service times; without loss of generality we may assume that the mean service time equals 1.

Consider the integral mean values

$$\xi_T = \frac{1}{T} \int_0^T C(t)dt, \ \eta_T = \frac{1}{T} \int_0^T M(t)dt.$$

Because of the ergodicity of the process $(M(t), N(t))$, with probability 1 there exist

$$\lim_{T \to \infty} \xi_T = EC(t) = \lambda,$$

$$\lim_{T \to \infty} \eta_T = EM(t) = \frac{\lambda}{1 - \lambda}(1 + \lambda\mu),$$

and therefore

$$\lim_{T \to \infty} \frac{\eta_T(1 - \xi_T) - \xi_T}{\xi_T^2} = \mu.$$

Thus if the interval of observation $(0, T)$ is long enough we can use

$$\zeta_T = \frac{\eta_T(1 - \xi_T) - \xi_T}{\xi_T^2}$$

as an estimator for the rate of retrials. But to use this estimator in practice we must know its statistical accuracy, i.e. we must know $\mathrm{Var}\zeta_T$. It is natural to consider asymptotic behavior of $\mathrm{Var}\zeta_T$ as $T \to \infty$.

Since $\zeta_T$ is a function of $\xi_T$ and $\eta_T$, the variance of $\zeta_T$ can be approximately calculated with the help of the following well known formula (Kendall, M.G. and Stuart, A. (1958) *The Advanced Theory of Statistics.* Vol.1, Hafner, New York):

$$\mathrm{Var} f(\xi, \eta) \;\simeq\; \left(\frac{\partial f(x, y)}{\partial \xi}\right)^2 \mathrm{Var}\xi + \left(\frac{\partial f(x, y)}{\partial \eta}\right)^2 \mathrm{Var}\eta$$

$$+ \; 2\frac{\partial f(x, y)}{\partial \xi} \frac{\partial f(x, y)}{\partial \eta} \mathrm{Cov}(\xi, \eta), \qquad (1.140)$$

where $x = \mathrm{E}\xi$, $y = \mathrm{E}\eta$.

In our case this equation becomes

$$\mathrm{Var}\zeta_T \;\simeq\; \frac{(\lambda - (2 - \lambda)M)^2}{\lambda^6}\mathrm{Var}\xi_T$$

$$+ \; \frac{(1 - \lambda)^2}{\lambda^4}\mathrm{Var}\eta_T$$

$$+ \; 2\frac{(1 - \lambda)(\lambda - (2 - \lambda)M)}{\lambda^5}\mathrm{Cov}(\xi_T, \eta_T),$$

where $M = \mathrm{E}M(t)$.

Thus the problem reduces to calculation of $\mathrm{Var}\xi_T$, $\mathrm{Var}\eta_T$ and $\mathrm{Cov}(\xi_T, \eta_T)$. This can be done with the help of the following general result.

**Statement 7** *Let $X_t$ be a stationary, uniformly geometrically ergodic Markov process with discrete state space $S$, rates of transition $q_{xy}$, stationary distribution $p_x$. Assume that we can observe processes $Y_t = f(X_t)$ and $Z_t = g(X_t)$, where $f$ and $g$ are some functions on the state space $S$, and denote*

$$\xi_T(f) = \frac{1}{T}\int_0^T f(X_t)dt, \quad \overline{f} = \sum_{x \in S} f(x)p_x,$$

$$\xi_T(g) = \frac{1}{T} \int_0^T g(X_t)dt, \quad \overline{g} = \sum_{x \in S} g(x)p_x.$$

Then as $T \to \infty$ we have:

$$\mathrm{Var}\xi_T(f) = \frac{2}{T} \sum_{y \in S} f(y)V_y^f + o\left(\frac{1}{T}\right), \qquad (1.141)$$

$$\mathrm{Var}\xi_T(g) = \frac{2}{T} \sum_{y \in S} g(y)V_y^g + o\left(\frac{1}{T}\right), \qquad (1.142)$$

$$\mathrm{Cov}(\xi_T(f), \xi_T(g)) = \frac{1}{T}\left(\sum_{y \in S} g(y)V_y^f + \sum_{y \in S} f(y)V_y^g\right) + o\left(\frac{1}{T}\right),$$
$$(1.143)$$

where the variables $V_y^f$ and $V_y^g$ can be found as a solution of equations

$$\sum_{x \in S} V_x^f q_{xy} = \overline{f}p_y - f(y)p_y, \qquad (1.144)$$

$$\sum_{x \in S} V_x^g q_{xy} = \overline{g}p_y - g(y)p_y, \qquad (1.145)$$

satisfying conditions

$$\sum_{x \in S} V_x^f = 0, \quad \sum_{x \in S} V_x^g = 0. \qquad (1.146)$$

**Theorem 1.25** As $T \to \infty$

$$\mathrm{Var}\zeta_T \simeq \frac{2}{\lambda^2(1-\lambda)^2 T}\Big((1-\lambda)^2 + \mu(4 - 6\lambda + 3\lambda^2)$$
$$+ \ \mu^2(3 - 6\lambda^2 + 5\lambda^3 - \lambda^4)\Big). \qquad (1.147)$$

*Proof.* First calculate $\mathrm{Var}\xi_T$, $\mathrm{Var}\eta_T$, $\mathrm{Cov}(\xi_T, \eta_T)$. In our case the state space is two-dimensional and the estimators $\xi_T$, $\eta_T$ can be thought of as $\xi_T(f)$, $\xi_T(g)$, where $f(m,n) = \delta(m)$, $g(m,n) = m$ ($\delta$ is the indicator function of positive integers). Correspondingly,

$$\mathrm{Var}\xi_T \simeq \frac{2}{T} \sum_{m=1}^{\infty} \sum_{n=0}^{\infty} V_{mn},$$

$$\mathrm{Var}\eta_T \simeq \frac{2}{T} \sum_{m=1}^{\infty} m \sum_{n=0}^{\infty} W_{mn},$$

$$\mathrm{Cov}(\xi_T, \eta_T) \simeq \frac{1}{T}\left(\sum_{m=1}^{\infty} m \sum_{n=0}^{\infty} V_{mn} + \sum_{m=1}^{\infty} \sum_{n=0}^{\infty} W_{mn}\right).$$

Here $V_{mn} = V_{mn}^f$, $W_{mn} = V_{mn}^g$. Thus the following equations hold:

$$-(1 + \lambda + n\mu)V_{0n} + \sum_{m=0}^{\infty} V_{mn} = \lambda s_{0n},$$

$$-(1 + \lambda + n\mu)V_{1n} + \lambda V_{0n} + (n+1)\mu V_{0n+1} = (\lambda - 1)s_{1n},$$

$$-(1 + \lambda + n\mu)V_{mn} + \lambda V_{m-1n-1} + n\mu V_{m-1n} = (\lambda - 1)s_{mn}, \ m > 1,$$

$$\sum_{m=0}^{\infty} \sum_{n=0}^{\infty} V_{mn} = 0,$$

$$-(1 + \lambda + n\mu)W_{0n} + \sum_{m=0}^{\infty} W_{mn} = M s_{0n},$$

$$-(1 + \lambda + n\mu)W_{1n} + \lambda W_{0n} + (n+1)\mu W_{0n+1} = (M - 1)s_{1n},$$

$$-(1+\lambda+n\mu)W_{mn}+\lambda W_{m-1n-1}+n\mu W_{m-1n} = (M-m)s_{mn}, \ m > 1,$$

$$\sum_{m=0}^{\infty} \sum_{n=0}^{\infty} W_{mn} = 0,$$

where, as we have defined in section 1.2, $s_{mn}$ is the stationary distribution of the process $(M(t), N(t))$.

Introducing generating functions

$$
\begin{aligned}
V(x, z) &= \sum_{m=0}^{\infty} \sum_{n=0}^{\infty} x^m z^n V_{mn}, \\
V_0(z) &= V(0, z) = \sum_{n=0}^{\infty} z^n V_{0n}, \\
W(x, z) &= \sum_{m=0}^{\infty} \sum_{n=0}^{\infty} x^m z^n W_{mn}, \\
W_0(z) &= W(0, z) = \sum_{n=0}^{\infty} z^n W_{0n},
\end{aligned}
$$

we get:

$$
\begin{aligned}
\mathrm{Var}\xi_T &\simeq -\frac{2}{T} V_0(1), \\
\mathrm{Var}\eta_T &\simeq \frac{2}{T} W_x'(1, 1), \\
\mathrm{Cov}(\xi_T, \eta_T) &\simeq \frac{1}{T} \left( V_x'(1, 1) - W_0(1) \right).
\end{aligned}
$$

Functions $V(x, z)$, $V_0(z)$, $W(x, z)$, $W_0(z)$ satisfy the following equations:

$$(\lambda x z - \lambda - 1)V(x, z) + V(1, z) + \mu z(x - 1)V_z'(x, z)$$
$$+\lambda x(1 - z)V_0(z) + \mu x(1 - z)V_0'(z)$$
$$= (\lambda - 1)s(x, z) + p_0(z), \tag{1.148}$$
$$-(1 + \lambda)V_0(z) - \mu z V_0'(z) + V(1, z) = \lambda p_0(z), \tag{1.149}$$
$$V(1, 1) = 0, \tag{1.150}$$
$$(\lambda x z - \lambda - 1)W(x, z) + W(1, z) + \mu z(x - 1)W_z'(x, z)$$
$$+\lambda x(1 - z)W_0(z) + \mu x(1 - z)W_0'(z)$$
$$= Ms(x, z) - x s_x'(x, z), \tag{1.151}$$
$$-(1 + \lambda)W_0(z) - \mu z W_0'(z) + W(1, z) = Mp_0(z), \tag{1.152}$$
$$W(1, 1) = 0, \tag{1.153}$$

where generating functions $p_0(z)$ and $s(x, z)$ were introduced in Section 1.2 as $\mathrm{E}\left(z^{N(t)}; C(t) = 0\right)$ and $\mathrm{E}\left(x^{M(t)} z^{N(t)}\right)$ respectively.

Differentiate equation (1.148) with respect to $x$ at the point $x = 1$, $z = 1$. Taking into account (1.150) we get:

$$-V_x'(1, 1) + \mu V_z'(1, 1) = (\lambda - 1)M. \tag{1.154}$$

Now put in (1.148) $x = 1$. Since

$$(\lambda - 1)s(1, z) + p_0(z) = \lambda(1 - z)p_1(z),$$

we get:

$$-\lambda V(1, z) + \lambda V_0(z) + \mu V_0'(z) = \lambda p_1(z).$$

Putting $z = 1$ in this equation we have

$$\lambda V_0(1) + \mu V_0'(1) = \lambda^2, \tag{1.155}$$

and differentiating with respect to $z$ at the point $z = 1$ we have:

$$-\lambda V_z'(1, 1) + \lambda V_0'(1) + \mu V_0''(1) = \lambda p_1'(1). \tag{1.156}$$

Similarly, from equation (1.149) we get:

$$-(1 + \lambda)V_0(1) - \mu V_0'(1) = \lambda(1 - \lambda), \tag{1.157}$$
$$V_z'(1, 1) - (1 + \lambda + \mu)V_0'(1) - \mu V_0''(1) = \lambda p_0'(1). \tag{1.158}$$

From (1.155) and (1.157) we can find $V_0(1)$ and $V_0'(1)$:

$$V_0(1) = -\lambda, \quad V_0'(1) = 2\lambda^2/\mu, \tag{1.159}$$

which yields, in particular, that

$$\text{Var}\xi_T \simeq \frac{2\lambda}{T}.$$

Since $p_0'(1) = \lambda^2/\mu$, $p_1'(1) = \lambda^2(\lambda + \mu)/(\mu(1 - \lambda))$, from (1.156), (1.158) we can find $V_z'(1, 1)$:

$$V_z'(1, 1) = \frac{\lambda^2(2 - \lambda)(1 + \mu)}{\mu(1 - \lambda)^2}.$$

Finally, from (1.154) we have:

$$V_x'(1, 1) = \frac{\lambda + \lambda^2(3 - 3\lambda + \lambda^2)\mu}{(1 - \lambda)^2}. \tag{1.160}$$

The second group of equations can be analyzed similarly.

Differentiate equation (1.151) with respect to $x$, $z$, $zz$ at the point $x = 1$, $z = 1$, Taking into account (1.153) we get:

$$W_x'(1, 1) = \mu W_z'(1, 1) + \text{Var}M(t), \tag{1.161}$$

$$\lambda W_0(1) + \mu W_0'(1) = \text{Cov}(M(t), N(t)), \tag{1.162}$$

$$\lambda W_z'(1, 1) - \lambda W_0'(1) - \mu W_0''(1) = -\frac{1}{2}\text{Cov}(M(t), N(t)(N(t) - 1)). \tag{1.163}$$

Besides, put $z = 1$ in equation (1.152):

$$-(1 + \lambda)W_0(1) - \mu W_0'(1) = \lambda(1 + \lambda\mu), \tag{1.164}$$

and differentiate (1.152) with respect to $z$ at the point $z = 1$:

$$-(1 + \lambda + \mu)W_0'(1) - \mu W_0''(1) + W_z'(1, 1) = \frac{\lambda^3(1 + \lambda\mu)}{(1 - \lambda)\mu}. \tag{1.165}$$

From (1.20), (1.162) and (1.164) we can find $W_0(1)$ and $\mu W_0'(1)$:

$$W_0(1) = -\frac{\lambda + \lambda^2(2 - \lambda^2)\mu}{(1 - \lambda)^2},$$

$$\mu W_0'(1) = \frac{\lambda^2(3 - \lambda) + \lambda^2(1 + 4\lambda - 2\lambda^2 - \lambda^3)\mu}{(1 - \lambda)^2}, \tag{1.166}$$

so that

$$\text{Cov}(\xi_T, \eta_T) \simeq \frac{\lambda}{(1 - \lambda)^2 T}(2 + \lambda(5 - 3\lambda)\mu).$$

Now from (1.163) and (1.165) eliminate $\mu W_0''(1)$:

$$(1 - \lambda)W_z'(1, 1) = (1 + \mu)W_0'(1) + \frac{\lambda^3}{(1 - \lambda)\mu}(1 + \lambda\mu)$$

Table 1.1 *Values of $\zeta_T$ for $T = 1000$ obtained by simulation of the M/M/1 retrial queue with $\lambda = 0.5$, $\mu = 1$*

| | | | | |
|---|---|---|---|---|
| 1.1721 | 0.4196 | 0.7882 | 0.9567 | 0.4943 |
| 1.7787 | 0.6157 | 1.0561 | 0.5120 | 0.6329 |
| 0.9817 | 0.6438 | 0.8325 | 0.6311 | 0.8615 |
| 1.5701 | 1.0854 | 0.5975 | 1.9268 | 0.8764 |
| 0.9671 | 1.3813 | 0.8303 | 1.2239 | 0.6404 |
| 1.0538 | 0.8301 | 0.7434 | 0.7181 | 1.0502 |
| 0.9327 | 0.9084 | 1.2142 | 0.8417 | 0.4476 |
| 1.0336 | 1.0769 | 0.4993 | 1.1884 | 0.5422 |
| 1.0034 | 0.4452 | 0.8322 | 0.7159 | 0.7928 |
| 1.1936 | 1.6180 | 0.6808 | 1.4565 | 1.2263 |

$$+ \quad \frac{1}{2} \text{Cov}(M(t), N(t)(N(t) - 1)).$$

Using (1.166) we can find $W'_z(1,1)$:

$$
\begin{aligned}
(1 - \lambda)^3 \mu W'_z(1,1) &= \lambda^2(3 - \lambda^2) + \lambda^2(4 + 3\lambda - \lambda^2 - 2\lambda^3)\mu \\
&+ \lambda^2(1 + 4\lambda - 2\lambda^2 - \lambda^3)\mu^2 \\
&+ \frac{\mu}{2} \text{Cov}(M(t), N(t)(N(t) - 1)).
\end{aligned}
$$

Finally from (1.161), (1.21), (1.22) we can find $W'_x(1,1)$:

$$
\begin{aligned}
(1 - \lambda)^4 W'_x(1,1) &= \lambda(1 + \lambda - 2\lambda^2 + \lambda^3) + \lambda^2(9 - 9\lambda + 3\lambda^2)\mu \\
&+ \lambda^2(3 + 6\lambda - 13\lambda^2 + 7\lambda^3 - \lambda^4)\mu^2,
\end{aligned}
$$

which yields that

$$
\begin{aligned}
\text{Var}\eta_T &\simeq \frac{2\lambda}{(1 - \lambda)^4 T} \left(1 + \lambda - 2\lambda^2 + \lambda^3 \right. \\
&+ \lambda\left(9 - 9\lambda + 3\lambda^2\right)\mu \\
&+ \left. \lambda\left(3 + 6\lambda - 13\lambda^2 + 7\lambda^3 - \lambda^4\right)\mu^2\right).
\end{aligned}
$$

Using equation (1.140) we get the final formula (1.147) for $\text{Var}\zeta_T$.

$\square$

We compared our approximation for $\text{Var}\zeta_T$ with results obtained by simulation. With this goal we simulated 50 times the model with parameters $\lambda = 0.5$, $\mu = 1$ for $T = 1000$ (in units of the mean holding times) and computed the estimator $\text{Var}\zeta_T$. The resulting values of $\zeta_T$ are shown in Table 1.1. The sample mean is 0.9304 (we recall that $\mu=1$) and the sample variance is 0.1204. Our formula (1.147) for $\mathrm{Var}\zeta_T$ gives $\mathrm{Var}\zeta_T \simeq 0.13$. Thus the simulation results are in good agreement with the approximation.

# CHAPTER 2

# The main multiserver model

## 2.1 Description of the model

Consider a group of $c$ fully available servers in which a Poisson flow of calls with rate $\lambda$ arrives. In the context of telecommunication engineering, these calls are referred to as primary calls. In other applications, the term primary customers is more appropriate.

If an arriving primary call finds some server free it immediately occupies a server and leaves the system after service. Otherwise, if all servers are engaged, it produces a source of repeated calls. Every such source after some delay produces repeated calls until after one or more attempts it finds a free server, in which case the source is eliminated and the call receives service and then leaves the system.

We assume that periods between successive retrials are exponentially distributed with parameter $\mu$, and service times are exponentially distributed with parameter $\nu$. Without loss of generality we may assume that $\nu = 1$. As usual, we suppose that interarrival periods, retrial times and service times are mutually independent.

The functioning of the system can be described by means of a bivariate process $(C(t), N(t))$, where $C(t)$ is the number of busy servers and $N(t)$ is the number of sources of repeated calls (queue length) at time $t$. Under the above assumptions the bivariate process $(C(t), N(t))$ is Markovian with the lattice semi-strip $S = \{0, 1, ..., c\} \times Z_+$ as the state space. Its infinitesimal transition rates $q_{(ij)(nm)}$ are given by:

1. for $0 \leq i \leq c - 1$

$$
q_{(ij)(nm)} = \begin{cases}
\lambda, & \text{if } (n, m) = (i + 1, j), \\
i, & \text{if } (n, m) = (i - 1, j), \\
j\mu, & \text{if } (n, m) = (i + 1, j - 1), \\
-(\lambda + i + j\mu), & \text{if } (n, m) = (i, j), \\
0 & \text{otherwise.}
\end{cases}
$$

2. for $i = c$

$$q_{(cj)(nm)} = \begin{cases} \lambda, & \text{if } (n,m) = (c, j+1), \\ c, & \text{if } (n,m) = (c-1, j), \\ -(\lambda + c), & \text{if } (n,m) = (c, j), \\ 0 & \text{otherwise.} \end{cases}$$

Random walks on the product of a finite set and the set of non-negative integers (in other terms, on a lattice semi-strip) arise in many applications. The best-known family of such walks was introduced in the 1970s by M. Neuts (Neuts, M.F. (1978) Markov chains with applications in queueing theory, which have a matrix geometric invariant probability vector. *Advances in Applied Probability*, **10**, No.1) and V. Malyshev (Malyshev, V.A. (1972) Homogeneous random walks on the product of a finite set and a half-line. In: *Probabilistic Methods of Research*, Moscow State University (in Russian)). The main assumption of their theories is the following condition of limited spacial homogeneity:

$$q_{(ij)(nm)} = q_{in}^{m-j}, \text{ if } j \geq 1.$$

This assumption allows extensive mathematical analysis of both stationary and transient behavior of the process. In contrast to this, for the retrial queue under consideration (as well as for other retrial queues) rates of transition from a point $(i, j)$ of the semi-strip $\{0, 1, ..., c\} \times Z_+$ depend on the second coordinate $j$. The main difficulties in analysis and the most interesting properties of retrial queues are connected with this fact.

From a practical point of view the most important characteristics of the quality of service to subscribers are:

- the stationary blocking probability $B = \lim_{t \to \infty} P\{C(t) = c\}$;

- the mean queue length in the steady state $N = \lim_{t \to \infty} EN(t)$;

- the stationary carried traffic (which is equal to the mean number of busy servers) $Y = \lim_{t \to \infty} EC(t)$;

- the mean waiting time $W$, which by Little's formula equals $\frac{N}{\lambda}$;

- the mean waiting time for customers which are really waiting for service (i.e. their first attempt was blocked) $W_B = \frac{W}{B}$.

## 2.2 Ergodicity

### 2.2.1 Sufficient conditions for ergodicity

Sufficient conditions for ergodicity of retrial queues can be obtained with the help of criteria based on mean drifts. The following theorem (Tweedie, R.L. (1975) Sufficient conditions for regularity, recurrence and ergodicity of Markov processes. *Math. Proceedings of the Cambridge Philosophical Society*, **78**, part 1) is the most convenient.

**Statement 8** *Let $X(t)$ be a Markov process with discrete state space $S$ and rates of transition $q_{sp}$, $s, p \in S$, $\sum_p q_{sp} = 0$. Assume that there exist*

1. *a function $\varphi(s)$, $s \in S$, which is bounded from below (this function is said to be a Lyapunov or test function);*

2. *a positive number $\varepsilon$ such that:*

   - *variables $y_s = \sum_{p \neq s} q_{sp}(\varphi(p) - \varphi(s)) < \infty$ for all $s \in S$;*
   - *$y_s \leq -\varepsilon$ for all $s \in S$ except perhaps a finite number of states.*

*Then the process $X(t)$ is regular and ergodic.*

For retrial queues a linear combination of coordinates of the vector Markov process which describes the functioning of a model usually can be used as the Lyapunov function $\varphi(s)$. Thus for the model under investigation we consider the following Lyapunov function:

$$\varphi(s) \equiv \varphi(i, j) = ai + j,$$

where $a$ is a parameter, which will be determined later on.

Then the mean drifts $y_s \equiv y_{ij}$ are given by:

$$y_{ij} = \begin{cases} \lambda a + j\mu(a - 1) + i \cdot (-a), & \text{if } 0 \leq i \leq c - 1, \\ \lambda \cdot 1 + c \cdot (-a), & \text{if } i = c. \end{cases}$$

Since for all $i = 0, 1, ..., c$ there exists

$$\lim_{j \to \infty} y_{ij} = L_i = \begin{cases} (a - 1) \cdot \infty, & \text{if } 0 \leq i \leq c - 1, \\ \lambda - ac, & \text{if } i = c. \end{cases}$$

the assumptions of Tweedie's theorem hold iff all variables $L_i$ are negative, i.e.

$$\begin{cases} a - 1 < 0, \\ \lambda - ac < 0. \end{cases}$$

These conditions represent a set of linear inequalities for the still unknown parameter $a$. Obviously, they can be written in the form

$$\frac{\lambda}{c} < a < 1.$$

Such an $a$ can be found iff the interval $(\frac{\lambda}{c}, 1)$ is not empty, i.e. iff $\lambda < c$. This is a sufficient condition for ergodicity of our model.

### 2.2.2 Necessary conditions for ergodicity

Necessary conditions for ergodicity of retrial queues are obtained quite simply from the fact that the mean number of busy servers $Y$ must be less than the total number of servers which are available to calls. In the steady state $Y$ is equal to the intensity of carried traffic and can usually be easily computed. Since in our model calls are not lost, the intensity of carried traffic is equal to the intensity of offered traffic, i.e. $\lambda$. Thus, the inequality $Y < c$ becomes $\lambda < c$. Therefore, in fact the condition $\lambda < c$ is necessary and sufficient for ergodicity of the model under consideration.

Non-ergodicity can be established also with the help of Lyapunov functions. The following theorem (Malyshev, V.A. and Menshikov, M.V. (1979) Ergodicity, continuity and analyticity of countable Markov chains. *Proceedings of the Moscow Math. Society*, **39**) is the most convenient.

**Statement 9** *Let $\{\zeta_n\}$ be a Markov chain with state space $S$ and one-step transition probabilities $r_{sp}$. Assume that there exist:*

1. *a nonnegative function $\varphi(s)$, $s \in S$, such that for some $d$ we have*

$$r_{sp} \neq 0 \implies |\varphi(s) - \varphi(p)| \leq d;$$

2. *a positive number $b$ such that:*

   - *set $A_b \equiv \{s \in S \,|\, \varphi(s) > b\} \neq \emptyset$;*
   - $\inf\limits_{s \in A_b} x_s \equiv \inf\limits_{s \in A_b} \mathrm{E}\left(\varphi(\zeta_{n+1}) - \varphi(s) \,|\, \zeta_n = s\right)$ *is nonnegative (or positive).*

*Then the chain $\{\zeta_n\}$ is nonergodic (correspondingly, transient).*

Consider now the main model. Let $\{\zeta_n\}$ be an embedded (at state transitions) Markov chain for the continuous time Markov process $(C(t), N(t))$. Its one step transition probabilities are given by

1. for $0 \leq i \leq c - 1$

$$r_{(ij)(nm)} = \begin{cases} \frac{\lambda}{\lambda + i + j\mu}, & \text{if } (n, m) = (i + 1, j), \\ \frac{i}{\lambda + i + j\mu}, & \text{if } (n, m) = (i - 1, j), \\ \frac{j\mu}{\lambda + i + j\mu}, & \text{if } (n, m) = (i + 1, j - 1), \\ 0 & \text{otherwise.} \end{cases}$$

2. for $i = c$

$$r_{(cj)(nm)} = \begin{cases} \frac{\lambda}{\lambda + c}, & \text{if } (n, m) = (c, j + 1), \\ \frac{c}{\lambda + c}, & \text{if } (n, m) = (c - 1, j), \\ 0 & \text{otherwise.} \end{cases}$$

Let $\lambda \geq c$. Consider the Lyapunov function $\varphi(i, j) = i + j$. Then:

1. If $r_{(ij)(kl)} \neq 0$ then $|i - k| \leq 1$, $|j - l| \leq 1$ and so

$$|\varphi(i, j) - \varphi(k, l)| = |i - k + j - l| \leq |i - k| + |j - l| \leq 2,$$

i.e. $d = 2$.

2. As the variable $b$ any positive number can be taken. Indeed,

(a) $A_b = \{(i, j) \mid i = 0, 1, ..., c; j > \max(b - i, 0)\} \neq \emptyset$;

(b)

$$x_{ij} = \begin{cases} \frac{\lambda - i}{\lambda + i + j\mu}, & \text{if } 0 \leq i \leq c - 1, \\ \frac{\lambda - c}{\lambda + c}, & \text{if } i = c. \end{cases}$$

Thus, for $\lambda \geq c$ variable $x_{ij}$ is always nonnegative and moreover $x_{ij} \geq 0$ for $(i, j) \in A_b$. Therefore, for $\lambda \geq c$ the process $(C(t), N(t))$ is nonergodic.

Now let $\lambda > c$. Consider the Lyapunov function $\varphi(i, j) = ai + j$, where $a \in (1, \frac{\lambda}{c})$. Then:

1. If $r_{(ij)(kl)} \neq 0$ then $|i - k| \leq 1$, $|j - l| \leq 1$ and so

$$|\varphi(i, j) - \varphi(k, l)| = |a(i - k) + j - l| \leq a|i - k| + |j - l| \leq a + 1,$$

i.e. $d = a + 1$.

2. The set $A_b$ is nonempty for any $b > 0$ and so we must choose $b$ in such a way that $\inf_{(i,j) \in A_b} x_{ij} > 0$. But for $i = c$

$$x_{cj} = \frac{\lambda - ac}{\lambda + c} > 0 \text{ for all } j.$$

and for $0 \leq i \leq c - 1$

$$\lim_{j \to \infty} x_{ij} = a - 1 > 0.$$

Thus, for any $i = 0, 1, ..., c$ there exist $N_i$ such that

$$x_{ij} > \frac{a - 1}{2} \text{ for all } N_i.$$

If

$$\varepsilon = \min \left\{ \frac{\lambda - ac}{\lambda + c}, \frac{a - 1}{2} \right\}$$

then $x_{ij} \geq \varepsilon$ for all $(i, j) \in S$ except for $(i, j)$ such that $i = 0, 1, ..., c - 1; 0 \leq j \leq N_i - 1$.

Let $b$ is large enough so that set $A_b$ does not contain these states. Then

$$x_{ij} \geq \varepsilon \text{ for all } (i, j) \in A_b,$$

i.e.

$$\inf_{(i,j) \in A_b} x_{ij} \geq \varepsilon > 0.$$

Therefore for $\lambda > c$ the chain $(C(t), N(t))$ is transient.

It would be interesting to compare conditions for ergodicity of the process $(C(t), N(t))$ and spatially homogeneous random walks. In both cases $\rho < 1$, where $\rho$ is 'traffic' (which must be introduced in the proper way), implies ergodicity. The condition $\rho > 1$ in both cases implies transience. The case $\rho = 1$ is more interesting. Spatially homogeneous random walks with $\rho = 1$ are null recurrent, whereas for the process $(C(t), N(t))$ the answer depends on the parameter $\mu$. As we have seen in the single-server case, for $\mu \geq 1$ the process $(C(t), N(t))$ is null recurrent and for $\mu < 1$ the process is transient.

From now on we will assume that $\lambda < c$ and the system is in the steady state.

## 2.3  Explicit formulas for the main performance characteristics

### 2.3.1  The case $c = 1$

As we have already shown, in Chapter 1, section 1.1, in the single-server case the joint distribution of the number of busy servers and the queue length

$$p_{0j} = \mathrm{P}\{C(t) = 0, N(t) = j\}$$

$$p_{1j} = \mathrm{P}\{C(t) = 1, N(t) = j\}$$

is given by the following explicit formulas:

$$p_{0j} = \frac{\lambda^j}{j!\mu^j} \prod_{k=0}^{j-1}(\lambda + k\mu) \cdot (1 - \lambda)^{\frac{\lambda}{\mu}+1},$$

$$p_{1j} = \frac{\lambda^{j+1}}{j!\mu^j} \prod_{k=1}^{j}(\lambda + k\mu) \cdot (1 - \lambda)^{\frac{\lambda}{\mu}+1},$$

and has the following partial generating functions:

$$p_0(z) \equiv \sum_{j=0}^{\infty} z^j p_{0j} = (1 - \lambda) \left( \frac{1 - \lambda}{1 - \lambda z} \right)^{\frac{\lambda}{\mu}},$$

$$p_1(z) \equiv \sum_{j=0}^{\infty} z^j p_{1j} = \lambda \left( \frac{1 - \lambda}{1 - \lambda z} \right)^{\frac{\lambda}{\mu}+1}.$$

In particular, the blocking probability $B \equiv \mathrm{P}\{C(t) = 1\}$ and the mean queue length $N \equiv \mathrm{E}N(t)$ are given by the formulas:

$$B = \lambda,$$

$$N = \frac{1 + \mu}{\mu} \cdot \frac{\lambda^2}{1 - \lambda}.$$

### 2.3.2 The case $c = 2$

Let

$$p_{0j} = \mathrm{P}\{C(t) = 0, N(t) = j\}$$
$$p_{1j} = \mathrm{P}\{C(t) = 1, N(t) = j\}$$
$$p_{2j} = \mathrm{P}\{C(t) = 2, N(t) = j\}$$

be the joint distribution of the number of busy servers and the number in orbit in the steady state. These probabilities satisfy the following set of Kolmogorov equations:

$$(\lambda + j\mu)p_{0j} = p_{1j}, \tag{2.1}$$
$$(\lambda + 1 + j\mu)p_{1j} = \lambda p_{0j} + (j + 1)\mu p_{0,j+1} + 2p_{2j}, \tag{2.2}$$
$$(\lambda + 2)p_{2j} = \lambda p_{1j} + (j + 1)\mu p_{1,j+1} + \lambda p_{2,j-1} \tag{2.3}$$

and the normalizing condition

$$\sum_{j=0}^{\infty} (p_{0j} + p_{1j} + p_{2j}) = 1. \tag{2.4}$$

Equation (2.2) in fact expresses $p_{1j}$ through $p_{0j}$. Using this, from (2.3) we can also express $p_{2j}$ through $p_{0j}$ :

$$2p_{2j} = [(\lambda + j\mu)^2 + j\mu]p_{0j} - (j+1)\mu p_{0,j+1}. \tag{2.5}$$

Substituting expressions (2.2) and (2.5) into (2.3) we get the following recursive relation for probabilities $p_{0j}$ :

$$\lambda[(\lambda + j\mu)^2 + j\mu]p_{0j} - (j+1)\mu[2 + 3\lambda + 2(j+1)\mu]p_{0,j+1}$$

$$= \lambda[(\lambda + (j-1)\mu)^2 + (j-1)\mu]p_{0,j-1} - j\mu[2 + 3\lambda + 2j\mu]p_{0j}.$$

This yields that for all $j$ we have:

$$\lambda[(\lambda + (j-1)\mu)^2 + (j-1)\mu]p_{0,j-1} - j\mu[2 + 3\lambda + 2j\mu]p_{0j} = 0,$$

or equivalently,

$$p_{0j} = \frac{\lambda}{j\mu} \cdot \frac{(\lambda + (j-1)\mu)^2 + (j-1)\mu}{2 + 3\lambda + 2j\mu} \cdot p_{0,j-1}.$$

This formula allows us to express all probabilities $p_{0j}$ through $p_{00}$:

$$p_{0j} = \frac{\lambda^j}{j!\mu^j} \cdot \prod_{k=0}^{j-1} \frac{(\lambda + k\mu)^2 + k\mu}{2 + 3\lambda + 2\mu + 2k\mu} \cdot p_{00}. \tag{2.6}$$

Now from (2.2) and (2.5) we can find probabilities $p_{1j}$ and $p_{2j}$ :

$$p_{1j} = (\lambda + j\mu)\frac{\lambda^j}{j!\mu^j} \cdot \prod_{k=0}^{j-1} \frac{(\lambda + k\mu)^2 + k\mu}{2 + 3\lambda + 2\mu + 2k\mu} \cdot p_{00},$$

$$p_{2j} = (1 + \lambda + (j+1)\mu)\frac{\lambda^j}{j!\mu^j} \cdot \prod_{k=0}^{j} \frac{(\lambda + k\mu)^2 + k\mu}{2 + 3\lambda + 2\mu + 2k\mu} \cdot p_{00}.$$

The initial probability $p_{00}$ can be obtained with the help of the normalizing condition (2.4) as

$$p_{00}^{-1} = \sum_{j=0}^{\infty} \frac{\lambda^j}{j!\mu^j} \prod_{k=0}^{j-1} \frac{(\lambda + k\mu)^2 + k\mu}{2 + 3\lambda + 2\mu + 2k\mu}$$

$$\times \left[ 1 + \lambda + j\mu + \frac{(1 + \lambda + (j+1)\mu)((\lambda + j\mu)^2 + j\mu)}{2 + 3\lambda + 2(j+1)\mu} \right].$$

In fact probability $p_{00}$, as well as generating functions $p_i(z) = \sum_{j=0}^{\infty} z^j p_{ij}$ and the main performance characteristics, can be expressed in terms of the hypergeometric functions

$$F(a, b, c; x) \equiv \sum_{j=0}^{\infty} \frac{x^j}{j!} \prod_{k=0}^{j-1} \frac{(a+k)(b+k)}{c+k}. \qquad (2.7)$$

For convenience of further references, recall some properties of these functions:

$$F'(a, b, c; x) = \frac{ab}{c} F(a+1, b+1, c+1; x), \qquad (2.8)$$

$$F''(a, b, c; x) = \frac{ab}{x(1-x)} \Big\{ F(a, b, c; x) \\ - \frac{c-(a+b+1)x}{c} F(a+1, b+1, c+1; x) \Big\}, \qquad (2.9)$$

$$F'(a+1, b+1, c+1; x) = \frac{c}{x(1-x)} \\ \times \Big\{ F(a, b, c; x) - \frac{c-(a+b+1)x}{c} F(a+1, b+1, c+1; x) \Big\}.$$

**Theorem 2.1** *For the M/M/2 retrial queue, the joint distribution of the number of busy servers and the number of sources of repeated calls in the steady state is given by the following partial generating functions:*

$$p_0(z) = F\left(a, b, c; \frac{\lambda z}{2}\right) \cdot p_{00}, \qquad (2.10)$$

$$p_1(z) = \Big\{ \lambda F\left(a, b, c; \frac{\lambda z}{2}\right) + \frac{\lambda^3}{2+3\lambda+2\mu} \\ \times F\left(a+1, b+1, c+1; \frac{\lambda z}{2}\right) \Big\} \cdot p_{00}, \qquad (2.11)$$

$$p_2(z) = \Big\{ \frac{\lambda^2}{2-\lambda z} F\left(a, b, c; \frac{\lambda z}{2}\right) + \frac{\lambda^3(\lambda z - 1)}{(2+3\lambda+2\mu)(2-\lambda z)} \\ \times F\left(a+1, b+1, c+1; \frac{\lambda z}{2}\right) \Big\} \cdot p_{00}, \qquad (2.12)$$

*where:*

$$a = \frac{2\lambda + 1 + \sqrt{4\lambda + 1}}{2\mu},$$

$$b = \frac{2\lambda + 1 - \sqrt{4\lambda + 1}}{2\mu}, \left.\right\} \qquad (2.13)$$

$$c = \frac{2 + 3\lambda + 2\mu}{2\mu},$$

$$p_{00} = \frac{2 - \lambda}{(2 + \lambda)F\left(a, b, c; \frac{\lambda}{2}\right) + \frac{\lambda^3}{2+3\lambda+2\mu}F\left(a+1, b+1, c+1; \frac{\lambda}{2}\right)}. \qquad (2.14)$$

*In particular, for the generating function of the queue length we have:*

$$p(z) = p_{00} \cdot \left\{ \left(1 + \lambda + \frac{\lambda^2}{2 - \lambda z}\right) F\left(a, b, c; \frac{\lambda z}{2}\right) + \frac{\lambda^3}{2 + 3\lambda + 2\mu} \right.$$
$$\left. \times \left(z - 1 + \frac{1}{2 - \lambda z}\right) F\left(a+1, b+1, c+1; \frac{\lambda z}{2}\right) \right\} \quad (2.15)$$

*Proof.* Using (2.6), generating function $p_0(z)$ can be written as

$$\sum_{j=0}^{\infty} \frac{(\lambda z)^j}{j! \mu^j} \prod_{k=0}^{j-1} \frac{k^2\mu^2 + k\mu(2\lambda + 1) + \lambda^2}{2 + 3\lambda + 2\mu + 2k\mu} p_{00}$$

$$= \sum_{j=0}^{\infty} \frac{(\lambda z/2)^j}{j!} \prod_{k=0}^{j-1} \frac{k^2 + k\frac{2\lambda+1}{\mu} + \frac{\lambda^2}{\mu^2}}{k + \frac{2+3\lambda+2\mu}{2\mu}} p_{00}$$

$$= \sum_{j=0}^{\infty} \frac{(\lambda z/2)^j}{j!} \prod_{k=0}^{j-1} \frac{\left(k + \frac{2\lambda+1+\sqrt{4\lambda+1}}{2\mu}\right)\left(k + \frac{2\lambda+1-\sqrt{4\lambda+1}}{2\mu}\right)}{k + \frac{2+3\lambda+2\mu}{2\mu}} p_{00}.$$

With the help of notations (2.13) we have:

$$p_0(z) = \sum_{j=0}^{\infty} \frac{(\lambda z/2)^j}{j!} \prod_{k=0}^{j-1} \frac{(a + k)(b + k)}{c + k} \cdot p_{00}.$$

Taking into account definition (2.7) of the hypergeometric function we get formula (2.10).

For the generating function $p_1(z)$, from (2.2) we have:

$$p_1(z) = \lambda p_0(z) + \mu z p_0'(z).$$

Replacing $p_0(z)$ in terms of the hypergeometric function we get:

$$p_1(z) = \lambda F\left(a, b, c; \frac{\lambda z}{2}\right) \cdot p_{00} + \mu z \frac{\lambda}{2} F'\left(a, b, c; \frac{\lambda z}{2}\right) \cdot p_{00}.$$

Using relation (2.8) for the derivative of the hypergeometric function gives:

$$p_1(z) = \left\{ \lambda F\left(a, b, c; \frac{\lambda z}{2}\right) \right.$$
$$\left. + \frac{\lambda \mu z}{2} \frac{ab}{c} F\left(a+1, b+1, c+1; \frac{\lambda z}{2}\right) \right\} \cdot p_{00}.$$

Taking into account the explicit formulas for variables $a, b, c$ we get (2.11).

Formula (2.12) can be obtained similarly from (2.5), (2.10), (2.8) and (2.9).

Equation (2.15) obviously follows from (2.10), (2.11) and (2.12).

Now relation (2.14) for the normalizing multiplier $p_{00}$ can be obtained from equation $p(1) = 1$. $\qquad\qquad\square$

From this theorem we immediately get the following explicit formulas for the main performance characteristics in terms of the ratio of two hypergeometric functions.

**Theorem 2.2** *For the M/M/2 retrial queue, the blocking probability and the mean queue length are given by*

$$B = \frac{\lambda^2 + (\lambda - 1)g}{2 + \lambda + g}, \tag{2.16}$$

$$N = \frac{1 + \mu}{\mu} \cdot \frac{\lambda^3 + (\lambda^2 - 2\lambda + 2)g}{(2 - \lambda)(2 + \lambda + g)},$$

*where*

$$g = \frac{\lambda^3}{2 + 3\lambda + 2\mu} \cdot \frac{F(a+1, b+1, c+1; \frac{\lambda}{2})}{F(a, b, c; \frac{\lambda}{2})}$$

*and the variables $a, b, c$ were defined in Theorem 2.1.*

### 2.3.3 General case

Let $p_{ij} = \mathrm{P}\{C(t) = i, N(t) = j\}$ be the joint distribution of the number of busy servers and the queue length in the steady state. These probabilities satisfy the following set of Kolmogorov equations:

$$(\lambda + i + j\mu)p_{ij} = \lambda p_{i-1,j} + (j+1)\mu p_{i-1,j+1}$$
$$+ (i+1)p_{i+1,j}, \text{ if } 0 \le i \le c-1, \tag{2.17}$$

$$(\lambda + c)p_{cj} = \lambda p_{c-1,j} + (j+1)\mu p_{c-1,j+1}$$
$$+ \lambda p_{c,j-1}, \text{ (case } i = c). \tag{2.18}$$

For generating functions

$$p_i(z) = \sum_{j=0}^{\infty} z^j p_{ij}, \ 0 \leq i \leq c$$

these equations become

$$(\lambda + i)p_i(z) + \mu z p_i'(z) = \lambda p_{i-1}(z) + \mu p_{i-1}'(z) + (i+1)p_{i+1}(z),$$
$$\text{if } 0 \leq i \leq c-1, \quad (2.19)$$
$$(\lambda + c)p_c(z) = \lambda p_{c-1}(z) + \mu p_{c-1}'(z) + \lambda z p_c(z),$$
$$(\text{case } i = c). \quad (2.20)$$

Now introduce the bivariate generating function

$$p(x, z) = \sum_{i=0}^{c} x^i p_i(z).$$

Then equations (2.19), (2.20) become:

$$\lambda(1-x)p(x,z) + \mu(z-x)p_z'(x,z) + (x-1)p_x'(x,z)$$

$$+\lambda x^c(x-z)p_c(z) + \mu x^c(x-z)p_c'(z) = 0. \quad (2.21)$$

Differentiating equation (2.21) with respect to $z, x, xx, xz, zz$ at the point $x = 1, z = 1$ we get the following equations:

$$\mu N - \lambda B - \mu N_c = 0, \quad (2.22)$$

$$\lambda + \mu N - Y - \lambda B - \mu N_c = 0,$$
$$\lambda Y + \mu p_{xz}'' - p_{xx}'' - \lambda c B - \mu c N_c = 0,$$
$$-\lambda N - \mu p_{zz}'' + (1 + \mu)p_{xz}'' + \lambda N_c - \mu c N_c - \lambda c B + \mu p_{czz}'' = 0,$$
$$\mu p_{zz}'' - \lambda N_c - \mu p_{czz}'' = 0,$$

where $N = EN(t) = p_z'(1,1)$, $B = P\{C(t) = c\} = p_c(1)$, $Y = EC(t) = p_x'(1,1)$, $N_c = E\{N(t); C(t) = c\} = p_c'(1)$. Eliminating from these equations variables $N_c, p_{xz}'', p_{zz}'', p_{czz}''$ and taking into account that $p_{xx}''(1,1) = \text{Var}C(t) + (EC(t))^2 - EC(t)$ we get:

$$Y = \lambda \quad (2.23)$$

$$N = \frac{1+\mu}{\mu} \cdot \frac{\lambda - \text{Var}C(t)}{c - \lambda} \quad (2.24)$$

Equation (2.24) can be rewritten in an equivalent form as

$$N = \frac{1+\mu}{\mu} \cdot \frac{\lambda + \lambda^2 - E\left(C(t)\right)^2}{c - \lambda}. \quad (2.25)$$

Equation (2.23) can be thought of as a variant of Little's formula and represents equality of the offered and carried traffic. Equation (2.24) is much more interesting. It gives a partial description of the dependence of the mean queue length upon the system parameters, and reduces calculation of the mean queue length to the calculation of the characteristics of the number of busy servers, which is a simpler problem.

Similarly, higher moments of the queue length can be expressed in terms of the distribution of the number of busy servers. For example, the second factorial moment $\Phi_2 \equiv \mathrm{E}\,(N(t) \cdot (N(t) - 1)) = p''_{zz}(1,1)$ is given by

$$\Phi_2 = \frac{1+2\mu}{2\mu^2(c-\lambda)^2}\left\{-(2+\mu)(c-\lambda)p'''_{xxx}(1,1)\right.$$

$$+ \left[\lambda(4+\mu)(c-\lambda) + 2c(1+\mu)(c-\lambda-1)\right]p''_{xx}(1,1)$$

$$\left. -2\lambda^2 c(1+\mu)(c-\lambda-1) - 2\lambda^3(c-\lambda)\right\}.$$

Another useful piece of information about the steady state distribution of the number of busy servers $p_i = \mathrm{P}\,(C(t) = i)$ can be obtained as follows.

Put in equations (2.19), (2.20) $z = 1$ :

$$\lambda p_i + \mu N_i - (i+1)p_{i+1} = \lambda p_{i-1} + \mu N_{i-1} - ip_i, \ 0 \leq i \leq c-1$$
$$\lambda p_{c-1} + \mu N_{c-1} - cp_c = 0,$$

where $N_i = \mathrm{E}\,(N(t); C(t) = i)$. These equations yield that

$$\lambda p_i + \mu N_i - (i+1)p_{i+1} = 0, \ 0 \leq i \leq c-1. \tag{2.26}$$

Denote the ratio $\frac{\mu N_i}{p_i} = \mathrm{E}\,(\mu N(t)|C(t) = i)$, which equals the rate of flow of repeated calls given that the number of busy servers equals $i$, as $r_i$. Then equation (2.26) can be rewritten as

$$p_{i+1} = \frac{\lambda + r_i}{i+1}p_i, \ 0 \leq i \leq c-1.$$

From this we recursively have:

$$p_i = \frac{(\lambda + r_{i-1})\ldots(\lambda + r_0)}{i!}p_0, \ 0 \leq i \leq c, \tag{2.27}$$

and from the normalizing condition $\sum_{k=0}^{c} p_k = 1$ :

$$p_0 = \left(\sum_{k=0}^{c} \frac{(\lambda + r_{k-1})\ldots(\lambda + r_0)}{k!}\right)^{-1}. \tag{2.28}$$

Thus the steady state distribution of the number of busy servers in the retrial queue is identical to the steady state distribution of the number of busy servers in the Erlang loss model with state dependent arrival rate $\Lambda_i = \lambda + r_i$. The extra load $r_i$ is formed by repeated calls.

Although parameters $r_i$, $0 \leq i \leq c - 1$, are unknown and thus equations (2.27), (2.28) do not give a closed form solution, these equations provide some insight into the problem and will be used later on.

It should be noted that equation (2.23) gives the following relation for the parameters $r_i$ :

$$\sum_{k=0}^{c} k \frac{(\lambda + r_{k-1}) \dots (\lambda + r_0)}{k!} = \lambda \sum_{k=0}^{c} \frac{(\lambda + r_{k-1}) \dots (\lambda + r_0)}{k!},$$

(2.29)

from which one of the $r_i$ can be eliminated.

## 2.4 Truncated model

### 2.4.1 Model description

In this model as opposed to the main model, the orbit size (i.e. the number of sources of repeated calls) is bounded by a given constant $M$. If the number of sources equals $M$ then the blocked calls are lost and have no influence on the functioning of the system. The stochastic dynamics of the system can be described by means of a bivariate process $(C^{(M)}(t), N^{(M)}(t))$, where $C^{(M)}(t)$ is the number of busy servers and $N^{(M)}(t)$ is the number of sources of repeated calls (queue length) at time $t$. Under the above assumptions the process $(C^{(M)}(t), N^{(M)}(t))$ is Markovian with the finite lattice semi-strip $S^{(M)} = \{0, 1, ..., c\} \times \{0, 1, ..., M\}$ as the state space. Its infinitesimal transition rates $q_{(ij)(nm)}^{(M)}$ are given by:

1. for $0 \leq i \leq c - 1$, $0 \leq j \leq M$

$$q_{(ij)(nm)}^{(M)} = \begin{cases} \lambda, & \text{if } (n,m) = (i+1, j), \\ i, & \text{if } (n,m) = (i-1, j), \\ j\mu, & \text{if } (n,m) = (i+1, j-1), \\ -(\lambda + i + j\mu), & \text{if } (n,m) = (i,j), \\ 0 & \text{otherwise.} \end{cases}$$

2. for $i = c$, $0 \le j \le M - 1$

$$q^{(M)}_{(cj)(nm)} = \begin{cases} \lambda, & \text{if } (n,m) = (c, j+1), \\ c, & \text{if } (n,m) = (c-1, j), \\ -(\lambda + c), & \text{if } (n,m) = (c,j), \\ 0 & \text{otherwise.} \end{cases}$$

3. for $i = c$, $j = M$

$$q^{(M)}_{(cM)(nm)} = \begin{cases} c, & \text{if } (n,m) = (c-1, M), \\ -c, & \text{if } (n,m) = (c, M), \\ 0 & \text{otherwise.} \end{cases}$$

Thus the rates $q^{(M)}_{(ij)(nm)}$ are the same as those of the main model except for the boundary state $i = c, j = M$.

The system $S^{(M)}$ is convenient since the corresponding set of Kolmogorov equations is finite and thus can be solved numerically by computer.

## 2.4.2 Ergodicity

Since the state space of the process $(C^{(M)}(t), N^{(M)}(t))$ is finite, the process is always ergodic. Its stationary distribution $p^{(M)}_{ij} = \mathrm{P}\{C^{(M)}(t) = i, N^{(M)}(t) = j\}$ may be found as a solution of the following set of linear equations:

$$\begin{aligned} (\lambda + i + j\mu)p^{(M)}_{ij} &= \lambda p^{(M)}_{i-1,j} + (j+1)\mu p^{(M)}_{i-1,j+1} \\ &+ (i+1)p^{(M)}_{i+1,j}, \\ &\quad 0 \le i \le c-1,\ 0 \le j \le M-1, \quad (2.30) \\ (\lambda + i + M\mu)p^{(M)}_{iM} &= \lambda p^{(M)}_{i-1,M} + (i+1)p^{(M)}_{i+1,M}, \\ &\quad 0 \le i \le c-1, \quad (2.31) \\ (\lambda + c)p^{(M)}_{cj} &= \lambda p^{(M)}_{c-1,j} + (j+1)\mu p^{(M)}_{c-1,j+1} \\ &+ \lambda p^{(M)}_{c,j-1},\ 0 \le j \le M-1, \quad (2.32) \\ cp^{(M)}_{cM} &= \lambda p^{(M)}_{c-1,M} + \lambda p^{(M)}_{c,M-1}, \quad (2.33) \end{aligned}$$

which satisfies the normalizing condition

$$\sum_{i=0}^{c}\sum_{j=0}^{M} p^{(M)}_{ij} = 1. \quad (2.34)$$

*2.4.3 Explicit formulas for the main performance characteristics*

For generating functions

$$p_i^{(M)}(z) = \sum_{j=0}^{M} z^j p_{ij}^{(M)}, \ 0 \le i \le c$$

equations (2.30)–(2.33) become

$$(\lambda + i)p_i^{(M)}(z) \ + \ \mu z \frac{dp_i^{(M)}(z)}{dz} = \lambda p_{i-1}^{(M)}(z) + \mu \frac{dp_{i-1}^{(M)}(z)}{dz}$$

$$+ \ p_{i+1}^{(M)}(z), \ 0 \le i \le c - 1, \tag{2.35}$$

$$(\lambda + c)p_c^{(M)}(z) \ - \ \lambda z^M p_{cM}^{(M)} = \lambda p_{c-1}^{(M)}(z) + \mu \frac{dp_{c-1}^{(M)}(z)}{dz}$$

$$+ \ \lambda z^{(M)} p_c(z) - \lambda z^{M+1} p_{cM}^{(M)}. \tag{2.36}$$

Now introduce the generating function

$$p^{(M)}(x, z) = \sum_{i=0}^{c} x^i p_i^{(M)}(z).$$

Then equations (2.35), (2.36) become:

$$\lambda(1 - x)p^{(M)}(x, z) + \mu(z - x)\frac{\partial p^{(M)}(x, z)}{\partial z}$$

$$+(x - 1)\frac{\partial p^{(M)}(x, z)}{\partial x} + \lambda x^c(x - z)p_c^{(M)}(z)$$

$$+\mu x^c(x - z)\frac{dp_c^{(M)}(z)}{dz} + \lambda z^M x^c(z - 1)p_{cM}^{(M)} = 0.$$

Differentiating this equation with respect to $z$, $x$, $xx$, $xz$, $zz$ at the point $x = 1, z = 1$ we get the following equations:

$$\mu N^{(M)} - \lambda B^{(M)} - \mu N_c^{(M)} + \lambda p_{cM}^{(M)} = 0,$$

$$\lambda + \mu N^{(M)} - Y^{(M)} - \lambda B^{(M)} - \mu N_c^{(M)} = 0,$$

$$\mu \frac{\partial^2 p^{(M)}(1,1)}{\partial z^2} - \lambda N_c^{(M)} - \mu \frac{d^2 p_c^{(M)}(1)}{dz^2} + \lambda M p_{cM}^{(M)} = 0,$$

$$-\lambda N^{(M)} - \mu \frac{\partial^2 p^{(M)}(1,1)}{\partial z^2} + (1 + \mu)\frac{\partial^2 p^{(M)}(1,1))}{\partial x \partial z}$$

$$+\lambda N_c^{(M)} - \mu c N_c^{(M)} - \lambda c B^{(M)} + \mu \frac{d^2 p_c^{(M)}(1)}{dz^2} + \lambda c p_{cM}^{(M)} = 0,$$

$$\lambda Y^{(M)} + \mu \frac{\partial^2 p^{(M)}(1,1)}{\partial x \partial z} - \frac{\partial^2 p^{(M)}(1,1)}{\partial x^2} - \lambda c B^{(M)} = \mu c N_c^{(M)},$$

where

$$N^{(M)} \equiv EN^{(M)}(t) = \frac{\partial p^{(M)}(1,1)}{\partial z},$$

$$B^{(M)} \equiv P\{C^{(M)}(t) = c\} = p_c^{(M)}(1),$$

$$Y^{(M)} \equiv EC^{(M)}(t) = \frac{\partial p^{(M)}(1,1)}{\partial x},$$

$$N_c^{(M)} \equiv E\{N^{(M)}(t); C^{(M)}(t) = c\} = \frac{dp_c^{(M)}(1)}{dz}.$$

Eliminating from these equations variables

$$N_c^{(M)}, \frac{\partial^2 p^{(M)}(1,1)}{\partial x \partial z}, \frac{\partial^2 p^{(M)}(1,1)}{\partial z^2}, \frac{d^2 p_c^{(M)}(1)}{dz^2}$$

and taking into account that

$$\frac{\partial^2 p^{(M)}(1,1)}{\partial x^2} = \text{Var}C^{(M)}(t) + \left(EC^{(M)}(t)\right)^2 - EC^{(M)}(t)$$

we get:

$$Y^{(M)} = \lambda - \lambda p_{cM}^{(M)} \tag{2.37}$$

$$N^{(M)} = \frac{1+\mu}{\mu} \cdot \frac{\lambda + \lambda^2 - E\left(C^{(M)}(t)\right)^2}{c - \lambda}$$
$$- \frac{\lambda}{\mu} \cdot \frac{(c+1+\lambda)(1+\mu) + M\mu}{c - \lambda} p_{cM}^{(M)} \tag{2.38}$$

Equation (2.37) can be thought of as a variant of Little's formula and represents a balance between offered, carried and lost traffic. Equation (2.38) is much more interesting. It gives a partial description of the dependence of the mean queue length upon the system parameters, and reduces calculation of the mean queue length to the calculation of the characteristics of the number of busy servers and the rate of lost traffic, which is a simpler problem.

## 2.4.4 Relation with the initial system

To state how the truncated model is related to the initial one we need some general notions and results from the theory of stochastic processes.

The first is the notion of stochastic comparability of random elements.

Let $(\Omega, \mathcal{F}, \mathrm{P})$ be a probability space and $(E, \mathcal{M})$ be a measurable space, where the set $E$ is partially ordered by means of some relation $\prec$. If $\xi^{(1)}{:}\Omega \to E$ and $\xi^{(2)}{:}\Omega \to E$ are two random elements, then $\xi^{(1)}$ is stochastically less than $\xi^{(2)}$ (notation $\xi^{(1)} \prec_{\mathrm{st}} \xi^{(2)}$) iff $\mathrm{P}(\xi^{(1)} \in A) \leq \mathrm{P}(\xi^{(2)} \in A)$ for any monotone subset $A$ of the state space $E$ (recall that subset $A$ of the state space $E$ is said to be a monotone set if $x \in A$ and $x \prec y$ imply that $y \in A$).

The most general results about stochastic comparability of random variables, vectors and processes were established in the monograph *Comparison Methods for Queues and Other Stochastic Models* by D. Stoyan (John Wiley, New York, 1983) and in a paper of Kamae, T., Krengel, U. and O'Brien, G.L. (1977) Stochastic inequalities on partially ordered spaces. *Ann.Probab.*, **5**, 899–912. We recall only two of them. The first describes the nature of the notion of stochastic comparability.

**Statement 10** *If $E$ is a complete separable metric space and the $\sigma$-algebra $\mathcal{M}$ is generated by the open sets, then $\xi^{(1)} \prec_{\mathrm{st}} \xi^{(2)}$ if and only if there exists a probability space $(\Omega^*, \mathcal{F}^*, \mathbf{P}^*)$ and random elements $\eta^{(1)}: \Omega^* \to E$ and $\eta^{(2)}: \Omega^* \to E$ such that:*

- *$\xi^{(1)}$ has the same distribution as $\eta^{(1)}$ and $\xi^{(2)}$ has the same distribution as $\eta^{(2)}$;*
- *$\eta^{(1)}(\omega) \prec \eta^{(2)}(\omega)$ for all outcomes $\omega \in \Omega^*$.*

The second statement gives a necessary and sufficient condition for stochastic comparability of Markov chains in terms of their rates of transitions (all of the continuous-time Markov chains considered below are assumed to be regular).

**Statement 11** *Let $\xi_t^{(1)}, \xi_t^{(2)}$ be two Markov processes with a countable partially ordered state space $E$ and infinitesimal characteristics $a_{x,y}^{(1)}, a_{x,y}^{(2)}$ respectively. For $A \subset E$ let $a^{(1)}(x, A) = \sum_{y \in A} a_{x,y}^{(1)}$ and $a^{(2)}(x, A) = \sum_{y \in A} a_{x,y}^{(2)}$. If*

- *$\xi_0^{(1)} \prec_{\mathrm{st}} \xi_0^{(2)}$;*
- *for any monotone subset $A \subset E$ and any $x, y \in E$ such that $x \prec y$ we have:*

$$a^{(1)}(x, A) \leq a^{(2)}(x, A), \; if \, x, y \notin A;$$
$$a^{(1)}(x, A) \geq a^{(2)}(x, A), \; if \, x, y \in A,$$

*then there exists a probability space $(\Omega, \mathcal{F}, P)$ and Markov chains $\eta_t^{(1)}, \eta_t^{(2)}$ on it, which coincide in distribution with $\xi_t^{(1)}, \xi_t^{(2)}$ respectively and for which $\eta_t^{(1)}(\omega) \prec \eta_t^{(2)}(\omega)$ for all $t \geq 0$ and all outcomes $\omega \in \Omega$.*

The main difficulty in application of this theorem is connected with the fact that in general it is very difficult to describe all monotone subsets of the state space and check the above inequalities for the rates of transitions directly. However, for some specific types of state space and stochastic processes these inequalities can be simplified. The most important for us is the case when the state space $(E, \prec)$ is a multidimensional integer lattice $Z^N$ (or its subset) with component-wise ordering of vectors $\mathbf{x} = (x_1, ..., x_N)$ and the processes $\xi_t^{(1)}, \xi_t^{(2)}$ are migration processes. Recall that a Markov random walk $\xi_t$ on the multidimensional integer lattice $Z^N$ (or its subset) is said to be a migration process if only the following transitions from a point $\mathbf{n} = (n_1, ..., n_N)$ of the state space are possible (below $\mathbf{e}_i$ is the $N$-dimensional vector whose $i$-th coordinate is 1 and the rest are equal to 0):

1. $\mathbf{n} + \mathbf{e}_i$ with rate $\lambda_{\mathbf{n}}^i$, $1 \leq i \leq N$;

2. $\mathbf{n} - \mathbf{e}_i$ with rate $\mu_{\mathbf{n}}^i$, $1 \leq i \leq N$;

3. $\mathbf{n} - \mathbf{e}_i + \mathbf{e}_j$ with rate $a_{\mathbf{n}}^{i,j}$, $1 \leq i, j \leq N$.

One-dimensional migration processes are the classical birth and death processes. Two-dimensional migration processes can be used to describe various retrial queues. For example, for the main model the process $(C(t), N(t))$ is a migration process with the following infinitesimal characteristics:

$$\lambda_{(n,m)}^1 = \begin{cases} \lambda, & \text{if } 0 \leq n \leq c-1, \\ 0, & \text{if } n = c; \end{cases}$$

$$\lambda_{(n,m)}^2 = \begin{cases} 0, & \text{if } 0 \leq n \leq c-1, \\ \lambda, & \text{if } n = c; \end{cases}$$

$$\mu_{(n,m)}^1 = n;$$

$$\mu_{(n,m)}^2 = 0;$$

$$a_{(n,m)}^{1,2} = 0;$$

$$a_{(n,m)}^{2,1} = \begin{cases} m\mu, & \text{if } 0 \leq n \leq c-1, \\ 0, & \text{if } n = c. \end{cases}$$

Similarly, the process $(C^{(M)}(t), N^{(M)}(t))$ for the truncated model is a migration process with the following infinitesimal characteristics:

$$\lambda^{1(M)}_{(n,m)} = \begin{cases} \lambda, & \text{if } 0 \leq n \leq c-1, \\ 0, & \text{if } n = c; \end{cases}$$

$$\lambda^{2(M)}_{(n,m)} = \begin{cases} 0, & \text{if } 0 \leq n \leq c-1, \\ \lambda, & \text{if } n = c, \ 0 \leq m \leq M-1, \\ 0, & \text{if } n = c, \ m = M; \end{cases}$$

$$\mu^{1(M)}_{(n,m)} = n;$$

$$\mu^{2(M)}_{(n,m)} = 0;$$

$$a^{1,2(M)}_{(n,m)} = 0;$$

$$a^{2,1(M)}_{(n,m)} = \begin{cases} m\mu, & \text{if } 0 \leq n \leq c-1, \\ 0, & \text{if } n = c. \end{cases}$$

For migration processes the above Statement 11 becomes the following Statement 12 (in fact we give a more general form of the statement proved by coupling arguments by Falin, G. (1988) Comparability of migration processes. *Theory of Probability and Its Applications*, **33**, No.2, 370–372).

**Statement 12** *Let $\xi^{(1)}_t$ and $\xi^{(2)}_t$ be $N$-dimensional migration processes with state spaces $A_1, A_2 \subset Z^N_+$ and infinitesimal characteristics $\lambda^{i(1)}_{\mathbf{n}}, \mu^{i(1)}_{\mathbf{n}}, a^{i,j(1)}_{\mathbf{n}}$ ($\mathbf{n} \in A_1$) and $\lambda^{i(2)}_{\mathbf{n}}, \mu^{i(2)}_{\mathbf{n}}, a^{i,j(2)}_{\mathbf{n}}$ ($\mathbf{n} \in A_2$) respectively. For any two vectors $\mathbf{n}, \mathbf{m} \in Z_+$ such that $\mathbf{n} \leq \mathbf{m}$, let $I \equiv I_{\mathbf{n},\mathbf{m}}$ denote the set $\{i | n_i = m_i\}$ and for $i \in I_{\mathbf{n},\mathbf{m}}$ let $J^{(1)}_i = J^{(1)}_{i,\mathbf{n},\mathbf{m}}$ and $J^{(2)}_i = J^{(2)}_{i,\mathbf{n},\mathbf{m}}$ denote sets $\left\{ j \in I \setminus \{i\} \Big| a^{j,i(1)}_{\mathbf{n}} > a^{j,i(2)}_{\mathbf{n}} \right\}$ and $\left\{ j \in I \setminus \{i\} \Big| a^{i,j(1)}_{\mathbf{n}} < a^{i,j(2)}_{\mathbf{n}} \right\}$ respectively. If*

* $\xi^{(1)}_0 \leq_{st} \xi^{(2)}_0$;

* *for any two vectors $\mathbf{n} \in A_1, \mathbf{m} \in A_2$ such that $\mathbf{n} \leq \mathbf{m}$ and all $i \in I_{\mathbf{n},\mathbf{m}}$ we have:*

$$\lambda^{i(1)}_{\mathbf{n}} + \sum_{j \in J^{(1)}_i \setminus I} a^{j,i(1)}_{\mathbf{n}} \leq \lambda^{i(2)}_{\mathbf{m}} + \sum_{j \in J^{(1)}_i \setminus I} a^{j,i(2)}_{\mathbf{m}};$$

$$\mu^{i(1)}_{\mathbf{n}} + \sum_{j \in J^{(2)}_i \setminus I} a^{i,j(1)}_{\mathbf{n}} \geq \mu^{i(2)}_{\mathbf{m}} + \sum_{j \in J^{(2)}_i \setminus I} a^{i,j(2)}_{\mathbf{m}},$$

*then there exist a probability space $(\Omega, \mathcal{F}, P)$ and a 'two-dimensional' Markov process $(\eta_t^{(1)}, \eta_t^{(2)})$ on it such that*

- *$\eta_t^{(1)}, \eta_t^{(2)}$ coincide in distribution with $\xi_t^{(1)}, \xi_t^{(2)}$ respectively;*
- *$\eta_t^{(1)}(\omega) \leq \eta_t^{(2)}(\omega)$ for all $t \geq 0$ and all outcomes $\omega \in \Omega$.*

In the two dimensional case this statement can be further simplified to

**Statement 13** *Let $\xi_t^{(1)}, \xi_t^{(2)}$ be two migration processes with state spaces $A_1, A_2 \subset Z_+^2$ and the infinitesimal characteristics $\lambda_{\mathbf{n}}^{i(1)}$, $\mu_{\mathbf{n}}^{i(1)}$, $a_{\mathbf{n}}^{i,j(1)}$ ($\mathbf{n} \in A_1$) and $\lambda_{\mathbf{n}}^{i(2)}$, $\mu_{\mathbf{n}}^{i(2)}$, $a_{\mathbf{n}}^{i,j(2)}$ ($\mathbf{n} \in A_2$) respectively. If*

- *$\xi_0^{(1)} \leq_{\mathrm{st}} \xi_0^{(2)}$;*
- *$\lambda_{\mathbf{n}}^{i(1)} \leq \lambda_{\mathbf{n}}^{i(2)}$; $i = 1, 2$; $\mathbf{n} \in A_1 \cap A_2$;*
- *$\mu_{\mathbf{n}}^{i(1)} \geq \mu_{\mathbf{n}}^{i(2)}$; $i = 1, 2$; $\mathbf{n} \in A_1 \cap A_2$;*
- *for any two vectors $\mathbf{n} \in A_1, \mathbf{m} \in A_2$ such that $n_i = m_i$ and $n_j \leq m_j$ ($i, j = 1, 2$; $i \neq j$) we have:*

$$\lambda_{\mathbf{n}}^{i(1)} + a_{\mathbf{n}}^{j,i(1)} \leq \lambda_{\mathbf{m}}^{i(2)} + a_{\mathbf{m}}^{j,i(2)};$$
$$\mu_{\mathbf{n}}^{i(1)} + a_{\mathbf{n}}^{i,j(1)} \geq \mu_{\mathbf{m}}^{i(2)} + a_{\mathbf{m}}^{i,j(2)},$$

*then there exist a probability space $(\Omega, \mathcal{F}, P)$ and a 'two-dimensional' Markov process $(\eta_t^{(1)}, \eta_t^{(2)})$ on it such that*

- *$\eta_t^{(1)}, \eta_t^{(2)}$ coincide in distribution with $\xi_t^{(1)}, \xi_t^{(2)}$ respectively;*
- *$\eta_t^{(1)}(\omega) \leq \eta_t^{(2)}(\omega)$ for all $t \geq 0$ and all outcomes $\omega \in \Omega$.*

Taking into account the description of both main and truncated retrial models as migration processes and applying Statement 13 we get the following main theorems.

**Theorem 2.3** *If*

$$(C^{(M)}(0), N^{(M)}(0)) \leq_{\mathrm{st}} (C(0), N(0)),$$

*then for all $t \geq 0$ we have:*

$$(C^{(M)}(t), N^{(M)}(t)) \leq_{\mathrm{st}} (C(t), N(t)).$$

*In particular, for the corresponding stationary distributions we get:*

$$\{p_{ij}^{(M)}\} \leq_{\mathrm{st}} \{p_{ij}\}.$$

**Theorem 2.4** *If*

$$(C^{(M)}(0), N^{(M)}(0)) \leq_{\text{st}} (C^{(M+1)}(0), N^{(M+1)}(0)),$$

*then for all $t \geq 0$ we have:*

$$(C^{(M)}(t), N^{(M)}(t)) \leq_{\text{st}} (C^{(M+1)}(t), N^{(M+1)}(t)).$$

*In particular, for the corresponding stationary distributions we get:*

$$\{p_{ij}^{(M)}\} \leq_{\text{st}} \{p_{ij}^{(M+1)}\}.$$

*2.4.5 An algorithm for numerical calculation of the stationary distribution in the truncated system*

The set (2.30)–(2.34) can be solved by a computer with the help of standard subroutines for solution of systems of linear equations. However, using the specific form of equations (2.30)–(2.33), it is possible to suggest a recursive algorithm for calculation of the probabilities $p_{ij}^{(M)}$.

Introduce new variables $r_{ij}^{(M)}$, $0 \leq i \leq c, 0 \leq j \leq M$, as follows:

$$r_{ij}^{(M)} = \frac{p_{ij}^{(M)}}{p_{0M}^{(M)}}.$$

Clearly, if we find variables $r_{ij}^{(M)}$ then we can calculate probabilities $p_{ij}^{(M)}$ as follows:

$$p_{ij}^{(M)} = \frac{r_{ij}^{(M)}}{\sum_{i=0}^{c} \sum_{j=0}^{M} r_{ij}^{(M)}}.$$

Variables $r_{ij}^{(M)}$ satisfy the following set of equations, which follow from equations (2.30)–(2.33) for probabilities $p_{ij}^{(M)}$:

$$r_{0M}^{(M)} = 1, \qquad\qquad\qquad\qquad\qquad\qquad (2.39)$$

$$(\lambda + i + j\mu)r_{ij}^{(M)} = \lambda r_{i-1,j}^{(M)} + (j+1)\mu r_{i-1,j+1}^{(M)} + (i+1)r_{i+1,j}^{(M)},$$
$$\text{if } 0 \leq i \leq c-1, \ 0 \leq j \leq M-1, \quad (2.40)$$

$$(\lambda + i + M\mu)r_{iM}^{(M)} = \lambda r_{i-1,M}^{(M)} + (i+1)r_{i+1,M}^{(M)},$$
$$\text{if } 0 \leq i \leq c-1, \ j = M, \qquad\qquad (2.41)$$

$$(\lambda + c)r_{cj}^{(M)} = \lambda r_{c-1,j}^{(M)} + (j+1)\mu r_{c-1,j+1}^{(M)} + \lambda r_{c,j-1}^{(M)},$$
$$\text{if } i = c, \ 0 \le j \le M - 1, \quad (2.42)$$

$$cr_{cM}^{(M)} = \lambda r_{c-1,M}^{(M)} + \lambda r_{c,M-1}^{(M)}. \quad (2.43)$$

Let us calculate variables $r_{ij}^{(M)}$ by groups, each of size $c+1$; first calculate $r_{0M}^{(M)}, ..., r_{cM}^{(M)}$, then $r_{0M-1}^{(M)}, ..., r_{cM-1}^{(M)}$ and so on, until we find $r_{00}^{(M)}, ..., r_{c0}^{(M)}$.

1. Put $j = M$.

1.2. To find the group $r_{0M}^{(M)}, ..., r_{cM}^{(M)}$ rewrite equation (2.41) as

$$r_{i+1,M}^{(M)} = \frac{(\lambda + i + M\mu)r_{iM}^{(M)} - \lambda r_{i-1,M}^{(M)}}{i+1}, \ 0 \le i \le c-1,$$

or equivalently

$$r_{i,M}^{(M)} = \frac{(\lambda + i - 1 + M\mu)r_{i-1,M}^{(M)} - \lambda r_{i-2,M}^{(M)}}{i}, \ 1 \le i \le c.$$

Since by (2.39) $r_{0M}^{(M)} = 1$, we can recursively calculate variables $r_{1M}^{(M)}, ..., r_{cM}^{(M)}$.

2. Put $j = j - 1$. Let us calculate variables $r_{0j}^{(M)}, ..., r_{cj}^{(M)}$.

2.1. The last variable, $r_{cj}^{(M)}$, can be found from equation (2.43) (if $j = M - 1$):

$$r_{cM-1}^{(M)} = \frac{cr_{cM}^{(M)} - \lambda r_{c-1,M}^{(M)}}{\lambda} \quad (2.44)$$

or from equation (2.42) with $j$ replaced by $j + 1$ (if $j < M - 1$):

$$r_{cj}^{(M)} = \frac{(\lambda + c)r_{c,j+1}^{(M)} - \lambda r_{c-1,j+1}^{(M)} - (j+2)\mu r_{c-1,j+2}^{(M)}}{\lambda}. \quad (2.45)$$

2.2. To find variables $r_{0j}^{(M)}, ..., r_{c-1j}^{(M)}$ let us use equation (2.40) for $i = 0, ..., c - 1$. This set of equations has the form

$$\alpha_i x_{i-1} + \beta_i x_i + \gamma_i x_{i+1} = \delta_i, \ 0 \le i \le c - 1, \quad (2.46)$$

where

$$x_i = r_{ij}^{(M)},$$
$$\alpha_i = -\lambda,$$
$$\beta_i = \lambda + i + j\mu,$$
$$\gamma_i = -(i+1),$$
$$\delta_i = (j+1)\mu r_{i-1,j+1}^{(M)},$$

and values

$$x_{-1} = 0, \quad x_c = r_{cj}^{(M)}$$

are known.

Such difference equations arise in numerical solutions of boundary value problem for second-order differential equations. The most effective computational algorithm for their solution is the so-called 'forward elimination, back substitution' algorithm (also known as 'Cholesky factorization') which can be found in many textbooks on numerical methods (see, for example, Godunov, S.K. and Ryabenkii, V.S. (1987) *Difference Schemes,* North-Holland, pp. 47–50). According to this algorithm one first calculates variables $B_i, D_i$, $0 \le i \le c - 1$, by recursive formulas

$$B_0 = \beta_0, \qquad D_0 = \delta_0,$$
$$B_i = \beta_i - \frac{\alpha_i \gamma_{i-1}}{B_{i-1}}, \qquad D_i = \delta_i - \frac{\alpha_i D_{i-1}}{B_{i-1}}, 1 \le i \le c - 1,$$

and then from equation

$$B_i x_i + \gamma_i x_{i+1} = D_i, \ 0 \le i \le c - 1,$$

recursively calculates (in reverse order) unknowns $x_{c-1}, \ldots, x_0$.

In our case this gives the following procedure:

- calculate variables $B_{ij}, D_{ij}, 0 \le i \le c-1$, (we add an additional index $j$ to indicate dependence on this variable) with the help of the equations

$$
\begin{aligned}
B_{0j} &= \lambda + j\mu, \\
B_{ij} &= \lambda + i + j\mu - \frac{\lambda i}{B_{i-1,j}}, \text{ for } 1 \le i \le c - 1; \\
D_{0j} &= 0, \\
D_{ij} &= (j+1)\mu r_{i-1,j+1}^{(M)} + \frac{\lambda D_{i-1,j}}{B_{i-1,j}}, \text{ for } 1 \le i \le c - 1.
\end{aligned}
\qquad (2.47)
$$

- then recursively calculate $r_{ij}^{(M)}, 0 \le i \le c - 1$, (in reverse order, starting with $r_{cj}^{(M)}$ known from step 2.1) with the help of the equation

$$r_{ij}^{(M)} = \frac{D_{ij} + (i+1)r_{i+1,j}^{(M)}}{B_{ij}}, \ i = c-1, c-2, ..., 1, 0.$$

3. Repeating step 2 while $j \ge 0$ (that is, successively for $j = M-2, M-3, \ldots, 0$) we find all variables $r_{ij}^{(M)}$.

Since $p_{ij}^{(M)} = r_{ij}^{(M)} \cdot p_{0M}^{(M)}$ , we have:

$$p_{0M}^{(M)} = \frac{1}{\sum\limits_{i=0}^{c} \sum\limits_{j=0}^{M} r_{ij}^{(M)}}.$$

Now we can calculate probabilities $p_{ij}^{(M)} = r_{ij}^{(M)} \cdot p_{0M}^{(M)}$ and the main probabilistic characteristics of the truncated system:

(a) The blocking probability

$$B^{(M)} \equiv \mathrm{P}\{C^{(M)}(t) = c\} = \sum_{j=0}^{M} r_{cj}^{(M)} \cdot p_{0M}^{(M)};$$

(b) The mean number of busy servers

$$Y^{(M)} \equiv \mathrm{E}C^{(M)}(t) = \sum_{i=0}^{c} \sum_{j=0}^{M} i r_{ij}^{(M)} \cdot p_{0M}^{(M)};$$

(c) The mean queue length

$$N^{(M)} \equiv \mathrm{E}N^{(M)}(t) = \sum_{i=0}^{c} \sum_{j=0}^{M} j r_{ij}^{(M)} \cdot p_{0M}^{(M)}.$$

The mean number of customers in the queue can be also calculated with the help of formula (2.38), where

$$\mathrm{E}\left(C^{(M)}(t)\right)^2 = \sum_{i=0}^{c} \sum_{j=0}^{M} i^2 r_{ij}^{(M)} \cdot p_{0M}^{(M)}.$$

To avoid subtraction in equations (2.47), introduce new variables $b_{ij} = B_{ij} - (\lambda + j\mu)$. Then recursive equations (2.47) become:

$$b_{0j} = 0,$$
$$b_{ij} = i\frac{j\mu + b_{i-1,j}}{\lambda + j\mu + b_{i-1,j}}, \ 1 \le i \le c-1;$$
$$D_{0j} = 0,$$
$$D_{ij} = (j+1)\mu r_{i-1,j+1}^{(M)} + \frac{\lambda D_{i-1,j}}{\lambda + j\mu + b_{i-1,j}}, \ 1 \le i \le c-1.$$

Correspondingly, unknowns $r_{ij}^{(M)}$ must be calculated from the relations

$$r_{ij}^{(M)} = \frac{D_{ij} + (i+1)r_{i+1,j}^{(M)}}{\lambda + j\mu + b_{ij}}, \ i = c-1, c-2, \dots, 1, 0.$$

For the same reason it would be useful to replace equations (2.44), (2.45) by equations which do not include subtractions. With this goal, sum the original Kolmogorov equations (2.30)–(2.33) with respect to $i = 0, 1, \ldots, c$. After some algebra we get

$$\lambda p_{cj}^{(M)} - (j+1)\mu \sum_{i=0}^{c-1} p_{i,j+1}^{(M)} = \lambda p_{c,j-1}^{(M)} - j\mu \sum_{i=0}^{c-1} p_{ij}^{(M)}, \ 0 \leq j \leq M-1;$$

$$\lambda p_{c,M-1}^{(M)} - M\mu \sum_{i=0}^{c-1} p_{iM}^{(M)} = 0.$$

Thus

$$\lambda p_{cj}^{(M)} - (j+1)\mu \sum_{i=0}^{c-1} p_{i,j+1}^{(M)} = 0, \ 0 \leq j \leq M-1, \qquad (2.48)$$

and correspondingly

$$\lambda r_{cj}^{(M)} = (j+1)\mu \sum_{i=0}^{c-1} r_{i,j+1}^{(M)}, \ 0 \leq j \leq M-1. \qquad (2.49)$$

This relation may be used instead of (2.44) and (2.45) to calculate $r_{cj}^{(M)}$ through known (from previous step) variables $r_{i,j+1}^{(M)}$.

To approximate the initial system under heavy traffic and/or low retrial rate $\mu$ the truncation limit $M$ is taken as very large, perhaps several thousand. If the number of servers $c$ is also large, the number of unknowns $r_{ij}^{(M)}$ can be several hundred thousands. Thus their storage in a computer memory may be a problem. To settle this problem note that for calculation of the current row $r_{ij}^{(M)}, 0 \leq i \leq c$, we need only the previous row $r_{ij+1}^{(M)}, 0 \leq i \leq c$, (and the value $r_{c-1j+2}^{(M)}$ in order to calculate $r_{cj}^{(M)}$, if equation (2.45) is used rather than equation (2.49)). Thus we may store in a computer memory only two rows of unknowns $r_{ij}^{(M)}$ – 'the current' $r_{ij}^{(M)}, 0 \leq i \leq c$, and the 'preceding' $r_{ij+1}^{(M)}, 0 \leq i \leq c$. Sums which are necessary for calculation of $p_{0M}^{(M)}, B^{(M)}, N^{(M)}$ (and any other characteristic) must be calculated successively after calculation of the new value of $r_{ij}^{(M)}$. As soon as the row $r_{ij}^{(M)}, 0 \leq i \leq c$, is calculated, it becomes 'preceding' and the row $r_{ij+1}^{(M)}, 0 \leq i \leq c$, is no longer needed. Using this approach requires $2(c+1)$ cells of computer memory independently of the value of $M$ (instead of $(M+1)(c+1)$ cells needed for a direct approach).

The following Pascal program calculates the joint stationary distribution $p_{ij}^{(M)}$ of the number of busy servers and the queue length, blocking probability $B^{(M)}$, the mean number of busy servers $Y^{(M)}$, the variance of the number of busy servers $\mathrm{Var}C^{(M)}(t)$ and the mean number of customers in the queue $N^{(M)}$.

```
Program retrial(Input,Output);
Uses Crt;
Var
i,j,c,M :integer;
lambda,mu,sum,bl,Y,V,N:extended;
r,p :array[0..20,0..100] of extended;
b,D : array[0..20] of extended;
Begin
writeln('input the number of servers '); read(c);
writeln('input the trancation limit '); read(M);
writeln('input the arrival rate '); read(lambda);
writeln('input the retrial rate '); read(mu);
r[0,M]:=1; r[1,M]:=lambda+M*mu;
for i:=2 to c do
r[i,M]:=((lambda+i-1+M*mu)*r[i-1,M]
-lambda*r[i-2,M])/i;
for j:=M-1 downto 0 do
begin
r[c,j]:=0;
for i:=0 to c-1 do r[c,j]:=r[c,j]+r[i,j+1];
r[c,j]:=(j+1)*mu*r[c,j]/lambda;
b[0]:=0; D[0]:=0;
for i:=1 to c-1 do
begin
b[i]:=i*(j*mu+b[i-1])/(lambda+j*mu+b[i-1]);
D[i]:=(j+1)*mu*r[i-1,j+1]
+lambda*D[i-1]/(lambda+j*mu+b[i-1]);
end;
for i:=c-1 downto 0 do
r[i,j]:=(D[i]+(i+1)*r[i+1,j])/(lambda+j*mu+b[i]);
end;
sum:=0; bl:=0; Y:=0; V:=0; N:=0;
for i:=0 to c do
for j:=0 to M do
```

```
begin
sum:=sum+r[i,j];
if i=c then bl:=bl+r[i,j];
Y:=Y+i*r[i,j];
V:=V+i*i*r[i,j];
N:=N+j*r[i,j];
end;
bl:=bl/sum; Y:=Y/sum; V:=V/sum; V:=V-Y*Y; N:=N/sum;
for i:=0 to c do
for j:=0 to M do
p[i,j]:=r[i,j]/sum;
writeln('blocking probability=',bl:6:4);
writeln('the mean number of busy servers=',Y:8:4);
writeln('the variance of
the number of busy servers=',V:8:4);
writeln('the mean number of sources=',N:8:4);
End.
```

## 2.5 Generalized truncated models

### 2.5.1 Models description

Obviously, if the truncation limit $M$ is large enough, then the truncated model which was described in the previous section may be used to approximate the original model. However, to get necessary accuracy sometimes one must take the variable $M$ very large.

In this section we describe other methods of numerical approximate analysis of retrial queues which are more effective as compared with the above direct truncation, in the sense that the truncation limits for these methods are much less.

The main basic idea of approximate calculation of an infinite system $S$ which cannot be calculated directly consists in replacing it by a 'calculable' system $S'$. The method of direct truncation takes as the approximating system $S'$ the finite system $S^{(M)}$ with a bounded number of sources of repeated calls. This system obviously is 'calculable' (as is any finite system). In contrast to this, under generalized truncation we are choosing as the approximating system a certain infinite system which will happen to be 'calculable'. The fact that we will approximate the initial (infinite) system by some infinite system will provide much better accuracy of approximation.

The simplest generalized truncated model (let us denote it as $\widetilde{S}^{(M)}$) may be described as follows. Assume, in contrast to the main model, that the intensity of repetition becomes equal to infinity as soon as the number of sources of repeated calls exceeds a level $M$. As a matter of fact this means that when the number of customers in orbit exceeds level $M$, then these customers form an ordinary queue, so that one customer from the queue will receive immediate service when the server becomes idle). The process $(\widetilde{C}^{(M)}(t), \widetilde{N}^{(M)}(t))$, where $\widetilde{C}^{(M)}(t)$ is the number of busy servers and $\widetilde{N}^{(M)}(t)$ is the queue length in this new 'truncated' system, is a Markov process with the state space $\{0, 1, ..., c-1\} \times \{0, 1, ..., M\} \cup \{c\} \times Z_+$. Its rates of transition $q_{(ij)(nm)}$ from a point $(i,j)$ of the state space are as follows:

(a) if $0 \leq i \leq c-1$, $0 \leq j \leq M$, then

$$q_{(ij)(nm)} = \begin{cases} \lambda, & \text{if } (n,m) = (i+1,j), \\ j\mu, & \text{if } (n,m) = (i+1,j-1), \\ i, & \text{if } (n,m) = (i-1,j). \\ -(\lambda + j\mu + i), & \text{if } (n,m) = (i,j), \\ 0, & \text{otherwise} \end{cases}$$

(b) if $i = c$, $0 \leq j \leq M$, then

$$q_{(cj)(nm)} = \begin{cases} \lambda, & \text{if } (n,m) = (c, j+1), \\ c, & \text{if } (n,m) = (c-1, j). \\ -(\lambda + c), & \text{if } (n,m) = (c,j), \\ 0, & \text{otherwise} \end{cases}$$

(c) if $i = c$, $j \geq M+1$, then

$$q_{(cj)(nm)} = \begin{cases} \lambda, & \text{if } (n,m) = (c, j+1), \\ c & \text{if } (n,m) = (c, j-1). \\ -(\lambda + c), & \text{if } (n,m) = (c,j), \\ 0, & \text{otherwise.} \end{cases}$$

Another way to reduce the initial retrial model to a numerically tractable model consists in the following. Assume in contrast to the main model that when the number of customers in orbit exceeds a level $M$, then only $M+1$ of them are allowed to perform retrials, whereas others wait until this overload period expires. Denote this model $\widehat{S}^{(M)}$.

The process $(\widehat{C}^{(M)}(t), \widehat{N}^{(M)}(t))$, where $\widehat{C}^{(M)}(t)$ is the number of busy servers and $\widehat{N}^{(M)}(t)$ is the queue length in this new 'truncated' system, is a Markov process with $\{0, 1, ..., c\} \times Z_+$ as the state space. Its rates of transition $q_{(ij)(nm)}$ from a point $(i, j)$ of the state space are

(a) if $0 \leq i \leq c - 1$, $0 \leq j \leq M$, then

$$q_{(ij)(nm)} = \begin{cases} \lambda, & \text{if } (n, m) = (i + 1, j), \\ j\mu, & \text{if } (n, m) = (i + 1, j - 1), \\ i, & \text{if } (n, m) = (i - 1, j). \\ -(\lambda + j\mu + i), & \text{if } (n, m) = (i, j), \\ 0, & \text{otherwise} \end{cases}$$

(b) if $i = c$, $0 \leq j \leq M$, then

$$q_{(cj)(nm)} = \begin{cases} \lambda, & \text{if } (n, m) = (c, j + 1), \\ c, & \text{if } (n, m) = (c - 1, j). \\ -(\lambda + c), & \text{if } (n, m) = (c, j), \\ 0, & \text{otherwise} \end{cases}$$

(c) if $0 \leq i \leq c - 1$, $j \geq M + 1$, then

$$q_{(ij)(nm)} = \begin{cases} \lambda, & \text{if } (n, m) = (i + 1, j), \\ (M + 1)\mu, & \text{if } (n, m) = (i + 1, j - 1), \\ i, & \text{if } (n, m) = (i - 1, j). \\ -(\lambda + (M + 1)\mu + i), & \text{if } (n, m) = (i, j), \\ 0, & \text{otherwise} \end{cases}$$

(d) if $i = c$, $j \geq M + 1$, then

$$q_{(cj)(nm)} = \begin{cases} \lambda, & \text{if } (n, m) = (c, j + 1), \\ c, & \text{if } (n, m) = (c - 1, j). \\ -(\lambda + c), & \text{if } (n, m) = (c, j), \\ 0, & \text{otherwise} \end{cases}$$

As a matter of fact both models (and the main model) can be described in the same framework. Namely, as opposite to the main retrial queue assume that the rate of retrial is a function of the total number of customers in orbit: $\mu = \mu_j$. Then

1. the case $\mu_j \equiv \mu$ corresponds to the main model;

2. the case

$$\mu_j = \begin{cases} \mu, & \text{if } 0 \leq j \leq M; \\ \\ \infty, & \text{if } j \geq M + 1 \end{cases}$$

corresponds to the first generalized truncated model $\widetilde{S}^{(M)}$;

3. the case

$$\mu_j = \begin{cases} \mu, & \text{if } 0 \leq j \leq M, \\[2mm] \dfrac{(M+1)\mu}{j}, & \text{if } j \geq M+1, \end{cases}$$

corresponds to the second generalized truncated model $\widehat{S}^{(M)}$.

An important feature of both generalized truncated models is that corresponding processes $(C(t), N(t))$ possess the property of limited spatial homogeneity with respect to the second coordinate, in the sense that for $j > M$ the rates of transitions from a point $(i, j)$ into a point $(n, m)$ of the state space depend on $j$ and $m$ only through $m - j$. Such chains are well studied in a monograph by Neuts, M. F. (1981) *Matrix-Geometric Solutions in Stochastic Models*, The Johns Hopkins University Press, Baltimore, MD. We recall the main definitions and results of this theory.

Let $\xi_t = (x_t, y_t)$ be a continuous time Markov chain with

$$S = \{(0,0), (1,0), \ldots, (c_0, 0)\} \cup \{0, 1, \ldots, c\} \times \{1, 2, \ldots\}$$

as the state space. Denote by $q_{(i,j)(n,m)}$ the rate of transition from the point $(i, j) \in S$ into the point $(n, m) \in S$ and by $\mathbf{Q}_{jm} = \left(q_{(i,j)(n,m)}\right)$ corresponding matrices. The matrix $\mathbf{Q}_{00}$ is a $(c_0 + 1) \times (c_0 + 1)$ matrix, matrices $\mathbf{Q}_{j0}$ for $j \geq 1$ are $(c + 1) \times (c_0 + 1)$ matrices, matrices $\mathbf{Q}_{0m}$ for $m \geq 1$ are $(c_0 + 1) \times (c + 1)$ matrices, matrices $\mathbf{Q}_{jm}$ for $j, m \geq 1$ are $(c + 1) \times (c + 1)$ matrices.

We assume that $\mathbf{Q}_{jm} = 0$ if $|j - m| > 1$. This means that the second coordinate of the process $\xi_t$ can change only by $\pm 1$; in most practical applications this assumption holds.

The chain $\xi_t$ is said to be limited spatial homogeneous if the rates $q_{(i,j)(n,m)}$ depend on $j$ and $m$ through $m - j$, probably except for $j, m \leq 1$,

$$q_{(i,j)(n,m)} = q_{in}^{(m-j)}, \text{ if } j \text{ or } m > 1.$$

Denote $(c+1) \times (c+1)$ matrices $\left(q_{i,n}^{(k)}\right)_{0 \leq i, n \leq c}$ by $\mathbf{Q}_k$, $k = -1, 0, +1$. Partitioning the state space $S$ as $S_0 \cup S_1 \cup S_2 \cup \ldots$, where $S_0 = \{(0,0), (1,0), \ldots, (c_0, 0)\}$ and $S_j = \{(0,j), (1,j), \ldots, (c,j)\}$ for $j \geq 1$, we may write the infinitesimal generator of the chain $\xi_t$ as follows:

$$\begin{pmatrix} \mathbf{Q}_{00} & \mathbf{Q}_{01} & \mathbf{O} & \mathbf{O} & \mathbf{O} & \mathbf{O} & \dots \\ \mathbf{Q}_{10} & \mathbf{Q}_{11} & \mathbf{Q}_1 & \mathbf{O} & \mathbf{O} & \mathbf{O} & \dots \\ \mathbf{O} & \mathbf{Q}_{-1} & \mathbf{Q}_0 & \mathbf{Q}_1 & \mathbf{O} & \mathbf{O} & \dots \\ \mathbf{O} & \mathbf{O} & \mathbf{Q}_{-1} & \mathbf{Q}_0 & \mathbf{Q}_1 & \mathbf{O} & \dots \\ \mathbf{O} & \mathbf{O} & \mathbf{O} & \mathbf{Q}_{-1} & \mathbf{Q}_0 & \mathbf{Q}_1 & \dots \\ \mathbf{O} & \mathbf{O} & \mathbf{O} & \mathbf{O} & \mathbf{Q}_{-1} & \mathbf{Q}_0 & \dots \\ \vdots & \vdots & \vdots & \vdots & \vdots & \vdots & \ddots \end{pmatrix} \qquad (2.50)$$

where $\mathbf{O}$ is the matrix with all elements equal to zero.

Such a chain is also known as a quasi-birth-and-death process. Let

$$\mathbf{Q} = \mathbf{Q}_{-1} + \mathbf{Q}_0 + \mathbf{Q}_{+1} = (q_{in})_{0 \le i, n \le c},$$

where

$$q_{in} = q_{in}^{(-1)} + q_{in}^{(0)} + q_{in}^{(+1)}.$$

Obviously $q_{in} \ge 0$ for $i \ne n$ and $\sum_{n=0}^c q_{in} = 0$. Thus the matrix $\mathbf{Q}$ can be thought of as the infinitesimal generator of a continuous time Markov chain with $\{0, 1, \dots, c\}$ as the state space. This chain is said to be an induced chain. We assume that it is irreducible and denote by $\pi_i$, $i = 0, 1, \dots, c$, its stationary distribution.

Denote by $d_i = \sum_{n=0}^c \left( q_{in}^{(+1)} - q_{in}^{(-1)} \right)$ the mean drift rate of the second coordinate of the chain $\xi_t$ when the initial state is $(i, j) \in S$ with $j \ge 2$. It is well known that the chain is ergodic iff the average mean drift $\sum_{i=0}^c \pi_i d_i$ is negative, in which case the stationary distribution $p_{ij}$ of the chain $\xi_t$ can be calculated as follows.

Partition the distribution $p_{ij}$ as $(\mathbf{p}_0, \mathbf{p}_1, \dots)$, where

$$\mathbf{p}_0 = (p_{00}, \dots, p_{c_0 0})$$

and

$$\mathbf{p}_j = (p_{0j}, \dots, p_{cj}) \text{ for } j \ge 1.$$

Then for $j \ge 1$

$$\mathbf{p}_j = \mathbf{p}_1 \cdot \mathbf{R}^{j-1},$$

where the $(c+1) \times (c+1)$ matrix $\mathbf{R}$ is the unique nonnegative solution with spectral radius less than 1 of the matrix equation

$$\mathbf{R}^2 \mathbf{Q}_{-1} + \mathbf{R} \mathbf{Q}_0 + \mathbf{Q}_{+1} = \mathbf{O}. \qquad (2.51)$$

The vectors $\mathbf{p}_0, \mathbf{p}_1$ can be found as a solution of the following finite set of linear equations:

$$\mathbf{p}_0 \mathbf{Q}_{00} + \mathbf{p}_1 \mathbf{Q}_{10} = \mathbf{0},$$

$$\mathbf{p}_0 \mathbf{Q}_{01} + \mathbf{p}_1 \left( \mathbf{Q}_{11} + \mathbf{R} \mathbf{Q}_{-1} \right) = \mathbf{0},$$

satisfying the normalizing condition

$$\mathbf{p}_0 \cdot \mathbf{e}_0 + \mathbf{p}_1 \left( \mathbf{I} - \mathbf{R} \right)^{-1} \cdot \mathbf{e} = 1. \tag{2.52}$$

Here $\mathbf{I}$ is the $(c+1) \times (c+1)$ identity matrix, $\mathbf{e}_0$ and $\mathbf{e}$ are column vectors from $R^{c_0+1}$ and $R^{c+1}$ respectively with all elements equal to 1.

Thus, Kolmogorov equations for the probabilities $p_{ij}$, which correspond to states $(i,j) \in S_0$, hold without change, whereas in equations which correspond to states $(i,j) \in S_1$ probabilities of states from $S_2$ are replaced according to relation $\mathbf{p}_2 = \mathbf{p}_1 \cdot \mathbf{R}$.

The matrix $\mathbf{R}$, which plays the key role in the analysis, can be effectively calculated as $\lim_{n \to \infty} \mathbf{R}_n$, where the sequence $\mathbf{R}_n$ is given by the following recursive formula:

$$\mathbf{R}_{n+1} = - \left( \mathbf{Q}_{+1} + \mathbf{R}_n^2 \mathbf{Q}_{-1} \right) \mathbf{Q}_0^{-1} \tag{2.53}$$

with the initial condition

$$\mathbf{R}_0 = \mathbf{O}.$$

## 2.5.2 Analysis of the model $\widetilde{S}^{(M)}$

Consider as the initial $c_0$ states the states $(i,j)$ with $0 \le i \le c, 0 \le j \le M - 1$ and $0 \le i \le c - 1, j = M$, arranged as follows:

$$(0,0), (1,0), \ldots, (c,0); \ldots; (0,M), (1,M), \ldots, (c-1,M),$$

and arrange the remaining states as follows: $(c,M), (c,M+1), \ldots$. Then the infinitesimal generator of the process $(\widetilde{C}^{(M)}(t), \widetilde{N}^{(M)}(t))$ can be written as follows:

$$\begin{pmatrix} \mathbf{Q}_{00} & (0,\ldots,0,\lambda)^T & 0 & 0 & \cdots \\ (0,\ldots,0,c) & -(c+\lambda) & \lambda & 0 & \cdots \\ 0 & c & -(c+\lambda) & \lambda & \cdots \\ 0 & 0 & c & -(c+\lambda) & \cdots \\ \vdots & \vdots & \vdots & \vdots & \ddots \end{pmatrix}$$

Application of the above general theory of Markov chains with modified matrix geometric invariant vector leads to the following procedure (in fact, all subsequent results can be easily obtained directly from Kolmogorov equations for the stationary distribution of the process $\left( \widetilde{C}^{(M)}(t), \widetilde{N}^{(M)}(t) \right)$).

1. The induced chain has the single state $c$ and thus its stationary distribution is $\pi_c = 1$.

2. The matrices $\mathbf{Q}_{-1}, \mathbf{Q}_0, \mathbf{Q}_1$ are $1 \times 1$ matrices, i.e. simply real numbers:

$$\mathbf{Q}_{-1} = c, \quad \mathbf{Q}_0 = -(c + \lambda), \quad \mathbf{Q}_1 = \lambda.$$

Therefore the mean drift $d_c = \lambda - c$. Thus the average mean drift $\sum d_i \pi_i = \lambda - c$, so that the condition $\lambda < c$ is necessary and sufficient for ergodicity of the process $\left( \widetilde{C}^{(M)}(t), \widetilde{N}^{(M)}(t) \right)$.

3. The matrix $\mathbf{R}$ is $1 \times 1$ matrix, i.e. simply a real number, satisfying the equation:

$$\mathbf{R}^2 \cdot c - \mathbf{R} \cdot (c + \lambda) + \lambda = 0.$$

This equation has two roots: 1 and $\frac{\lambda}{c}$. The second root is less than 1 and so $\mathbf{R} = \frac{\lambda}{c}$. Thus, for $j \geq M$

$$\widetilde{p}_{cj}^{(M)} = \widetilde{p}_{cM}^{(M)} \cdot \left( \frac{\lambda}{c} \right)^{j-M} \tag{2.54}$$

4. Therefore, for probabilities $\widetilde{p}_{ij}^{(M)}$, $0 \leq i \leq c$, $0 \leq j \leq M$, the following equations hold:

$$
\begin{aligned}
(\lambda + i + j\mu)\widetilde{p}_{ij}^{(M)} &= \lambda \widetilde{p}_{i-1,j}^{(M)} + (j+1)\mu \widetilde{p}_{i-1,j+1}^{(M)} \\
&+ (i+1)\widetilde{p}_{i+1,j}^{(M)}, \\
&\quad 0 \leq i \leq c-1, \ 0 \leq j \leq M-1, \tag{2.55} \\
(\lambda + c)\widetilde{p}_{cj}^{(M)} &= \lambda \widetilde{p}_{c-1,j}^{(M)} + (j+1)\mu \widetilde{p}_{c-1,j+1}^{(M)} \\
&+ \lambda \widetilde{p}_{c,j-1}^{(M)}, \\
&\quad 0 \leq j \leq M-1, \tag{2.56} \\
(\lambda + i + M\mu)\widetilde{p}_{iM}^{(M)} &= \lambda \widetilde{p}_{i-1,M}^{(M)} + (i+1)\widetilde{p}_{i+1,M}^{(M)}, \\
&\quad 0 \leq i \leq c-1, \tag{2.57} \\
c\widetilde{p}_{cM}^{(M)} &= \lambda \widetilde{p}_{c-1,M}^{(M)} + \lambda \widetilde{p}_{c,M-1}^{(M)}, \tag{2.58}
\end{aligned}
$$

with the normalizing condition

$$\sum_{i=0}^{c} \sum_{j=0}^{M} \widetilde{p}_{ij}^{(M)} + \widetilde{p}_{cM}^{(M)} \cdot \frac{\lambda}{c - \lambda} = 1. \tag{2.59}$$

It is worth noting that equations (2.55)–(2.58) are identical to the Kolmogorov equations for stationary probabilities $p_{ij}^{(M)}$ of the ordinary truncated model. Thus for $0 \leq i \leq c, 0 \leq j \leq M$ probabilities $\widetilde{p}_{ij}^{(M)}$ coincide up to a normalizing constant with $p_{ij}^{(M)}$ :

$$\widetilde{p}_{ij}^{(M)} = \text{Const} \cdot p_{ij}^{(M)}, \ 0 \leq i \leq c, \ 0 \leq j \leq M. \qquad (2.60)$$

Summing (2.60) with respect to $0 \leq i \leq c, \ 0 \leq j \leq M$ and using (2.59) we get:

$$\text{Const} = \sum_{i=0}^{c} \sum_{j=0}^{M} \widetilde{p}_{ij}^{(M)} = 1 - \frac{\lambda}{c-\lambda} \cdot \widetilde{p}_{cM}^{(M)}. \qquad (2.61)$$

Now from (2.60) we have:

$$\widetilde{p}_{cM}^{(M)} = \frac{p_{cM}^{(M)}}{1 + \frac{\lambda}{c-\lambda} p_{cM}^{(M)}},$$

$$\text{Const} = \frac{1}{1 + \frac{\lambda}{c-\lambda} p_{cM}^{(M)}}.$$

This allows us to connect stationary performance characteristics of truncated models $S^{(M)}$, $\widetilde{S}^{(M)}$. In particular, the blocking probability $\widetilde{B}^{(M)}$ in the system $\widetilde{S}^{(M)}$ is given by:

$$\widetilde{B}^{(M)} = \frac{B^{(M)} + \frac{\lambda}{c-\lambda} p_{cM}^{(M)}}{1 + \frac{\lambda}{c-\lambda} p_{cM}^{(M)}},$$

where $B^{(M)}$ is the blocking probability in the system $S^{(M)}$, and the mean number of sources of repeated calls $\widetilde{N}^{(M)}$ in the system $\widetilde{S}^{(M)}$ is given by:

$$\widetilde{N}^{(M)} = \frac{N^{(M)} + \frac{(M+1)c-M\lambda}{(c-\lambda)^2} \lambda p_{cM}^{(M)}}{1 + \frac{\lambda}{c-\lambda} p_{cM}^{(M)}},$$

where $N^{(M)}$ is the mean number of sources of repeated calls in the system $S^{(M)}$.

## 2.5.3 Analysis of the model $\widehat{S}^{(M)}$

Partition the state space $S = \{0, 1, \ldots, c\} \times Z_+$ of the process $\left(\widehat{C}^{(M)}(t), \widehat{N}^{(M)}(t)\right)$ as $S_0 \cup S_1 \cup \ldots$, where subset $S_j$ is formed by states $\{(0, j), (1, j), \ldots, (c, j)\}$, and consider as the initial states the $(c+1) \times M$ states $(i, j)$ for $0 \leq i \leq c, 0 \leq j \leq M-1$. To describe

the infinitesimal generator of the process $\left(\widehat{C}^{(M)}(t), \widehat{N}^{(M)}(t)\right)$ in
the form (2.50) introduce the following $(c+1) \times (c+1)$ matrices:

$$\mathbf{Q}_1 = \begin{pmatrix} 0 & 0 & 0 & \ldots & 0 & 0 & 0 \\ 0 & 0 & 0 & \ldots & 0 & 0 & 0 \\ 0 & 0 & 0 & \ldots & 0 & 0 & 0 \\ \vdots & \vdots & \vdots & \ddots & \vdots & \vdots & \vdots \\ 0 & 0 & 0 & \ldots & 0 & 0 & 0 \\ 0 & 0 & 0 & \ldots & 0 & 0 & 0 \\ 0 & 0 & 0 & \ldots & 0 & 0 & \lambda \end{pmatrix}$$

$$\mathbf{B}_j = \begin{pmatrix} 0 & j\mu & 0 & \ldots & 0 & 0 & 0 \\ 0 & 0 & j\mu & \ldots & 0 & 0 & 0 \\ 0 & 0 & 0 & \ldots & 0 & 0 & 0 \\ \vdots & \vdots & \vdots & \ddots & \vdots & \vdots & \vdots \\ 0 & 0 & 0 & \ldots & 0 & j\mu & 0 \\ 0 & 0 & 0 & \ldots & 0 & 0 & j\mu \\ 0 & 0 & 0 & \ldots & 0 & 0 & 0 \end{pmatrix}$$

$$\mathbf{A}_j = \begin{pmatrix} q_{0j} & \lambda & 0 & \ldots & 0 & 0 & 0 \\ 1 & q_{1j} & \lambda & \ldots & 0 & 0 & 0 \\ 0 & 2 & q_{2j} & \ldots & 0 & 0 & 0 \\ \vdots & \vdots & \vdots & \ddots & \vdots & \vdots & \vdots \\ 0 & 0 & 0 & \ldots & & \lambda & 0 \\ 0 & 0 & 0 & \ldots & c-1 & q_{c-1,j} & \lambda \\ 0 & 0 & 0 & \ldots & 0 & c & q_{c,j} \end{pmatrix},$$

where $q_{ij} = -(\lambda + i + j\mu)$ if $0 \le i \le c-1$, and $q_{cj} = -(\lambda + c)$.

Then we can write the infinitesimal generator of the process
$\left(\widehat{C}^{(M)}(t), \widehat{N}^{(M)}(t)\right)$ in the form (2.50) with

$$\mathbf{Q}_{00} = \begin{pmatrix} \mathbf{A}_0 & \mathbf{Q}_1 & \mathbf{O} & \ldots & \mathbf{O} & \mathbf{O} & \mathbf{O} \\ \mathbf{B}_1 & \mathbf{A}_1 & \mathbf{Q}_1 & \ldots & \mathbf{O} & \mathbf{O} & \mathbf{O} \\ \mathbf{O} & \mathbf{B}_2 & \mathbf{A}_2 & \ldots & \mathbf{O} & \mathbf{O} & \mathbf{O} \\ \vdots & \vdots & \vdots & \ddots & \vdots & \vdots & \vdots \\ \mathbf{O} & \mathbf{O} & \mathbf{O} & \ldots & \mathbf{A}_{M-3} & \mathbf{Q}_1 & \mathbf{O} \\ \mathbf{O} & \mathbf{O} & \mathbf{O} & \ldots & \mathbf{B}_{M-2} & \mathbf{A}_{M-2} & \mathbf{Q}_1 \\ \mathbf{O} & \mathbf{O} & \mathbf{O} & \ldots & \mathbf{O} & \mathbf{B}_{M-1} & \mathbf{A}_{M-1} \end{pmatrix},$$

$$\mathbf{Q}_{01} = \begin{pmatrix} \mathbf{O} \\ \mathbf{O} \\ \mathbf{O} \\ \vdots \\ \mathbf{O} \\ \mathbf{O} \\ \mathbf{Q}_1 \end{pmatrix},$$

$$\mathbf{Q}_{10} = (\ \mathbf{O} \quad \mathbf{O} \quad \mathbf{O} \quad \ldots \quad \mathbf{O} \quad \mathbf{O} \quad \mathbf{B}_M\ ),$$

$$\mathbf{Q}_{11} = \mathbf{A}_M,$$

$$\mathbf{Q}_0 = \mathbf{A}_{M+1},$$

$$\mathbf{Q}_{-1} = \mathbf{B}_{M+1},$$

where $\mathbf{O}$ is the $(c+1) \times (c+1)$ matrix with all elements equal to zero, and matrix $\mathbf{Q}_1$ was introduced earlier.

Application of the above general theory of Markov chains with modified matrix geometric invariant vector leads to the following procedure.

1. The induced chain has the state space $\{0, 1, \ldots, c\}$. Its infinitesimal generator is the following tri-diagonal matrix:

$$\mathbf{Q} = \begin{pmatrix} -\Lambda_M & \Lambda_M & \cdots & 0 & 0 \\ 1 & -(1+\Lambda_M) & \cdots & 0 & 0 \\ \vdots & \vdots & \ddots & \vdots & \vdots \\ 0 & 0 & \cdots & -(c-1+\Lambda_M) & \Lambda_M \\ 0 & 0 & \cdots & c & -c \end{pmatrix},$$

where $\Lambda_M = \lambda + (M+1)\mu$. It is the generator of the process of the number of busy servers in the Erlang loss model $M/M/c/0$ which serves a Poisson input flow with rate $\Lambda_M$. Thus, the stationary distribution of the induced chain is

$$\pi_i = \frac{\Lambda_M^i}{i!} \left/ \sum_{k=0}^{c} \frac{\Lambda_M^k}{k!} \right., \ 0 \le i \le c.$$

2. The vector of mean drifts $\mathbf{d} = (\mathbf{Q}_{+1} - \mathbf{Q}_{-1})\mathbf{e}$ is

$$(-(M+1)\mu, -(M+1)\mu, \ldots, -(M+1)\mu, \lambda).$$

Thus the average mean drift is equal to $-(M+1)\mu(1-\pi_c) + \lambda\pi_c$, so that the process $\left(\widehat{C}^{(M)}(t), \widehat{N}^{(M)}(t)\right)$ is ergodic iff

$$\lambda < \frac{(M+1)\mu(1-\pi_c)}{\pi_c}. \tag{2.62}$$

As $M \to \infty$, the right-hand side of this inequality tends to $c$. Thus for any given $\lambda \in (0, c)$ (when the initial retrial queue is ergodic) one should take $M$ large enough to guarantee that relation (2.62) holds.

3. To calculate the matrix $\mathbf{R}$ one may use recursive procedure (2.53). Note that since the matrix $\mathbf{Q}_0$ is tri-diagonal the product $\left(\mathbf{Q}_{+1} + \mathbf{R}_n^2 \mathbf{Q}_{-1}\right) \mathbf{Q}_0^{-1}$ can be easily calculated with the 'forward elimination, back substitution' algorithm without calculation of the inverse matrix $\mathbf{Q}_0^{-1}$.

4. Now one can write down a closed set of linear equations for probabilities $\widehat{p}_{ij}^{(M)}$, $0 \le i \le c$, $0 \le j \le M$ :

$$
\begin{aligned}
(\lambda + i + j\mu)\widehat{p}_{ij}^{(M)} &= \lambda\widehat{p}_{i-1,j}^{(M)} + (j+1)\mu\widehat{p}_{i-1,j+1}^{(M)} \\
&+ (i+1)\widehat{p}_{i+1,j}^{(M)}, \\
&\quad 0 \le i \le c-1,\ 0 \le j \le M-1, \\
(\lambda + c)\widehat{p}_{cj}^{(M)} &= \lambda\widehat{p}_{c-1,j}^{(M)} + (j+1)\mu\widehat{p}_{c-1,j+1}^{(M)} \\
&+ \lambda\widehat{p}_{c,j-1}^{(M)},\ 0 \le j \le M-1, \\
(\lambda + i + M\mu)\widehat{p}_{iM}^{(M)} &= \lambda\widehat{p}_{i-1,M}^{(M)} + (i+1)\widehat{p}_{i+1,M}^{(M)} \\
&+ (M+1)\mu\sum_{k=0}^{c}\widehat{p}_{kM}^{(M)}r_{k,i-1},\ 0 \le i \le c-1, \\
(\lambda + c)\widehat{p}_{cM}^{(M)} &= \lambda\widehat{p}_{c-1,M}^{(M)} + \lambda\widehat{p}_{c,M-1}^{(M)} \\
&+ (M+1)\mu\sum_{k=0}^{c}\widehat{p}_{kM}^{(M)}r_{k,c-1},
\end{aligned}
$$

where $r_{k,i}$ are elements of the matrix $\mathbf{R}$.

## 2.6 Numerical calculation of the stationary characteristics of the main model

### 2.6.1 Introduction

Since we have no explicit analytical formulas for the main performance characteristics of multiserver retrial queues, the only way to obtain precise numerical data is numerical solution of the Kolmogorov equations for stationary probabilities. But since this system of equations is infinite it cannot be solved directly even by a computer. Transformations which reduce this set of equations to

Table 2.1 *Dependence of the blocking probability and the mean queue length on the truncation level $M$ ($c = 5, \lambda = 4, \mu = 20$.)*

| $M$ | $B^{(M)}$ | $N^{(M)}$ |
|---|---|---|
| 0 | 0.1991 | 0 |
| 5 | 0.4539 | 0.9797 |
| 10 | 0.5102 | 1.7057 |
| 15 | 0.5268 | 2.0893 |
| 20 | 0.5321 | 2.2687 |
| 25 | 0.5338 | 2.3456 |
| 30 | 0.5344 | 2.3773 |
| $\infty$ | 0.5347 | 2.3974 |

a solution of some finite problem are not available in the general case. Therefore we need a method of approximate numerical solution of this system. At present the most frequently used such method is one due to Wilkinson, who proposed replacement of the initial queueing system where the number of sources of repeated calls is unbounded, by the truncated system where the number of sources of repeated calls is bounded by some sufficiently large constant $M$. The truncated system is convenient since for it the set of Kolmogorov equations is finite, and thus can be solved numerically by computer.

As one may expect based on intuitive reasonings, as the truncation limit $M$ tends to infinity, performance characteristics of the truncated system converge to the corresponding characteristics of the initial system. This is clear from Table 2.1 where we give blocking probability and mean queue length for various values of $M$ in the case $c = 5, \lambda = 4, \mu = 20$.

It should be noted that the blocking probability $B^{(M)}$ converges to the blocking probability $B$ in the initial main model faster than the queue length $N^{(M)}$ converges to $N$. We may expect that other characteristics of the number of busy servers converge faster than characteristics of the queue length. Taking this into account we may use formula (2.24) to approximate the mean queue length:

$$N^{(M)}_{\text{appr}} = \frac{1 + \mu}{\mu} \cdot \frac{\lambda - \text{Var} C^{(M)}(t)}{c - \lambda}. \qquad (2.63)$$

In Table 2.2 we give both approximations for the mean queue length for various values of $M$ in the case $c = 5, \lambda = 4, \mu = 20$.

Table 2.2 *Approximation of the mean queue length in the main model with the help of the truncated model by means of $N^{(M)}$ and equation (2.63) ($c = 5, \lambda = 4, \mu = 20$.)*

| $M$ | $N^{(M)}_{\text{appr}}$ | $N^{(M)}$ |
|---|---|---|
| 0 | 2.3379 | 0 |
| 5 | 2.2870 | 0.9797 |
| 10 | 2.3583 | 1.7057 |
| 15 | 2.3844 | 2.0893 |
| 20 | 2.3931 | 2.2683 |
| 25 | 2.3960 | 2.3456 |
| 30 | 2.3970 | 2.3773 |
| $\infty$ | 2.3974 | 2.3974 |

Table 2.3 *Dependence of the blocking probability and the mean queue length in the generalized truncated model on the truncation level $M$ ($c = 5, \lambda = 4, \mu = 20$.)*

| $M$ | $\widetilde{B}^{(M)}$ | $B^{(M)}$ | $\widetilde{N}^{(M)}$ | $N^{(M)}$ | $\widetilde{N}^{(M)}_{\text{appr}}$ |
|---|---|---|---|---|---|
| 0 | 0.5541 | 0.1991 | 2.2165 | 0 | 2.3273 |
| 5 | 0.5370 | 0.4539 | 2.3516 | 0.9797 | 2.3896 |
| 10 | 0.5352 | 0.5102 | 2.3831 | 1.7057 | 2.3958 |
| 15 | 0.5348 | 0.5268 | 2.3928 | 2.0893 | 2.3970 |
| 20 | 0.5347 | 0.5321 | 2.3959 | 2.2687 | 2.3973 |
| 25 | 0.5347 | 0.5338 | 2.3969 | 2.3456 | 2.3974 |
| 30 | 0.5347 | 0.5344 | 2.3973 | 2.3773 | 2.3974 |
| $\infty$ | 0.5347 | 0.5347 | 2.3974 | 2.3974 | 2.3974 |

As one can see $N^{(M)}_{\text{appr}}$ converges to $N$ faster than $N^{(M)}$.

Now consider approximation of the performance characteristics of the main model with the help of the generalized truncated model. In Table 2.3 we give blocking probability $\widetilde{B}^{(M)}$ and mean queue length $\widetilde{N}^{(M)}$ in the generalized truncated model for various values of $M$ in the case $c = 5, \lambda = 4, \mu = 20$. For convenience of comparison we give also results from Table 2.1 for the ordinary truncated model.

As comparison of numerical data shows, we have a much higher rate of convergence with the generalized truncated system than for ordinary truncated system.

Taking into account the above remarks concerning use of the semi-explicit formula (2.24) we may approximate $N$ as follows:

$$\widetilde{N}_{\text{appr}}^{(M)} = \frac{1+\mu}{\mu} \cdot \frac{\lambda - \text{Var}\widetilde{C}^{(M)}(t)}{c - \lambda}. \qquad (2.64)$$

Values of variable $\widetilde{N}_{\text{appr}}^{(M)}$ are also given in Table 2.3. As we can see $\widetilde{N}_{\text{appr}}^{(M)}$ provides the best approximation for the queue length $N$ in the main model.

It should be noted that in this approximation the initial step (i.e. the generalized truncated system with the level of truncation $M = 0$ ) is the ordinary $M/M/c$ queue, which has been used for the purpose of approximations of characteristics of retrial queues for a long time. The usual argument in favour of this approximation is that as $\mu \to \infty$ the retrial queue can be viewed as the corresponding $M/M/c$ system with queueing of blocked calls. This approach in fact means that we replace a function of $\mu$ by its limit as $\mu \to \infty$. But this cannot explain the surprisingly high precision of such a replacement. A much better explanation is provided by viewing the standard $M/M/c$ system with queueing of blocked calls as the first in the sequence of systems rapidly converging to the initial system.

From the mathematical point of view the following questions should be investigated rigorously:

(1) Would stationary characteristics of the truncated systems converge as the truncation limit $M \to \infty$ to the corresponding stationary characteristics of the original system?

(2) If the answer to Question 1 is affirmative, how much is the error of this approximation?

*2.6.2 Convergence of the 'truncated' solution to the initial*

Let us denote the sum $\sum\limits_{n=i}^{\infty} \sum\limits_{m=j}^{\infty} p_{nm}$ by $\overline{p}_{ij}$. The inequality

$$\{p_{ij}^{(M)}\} \leq_{\text{st}} \{p_{ij}^{(M+1)}\} \leq_{\text{st}} \{p_{ij}\}$$

implies in particular that

$$\overline{p}_{ij}^{(M)} \leq \overline{p}_{ij}^{(M+1)} \leq \overline{p}_{ij},$$

i.e. monotonicity and boundedness of the sequence $\{\overline{p}_{ij}^{(M)}\}$. Thus we can guarantee existence of the limit $a_{ij} = \lim\limits_{M \to \infty} \overline{p}_{ij}^{(M)}$ and the inequality $a_{ij} \le \overline{p}_{ij}$. But

$$p_{ij}^{(M)} = \overline{p}_{ij}^{(M)} + \overline{p}_{i+1j+1}^{(M)} - \overline{p}_{i+1j}^{(M)} - \overline{p}_{ij+1}^{(M)}.$$

Hence, there exists $b_{ij} = \lim\limits_{M \to \infty} p_{ij}^{(M)} = a_{ij} + a_{i+1j+1} - a_{i+1j} - a_{ij+1}$ and (since $\overline{a}_{ij} \le \overline{\overline{p}}_{ij} < \infty$) $\overline{b}_{ij} = a_{ij}$.

Taking limits of both sides in the Kolmogorov equations for the probabilities $p_{ij}^{(M)}$ we get that the variables $b_{ij}$ satisfy the set of Kolmogorov equations for the process $(C(t), N(t))$. Besides,

$$\sum_{i=0}^{c} \sum_{j=0}^{\infty} b_{ij} = \overline{b}_{00} = a_{00} = \lim_{M \to \infty} \overline{p}_{00}^{(M)} = \lim_{M \to \infty} 1 = 1,$$

i.e. variables $b_{ij}$ satisfy the normalizing condition as well, so that $b_{ij} = p_{ij}$.

Using the general theorems about convergence of probability measures we can guarantee convergence of the stationary distribution of the number of busy servers and so, applying equation (2.24), we have that $N^{(M)} \to N$.

Using formulas which relate characteristics of the generalized truncated model $\widetilde{S}^{(M)}$ with the corresponding characteristics of the truncated model $S^{(M)}$ we can guarantee the convergence of the characteristics of the model $\widetilde{S}^{(M)}$ to the corresponding characteristics of the initial model.

### 2.6.3 Error estimate

*The blocking probability*

Since $C^{(M)}(t) \le_{\text{st}} C(t)$, we have (by the definition of stochastic ordering) that

$$P(C^{(M)}(t) \ge n) \le P(C(t) \ge n), \ 0 \le n \le c.$$

In the case $n = c$ this inequality becomes $B^{(M)} \le B$, i.e.

$$B - B^{(M)} \ge 0.$$

To get an estimate for $B - B^{(M)}$ from above we will use relations (2.23) for $Y$ and (2.37) for $Y^{(M)}$. Since

$$Y = \sum_{n=1}^{c} P(C(t) \geq n), \ Y^{(M)} = \sum_{n=1}^{c} P(C^{(M)}(t) \geq n),$$

we have:

$$
\begin{aligned}
B - B^{(M)} &= P(C(t) \geq c) - P(C^{(M)}(t) \geq c) \\
&\leq \sum_{n=1}^{c} \left( P(C(t) \geq n) - P(C^{(M)}(t) \geq n) \right) \\
&= Y - Y^{(M)} \\
&= \lambda p_{cM}^{(M)}.
\end{aligned}
$$

The probability $p_{cM}^{(M)}$ can be considered as known since it is calculated numerically. Below we will give an estimate of this probability in terms of the system parameters.

### The mean queue length

An error estimate in approximating the mean queue length $N$ depends on the method of approximation. If $N$ is approximated with the help of $N^{(M)}$ then from equations (2.25) and (2.38) we have:

$$
\begin{aligned}
N - N^{(M)} &= \frac{1+\mu}{\mu} \cdot \frac{\left( c+1+\lambda + M\frac{\mu}{1+\mu} \right) \lambda p_{cM}^{(M)}}{c - \lambda} \\
&- \frac{1+\mu}{\mu} \cdot \frac{E\left(C(t)\right)^2 - E\left(C^{(M)}(t)\right)^2}{c - \lambda}.
\end{aligned}
$$

Since

$$
\begin{aligned}
& E\left(C(t)\right)^2 - E\left(C^{(M)}(t)\right)^2 \\
&= \sum_{n=1}^{c} (2n - 1) \left( P(C(t) \geq n) - P(C^{(M)}(t) \geq n) \right),
\end{aligned}
$$

and all terms in this sum are nonnegative, we get:

$$
\begin{aligned}
E\left(C(t)\right)^2 - E\left(C^{(M)}(t)\right)^2 &\leq (2c - 1) \sum_{n=1}^{c} P(C(t) \geq n) \\
&- (2c - 1) \sum_{n=1}^{c} P(C^{(M)}(t) \geq n)
\end{aligned}
$$

$$= (2c-1)\left(Y - Y^{(M)}\right)$$

$$= (2c-1)\lambda p_{cM}^{(M)},$$

and

$$\mathrm{E}\left(C(t)\right)^2 - \mathrm{E}\left(C^{(M)}(t)\right)^2 \geq \sum_{n=1}^{c} \mathrm{P}(C(t) \geq n)$$

$$- \sum_{n=1}^{c} \mathrm{P}(C^{(M)}(t) \geq n)$$

$$= Y - Y^{(M)}$$

$$= \lambda p_{cM}^{(M)}.$$

Thus

$$\frac{1+\mu}{\mu} \cdot \frac{\lambda - c + 2 + M\frac{\mu}{1+\mu}}{c - \lambda} \lambda p_{cM}^{(M)} \leq N - N^{(M)}$$

$$\leq \frac{1+\mu}{\mu} \cdot \frac{\lambda + c + M\frac{\mu}{1+\mu}}{c - \lambda} \lambda p_{cM}^{(M)}.$$

If $N$ is approximated with the help of equation (2.63) then

$$N - N_{\mathrm{appr}}^{(M)}$$

$$= \frac{1+\mu}{\mu} \frac{\mathrm{Var}C^{(M)}(t) - \mathrm{Var}C(t)}{c - \lambda}$$

$$= \frac{1+\mu}{\mu} \frac{\mathrm{E}\left(C^{(M)}(t)\right)^2 - \left(\mathrm{E}C^{(M)}(t)\right)^2 - \mathrm{E}\left(C(t)\right)^2 + \left(\mathrm{E}C(t)\right)^2}{c - \lambda}$$

$$= \frac{1+\mu}{\mu} \frac{\mathrm{E}\left(C^{(M)}(t)\right)^2 - \left(\lambda - \lambda p_{cM}^{(M)}\right)^2 - \mathrm{E}\left(C(t)\right)^2 + \lambda^2}{c - \lambda}.$$

We have already shown that

$$\lambda p_{cM}^{(M)} \leq \mathrm{E}\left(C(t)\right)^2 - \mathrm{E}\left(C^{(M)}(t)\right)^2 \leq (2c-1)\lambda p_{cM}^{(M)}.$$

Thus

$$\frac{1+\mu}{\mu} \cdot \frac{-2c + 2\lambda + 1 - \lambda p_{cM}^{(M)}}{c - \lambda} \lambda p_{cM}^{(M)} \leq N - N_{\mathrm{appr}}^{(M)}$$

$$\leq \frac{1+\mu}{\mu} \cdot \frac{2\lambda - 1 - \lambda p_{cM}^{(M)}}{c - \lambda} \lambda p_{cM}^{(M)}.$$

*The boundary probability $p_{cM}^{(M)}$*

Using relation (2.48) we can transform equation (2.32) as follows:

$$
\begin{aligned}
(\lambda + c)p_{cj}^{(M)} &= \lambda p_{c-1,j}^{(M)} + (j+1)\mu p_{c-1,j+1}^{(M)} + \lambda p_{c,j-1}^{(M)} \\
&\leq \lambda \sum_{i=0}^{c-1} p_{ij}^{(M)} + (j+1)\mu \sum_{i=0}^{c-1} p_{i,j+1}^{(M)} + \lambda p_{c,j-1}^{(M)} \\
&= \lambda \frac{\lambda}{j\mu} p_{c,j-1}^{(M)} + \lambda p_{cj}^{(M)} + \lambda p_{c,j-1}^{(M)},
\end{aligned}
$$

i.e.

$$
cp_{cj}^{(M)} \leq \lambda \frac{\lambda + j\mu}{j\mu} p_{c,j-1}^{(M)}, \text{ for } 1 \leq j \leq M-1. \tag{2.65}
$$

Similarly from equation (2.33) one can show that (2.65) holds for $j = M$ as well.

Now from (2.65) we have recursively:

$$
p_{cj}^{(M)} \geq \frac{\frac{(\lambda/c)^j}{j!\mu^j} \prod_{i=1}^{j}(\lambda + i\mu)}{\frac{(\lambda/c)^M}{M!\mu^M} \prod_{i=1}^{M}(\lambda + i\mu)} \cdot p_{cM}^{(M)}, \text{ for } 0 \leq j \leq M.
$$

Summing these inequalities with respect to $j = 0, \ldots, M$, we get:

$$
B^{(M)} \geq \frac{\sum_{j=0}^{M} \frac{(\lambda/c)^j}{j!\mu^j} \prod_{i=1}^{j}(\lambda + i\mu)}{\frac{(\lambda/c)^M}{M!\mu^M} \prod_{i=1}^{M}(\lambda + i\mu)} \cdot p_{cM}^{(M)}. \tag{2.66}
$$

On the other hand, since

$$
EC^{(M)}(t) = \lambda - \lambda p_{cM}^{(M)}
$$

and

$$
EC^{(M)}(t) \geq cB^{(M)},
$$

we can guarantee that

$$
cB^{(M)} \leq \lambda - \lambda p_{cM}^{(M)}. \tag{2.67}
$$

From (2.66) and (2.67) we have the following inequality for $p_{cM}^{(M)}$:

$$\lambda - \lambda p_{cM}^{(M)} \geq \frac{c \sum_{j=0}^{M} \frac{(\lambda/c)^j}{j! \mu^j} \prod_{i=1}^{j} (\lambda + i\mu)}{\frac{(\lambda/c)^M}{M! \mu^M} \prod_{i=1}^{M} (\lambda + i\mu)} \cdot p_{cM}^{(M)}.$$

Solving this inequality for $p_{cM}^{(M)}$ we finally get the following estimate for $p_{cM}^{(M)}$:

$$p_{cM}^{(M)} \leq \frac{\frac{(\lambda/c)^M}{M! \mu^M} \prod_{i=0}^{M} (\lambda + i\mu)}{\sum_{j=0}^{M} \frac{(\lambda/c)^j}{j! \mu^j} (c + \lambda + j\mu) \prod_{i=0}^{j-1} (\lambda + i\mu)}.$$

It should be noted that in the single-server case all the above inequalities become exact equalities. In this sense the above bounds cannot be improved.

### 2.6.4 Some numerical results

Quality of service to subscribers in queueing systems with repeated attempts is characterized by several performance measures. $W$, the mean waiting time in the steady state, is often considered to be the most important of these performance measures. However, $W$ is an average over all primary calls, including those calls which receive immediate service and really do not wait at all. A fuller under-standing of the waiting time process can be obtained by observing the blocking probability $B$ (the probability that the waiting time is positive) and the conditional mean waiting time $W_B = \frac{W}{B}$ given that the waiting time is positive. Table 2.4 shows dependence of these characteristics on $\lambda$ and $\mu$ for the main model with $c = 10$ servers.

One could observe from Table 2.4 that variables $W_B$ and $W$ decrease as $\mu$ increases, whereas the blocking probability $B$ increases as $\mu$ increases. Thus, impact of the rate of retrials on the quality of service of customers depends on what is considered as the objective function.

In communication engineering, $B$ is usually taken as the main performance measure. Correspondingly, one of the most important problems of design of real systems with repeated demands consists

Table 2.4 *Dependence of the blocking probability B, the conditional mean waiting time $W_B$ and the mean waiting time W on $\lambda$ and $\mu$ for the main model with $c = 10$ servers*

|            |       | $\mu = 0.1$ | $\mu = 1$ | $\mu = 5$ | $\mu = 10$ | $\mu = 20$ |
|------------|-------|-------------|-----------|-----------|------------|------------|
|            | $B$   | 3.8E-5      | 4.0E-5    | 4.2E-5    | 4.4E-5     | 4.5E-5     |
| $\lambda = 2$ | $W_B$ | 10.1309     | 1.1323    | 0.3321    | 0.2307     | 0.1790     |
|            | $W$   | 3.9E-4      | 4.5E-5    | 1.4E-5    | 1.0E-5     | 8.1E-6     |
|            | $B$   | 0.0056      | 0.0061    | 0.0071    | 0.0076     | 0.0081     |
| $\lambda = 4$ | $W_B$ | 10.2481     | 1.2003    | 0.3885    | 0.2825     | 0.2271     |
|            | $W$   | 0.0571      | 0.0073    | 0.0028    | 0.0022     | 0.0018     |
|            | $B$   | 0.0558      | 0.0647    | 0.0795    | 0.0862     | 0.0916     |
| $\lambda = 6$ | $W_B$ | 10.8924     | 1.3731    | 0.5036    | 0.3854     | 0.3219     |
|            | $W$   | 0.6075      | 0.0889    | 0.0401    | 0.0332     | 0.0295     |
|            | $B$   | 0.2424      | 0.2792    | 0.3355    | 0.3589     | 0.3775     |
| $\lambda = 8$ | $W_B$ | 13.7620     | 1.9489    | 0.8422    | 0.6856     | 0.5998     |
|            | $W$   | 3.3362      | 0.5442    | 0.2825    | 0.2460     | 0.2264     |

of determining for given offered traffic $\lambda$ and retrial rate $\mu$ the number of servers $c = c(\lambda, \mu, B)$ which guarantees a preassigned value of blocking probability $B$. The dependence of $c(\lambda, \mu, B)$ on $\lambda, \mu$ and $B$ is illustrated by Table 2.5.

## 2.7 Limit theorems

### 2.7.1 High rate of retrials

In real situations, many subscribers who get a busy signal almost immediately repeat their calls. Therefore, an investigation of the asymptotic behaviour of performance characteristics of retrial systems is of special interest for practical applications.

As $\mu \to \infty$ the stationary distribution of states of any retrial queue $p_x = p_x(\mu)$, where $x$ is a point of the state space of the corresponding stochastic process, usually has a limit $p_x(\infty)$ which is the stationary distribution of states of some 'limit' system. From intuitive considerations it is easy to determine the structure of this limit system and the limit distribution $p_x(\infty)$. Say, for the main model the 'limit' system is the standard $M/M/c/\infty$ queueing model. Of course, the equality $\lim_{\mu \to \infty} p_x(\mu) = p_x(\infty)$ must be proved

Table 2.5 *Dependence of the number of servers which guarantees a pre-assigned value of blocking probability B on* $\lambda, \mu$ *and* $B$

|  |  | $\mu = 0.1$ | $\mu = 1$ | $\mu = 5$ | $\mu = 10$ | $\mu = 20$ |
|---|---|---|---|---|---|---|
|  | $B = 0.1\%$ | 21 | 21 | 22 | 22 | 22 |
| $\lambda = 10$ | $B = 1\%$ | 18 | 18 | 19 | 19 | 19 |
|  | $B = 5\%$ | 15 | 16 | 16 | 16 | 16 |
|  | $B = 0.1\%$ | 71 | 72 | 73 | 73 | 73 |
| $\lambda = 50$ | $B = 1\%$ | 64 | 65 | 66 | 67 | 67 |
|  | $B = 5\%$ | 59 | 60 | 61 | 62 | 62 |
|  | $B = 0.1\%$ | 129 | 129 | 130 | 131 | 132 |
| $\lambda = 100$ | $B = 1\%$ | 119 | 120 | 121 | 122 | 123 |
|  | $B = 5\%$ | 111 | 112 | 114 | 115 | 116 |
|  | $B = 0.1\%$ | 238 | 239 | 241 | 242 | 243 |
| $\lambda = 200$ | $B = 1\%$ | 224 | 225 | 228 | 230 | 231 |
|  | $B = 5\%$ | 212 | 215 | 218 | 220 | 221 |

as a pure mathematical result. But the problem above all is to obtain the next term of the asymptotic expansion of $p_x(\mu)$ in a power series in the mean times between successive retrials $1/\mu$:

$$p_x(\mu) = p_x(\infty) + \frac{1}{\mu} a_x + ...$$

The term $a_x$ describes the influence of repeated calls and hence is of special interest.

Asymptotic analysis of retrial queues under high rate of retrials can be performed with the help of classical perturbation analysis. In the most general form this can be described as follows.

Let the vector of interest $\mathbf{p}_\varepsilon$ be the unique solution to the equation

$$\mathbf{p}_\varepsilon L_\varepsilon = \mathbf{a}, \tag{2.68}$$

where $L_\varepsilon$ is a linear mapping of corresponding linear spaces. Assume that $L_\varepsilon$ can be decomposed into a sum $L_\varepsilon = L_0 + P_\varepsilon$, where the linear operator $L_0$ corresponds to the unperturbed problem $\mathbf{p}_0 L_0 = \mathbf{a}$ and $P_\varepsilon$ is a perturbation. Suppose that the solution $\mathbf{p}_0$ of the unperturbed problem is unique and can be explicitly obtained as $\mathbf{p}_0 = \mathbf{a} L_0^{-1}$. Rewrite the equation (2.68) for $\mathbf{p}_\varepsilon$ as

$$\mathbf{p}_\varepsilon L_0 = \mathbf{a} - \mathbf{p}_\varepsilon P_\varepsilon,$$

so that

$$\begin{aligned} \mathbf{p}_\varepsilon &= (\mathbf{a} - \mathbf{p}_\varepsilon P_\varepsilon)\, L_0^{-1} \\ &= \mathbf{p}_0 - \mathbf{p}_\varepsilon P_\varepsilon L_0^{-1}. \end{aligned}$$

Thus, $\mathbf{p}_0$ can be considered as an approximation of $\mathbf{p}_\varepsilon$ if $\mathbf{p}_\varepsilon P_\varepsilon L_0^{-1}$ is 'small'.

In our case $\varepsilon = 1/\mu$, $\mathbf{p}_\varepsilon$ is the stationary distribution of the Markov process $(C(t), N(t))$, operator $L_\varepsilon$ will be the slightly modified (to include the normalizing condition) infinitesimal generator of Markov process $(C(t), N(t))$. Thus the set of equations (2.68) is simply the set of Kolmogorov equations for the stationary distribution (plus the normalizing condition) and the decomposition $L_\varepsilon = L_0 + P_\varepsilon$ means transformation of these Kolmogorov equations to the Kolmogorov equations for the standard $M/M/c/\infty$ queueing model.

**Theorem 2.5** *As $\mu \to \infty$, the stationary blocking probability $B$ and the stationary mean queue length $N$ are*

$$\begin{aligned} B &= B(\infty) \\ &+ \frac{c - 1 - \lambda + B(\infty)}{\mu} \ln\left(1 - \frac{\lambda}{c}\right) B(\infty) \\ &+ o\left(\frac{1}{\mu}\right), \end{aligned} \tag{2.69}$$

$$\begin{aligned} N &= N(\infty) \\ &+ \frac{\lambda - (c^2 - c - 2\lambda c + 2\lambda + \lambda^2 - \lambda B(\infty))\ln\left(1 - \frac{\lambda}{c}\right)}{\lambda \mu} N(\infty) \\ &+ o\left(\frac{1}{\mu}\right), \end{aligned} \tag{2.70}$$

*where*

$$B(\infty) = \frac{\frac{\lambda^c}{(c-1)!}}{(c - \lambda) \sum\limits_{i=0}^{c-1} \frac{\lambda^i}{i!} + \frac{\lambda^c}{(c-1)!}},$$

$$N(\infty) = \frac{\lambda}{c - \lambda} B(\infty)$$

*are the stationary blocking probability and the mean queue length respectively in the standard $M/M/c/\infty$ queueing system.*

*Proof.* We start the proof with two lemmas which give the asymptotics of the stationary distribution $p_{ij}$.

**Lemma 2.1** *As $\mu \to \infty$ the stationary joint distribution $\{p_{ij}\}$ of the number of busy servers $C(t)$ and the queue length $N(t)$ in the main retrial model converges to the corresponding distribution $\{p_{ij}(\infty)\}$ in the corresponding standard $M/M/c/\infty$ queueing system:*

$$\lim_{\mu \to \infty} p_{ij} = p_{ij}(\infty),$$

*where*

$$
p_{ij}(\infty) = \begin{cases}
\left[ \sum_{i=0}^{c-1} \frac{\lambda^i}{i!} + \frac{\lambda^c}{(c-1)!(c-\lambda)} \right]^{-1}, & \text{if } i = 0,\ j = 0, \\[2ex]
\frac{\lambda^i}{i!} p_{00}(\infty), & \text{if } 0 \le i \le c-1,\ j = 0, \\[2ex]
0, & \text{if } 0 \le i \le c-1,\ j \ge 1, \\[2ex]
\frac{\lambda^c}{c!} \cdot \left( \frac{\lambda}{c} \right)^j p_{00}(\infty), & \text{if } i = c,\ j \ge 0.
\end{cases}
\tag{2.71}
$$

*Proof.* The proof is based on the Kolmogorov equations (2.17–2.18) for the stationary distribution $p_{ij}$. For convenience we repeat these equations here:

$$(\lambda + i + j\mu)p_{ij} = \lambda p_{i-1,j} + (j+1)\mu p_{i-1,j+1} + (i+1)p_{i+1,j},$$
$$0 \le i \le c-1, \tag{2.72}$$

$$(\lambda + c)p_{cj} = \lambda p_{c-1,j} + (j+1)\mu p_{c-1,j+1} + \lambda p_{c,j-1}, \quad (\text{case } i = c), \tag{2.73}$$

The proof of the lemma consists of four stages. First we prove relation (2.71) for $i = 0, ..., c-1;\ j \ge 1$, then for $i = 0,\ j = 0$, then for $i = 1, ..., c-1,\ j = 0$ and finally for $i = 0,\ j > 0$.

**1.** Let $p_i = \mathrm{P}(C(t) = i)$, $N_i = \mathrm{E}(N(t); C(t) = i)$. Summing equations (2.72), (2.73) with respect to $j$ we get that

$$N_i = \frac{(i+1)p_{i+1} - \lambda p_i}{\mu} \quad \text{for } i = 0, ..., c-1.$$

The numerator of the fraction in the right-hand side of this equation is bounded since $p_i$, $p_{i+1}$ are probabilities. Thus, as $\mu \to \infty$ there exists $\lim_{\mu \to \infty} N_i = 0$, $i = 0, ..., c-1$. Because for $j \ge 1$ inequality $p_{ij} \le \frac{N_i}{j}$ holds, we can guarantee that there exists

$$\lim_{\mu \to \infty} p_{ij} = 0 = p_{ij}(\infty) \text{ for } 0 \le i \le c-1,\ j \ge 1.$$

Moreover, taking into account this relation, we can easily establish from (2.72) by induction with respect to $i$ that

$$\lim_{\mu \to \infty} \mu p_{ij} = 0 \text{ for } 0 \leq i \leq c - 2, \ j \geq 1. \tag{2.74}$$

**2.** Consider relations

$$\sum_{i=0}^{c} p_i = 1, \quad \sum_{i=0}^{c} i p_i = \lambda.$$

The first is another form of the normalizing condition, and the second is another form of equation (2.23) for the mean number of busy servers in the steady state. Eliminate from these relations probability $p_c$ and replace all probabilities $p_i$ for $i = 0, ..., c - 1$ as a sum of two terms: $p_{i0}$ and $\sum_{j=1}^{\infty} p_{ij}$ :

$$c - \lambda = \sum_{i=0}^{c-1} (c - i) p_{i0} + \sum_{i=0}^{c-1} (c - i) \sum_{j=1}^{\infty} p_{ij}. \tag{2.75}$$

Further, using the main equation (2.72) for $j = 0$ express $p_{i0}$ in terms of $p_{00}$ and $p_{k1}$ :

$$p_{i0} = \frac{\lambda^i}{i!} p_{00} - \mu \frac{\lambda^i}{i!} \sum_{l=1}^{i} \frac{(l-1)!}{\lambda^l} \sum_{k=0}^{l-2} p_{k1}, \ i = 0, ..., c. \tag{2.76}$$

Substitute this relation for $p_{i0}$ into (2.75), and express from the resulting formula probability $p_{00}$ :

$$p_{00} = \frac{c - \lambda - \sum\limits_{i=0}^{c-1} (c - i) \sum\limits_{j=1}^{\infty} p_{ij} + \mu \sum\limits_{i=0}^{c-1} (c - i) \frac{\lambda^i}{i!} \sum\limits_{l=1}^{i} \frac{(l-1)!}{\lambda^l} \sum\limits_{k=0}^{l-2} p_{k1}}{\sum\limits_{i=0}^{c-1} (c - i) \frac{\lambda^i}{i!}}.$$

$$\tag{2.77}$$

The numerator of the fraction in the right-hand side of this relation contains unknown probabilities $p_{ij}$. However, as $\mu \to \infty$ they tend to zero. Indeed, for the first sum we have:

$$0 < \sum_{i=0}^{c-1} (c - i) \sum_{j=1}^{\infty} p_{ij} < \sum_{i=0}^{c-1} (c - i) N_i$$

$$= \frac{1}{\mu} \sum_{i=0}^{c-1} (c - i)((i + 1) p_{i+1} - \lambda p_i) = \frac{\lambda - \text{Var} C(t)}{\mu} < \frac{\lambda}{\mu}.$$

The second sum contains only a finite number of terms $\mu p_{k1}$ for

$k = 0, ..., c - 3$. By (2.74), their limits as $\mu \to \infty$ are zero. Thus, (2.77) implies that there exists

$$\lim_{\mu\to\infty} p_{00} = \frac{c - \lambda}{\sum\limits_{i=0}^{c-1}(c - i)\frac{\lambda^i}{i!}} = p_{00}(\infty).$$

**3.** Using this relation we get from (2.76) that there exists

$$\lim_{\mu\to\infty} p_{i0} = \frac{\lambda^i}{i!}p_{00}(\infty) = p_{i0}(\infty), \quad i = 0, ..., c.$$

**4.** Eliminate term $(j + 1)\mu p_{c-1,j+1}$ from equation (2.72) for $i = c - 1$, $j = j + 1$ and equation (2.73):

$$cp_{c,j+1} = (\lambda + c)p_{cj} - \lambda p_{c,j-1}$$

$$-\lambda p_{c-1,j} - \lambda p_{c-2,j+1} + (\lambda + c - 1)p_{c-1,j+1} \quad (2.78)$$

$$-(j + 2)\mu p_{c-2,j+2}, \quad j \geq 1.$$

For $j = 0$ existence and value of $\lim_{\mu\to\infty} p_{cj}$ are already established. This allows us to prove by induction with respect to $j$ that $\lim_{\mu\to\infty} p_{cj}$ exists for all $j$ and is equal to $p_{cj}(\infty)$. □

**Lemma 2.2** *There exist* $a_{ij} = \lim_{\mu\to\infty} \mu(p_{ij} - p_{ij}(\infty))$ *and*

$$a_{ij} = \begin{cases} \dfrac{\lambda^i}{i!}\dfrac{\lambda^c}{(c-1)!}\ln\left(1 - \dfrac{\lambda}{c}\right)\dfrac{p_{00}^2(\infty)}{c-\lambda}, \\[2mm] \quad \text{if } 0 \leq i \leq c - 1, \; j = 0, \\[2mm] 0, \quad \text{if } 0 \leq i \leq c - 2, \; j \geq 1, \\[2mm] \dfrac{1}{j}\dfrac{\lambda^j}{c^{j-1}}\dfrac{\lambda^c}{c!}p_{00}(\infty), \\[2mm] \quad \text{if } i = c - 1, \; j \geq 1, \\[2mm] \dfrac{\lambda^c}{c!}\left(\dfrac{\lambda}{c}\right)^j\left\{\dfrac{\lambda^c}{(c-1)!}\ln\left(1 - \dfrac{\lambda}{c}\right)\dfrac{p_{00}(\infty)}{c-\lambda}\right. \\[2mm] \left. + \dfrac{\lambda}{c}\left(\sum\limits_{i=1}^{j}\dfrac{1}{i} - \dfrac{c-1}{j+1}\right)\right\}p_{00}(\infty), \\[2mm] \quad \text{if } i = c, \; j \geq 0. \end{cases} \quad (2.79)$$

*Proof.* **1.** If $0 \leq i \leq c-2$, $j \geq 1$, then as a matter of fact the limits (2.79) were calculated in a proof of Lemma 2.1 and equality (2.74) gives that $a_{ij} = 0$.

**2.** To calculate $a_{c-1,j}$ for $j \geq 1$ replace in equation (2.73) index $j$ by $j+1$ and rewrite the obtained equation as

$$\mu p_{c-1,j} = \frac{1}{j} \left( (\lambda + c) p_{c,j-1} - \lambda p_{c-1,j-1} - \lambda p_{c,j-2} \right), \; j \geq 1.$$

We have already proved in Lemma 2.1 that all probabilities in the right-hand side of this relation have limits and we know the values of these limits. Thus, we can guarantee that for $j \geq 1$ there exists $a_{c-1,j} = \lim_{\mu \to \infty} \mu p_{c-1,j}$ and

$$a_{c-1,j} = \frac{1}{j} \frac{\lambda^j}{c^{j-1}} \frac{\lambda^c}{c!} p_{00}(\infty). \tag{2.80}$$

**3.** Now let $i = 0, j = 0$. It follows from (2.77) that

$$\mu(p_{00} - p_{00}(\infty)) = \left[ \mu^2 \sum_{i=0}^{c-1} (c-i) \frac{\lambda^i}{i!} \sum_{l=1}^{i} \frac{(l-1)!}{\lambda^l} \sum_{k=0}^{l-2} p_{k1} \right.$$
$$\left. - \sum_{i=0}^{c-1} (c-i) \sum_{j=1}^{\infty} p_{ij} \right] \left[ \sum_{i=0}^{c-1} (c-i) \frac{\lambda^i}{i!} \right]^{-1} . \tag{2.81}$$

The right-hand side contains two types of unknown terms: probabilities $p_{k1}$ for $k \leq l - 2 \leq i - 2 \leq c - 3$ with coefficients $\mu^2$, and sums $\sum_{j=1}^{\infty} p_{ij}, 0 \leq i \leq c - 1$, with coefficients $\mu$.

To find $\lim_{\mu \to \infty} \mu^2 p_{k1}$ consider equation (2.72) with $i = k, j \geq 1$ :

$$\mu^2 p_{kj} = \frac{\mu}{\lambda + k + j\mu} \left\{ \lambda \mu p_{k-1,j} + (j+1)\mu^2 p_{k-1,j+1} \right.$$
$$+ \left. (k+1)\mu p_{k+1,j} \right\}, \; 0 \leq k \leq c - 1. \tag{2.82}$$

Putting here $k = 0$ we get:

$$\mu^2 p_{0j} = \frac{\mu}{\lambda + j\mu} \mu p_{1,j}.$$

But we already know that $\lim_{\mu \to \infty} \mu p_{1j} = 0$. Thus $\lim_{\mu \to \infty} \mu^2 p_{01} = 0$. Now by induction on $k$ it is easy to show that

$$\lim_{\mu \to \infty} \mu^2 p_{k1} = 0, \; \text{if } 0 \leq k \leq c - 3, j \geq 1. \tag{2.83}$$

Indeed, the first and the third terms in the right-hand side of equa-

tion (2.82) tend to zero by (2.74) (since $k + 1 \leq c - 2$). The second
term tends to zero by assumption of induction.

Thus, the first term in the numerator in the right-hand side of
(2.81) as $\mu \to \infty$ has limit zero.

Now calculate the limit of the second term in the right-hand side
of (2.81) as $\mu \to \infty$, i.e. $\lim_{\mu \to \infty} \mu \sum_{j=1}^{\infty} p_{ij}$ for $0 \leq i \leq c - 1$.

The simplest way to find this limit consists in exchanging the
order of lim and sum and applying the already known $\lim_{\mu \to \infty} \mu p_{ij}$
for $0 \leq i \leq c - 1$, $j \geq 1$. However this requires proof of uniform
convergence of the series $\sum_{j=1}^{\infty} \mu p_{ij}$, and so we will use another
way. We have:

$$\mu \sum_{j=1}^{\infty} p_{ij} \leq \mu \sum_{j=1}^{\infty} j p_{ij} = \mu N_i = (i+1)p_{i+1} - \lambda p_i$$

$$= (i+1)p_{i+1,0} - \lambda p_{i0} + (i+1)\sum_{j=1}^{\infty} p_{i+1,j} - \lambda \sum_{j=1}^{\infty} p_{ij}$$

$$\leq (i+1)p_{i+1,0} - \lambda p_{i0} + (i+1)N_{i+1}$$

$$= (i+1)p_{i+1,0} - \lambda p_{i0} + (i+1)\frac{(i+2)p_{i+2} - \lambda p_{i+1}}{\mu}.$$

Thus from Lemma 2.1 we get:

$$\lim_{\mu \to \infty} \mu \sum_{j=1}^{\infty} p_{ij} = 0 \text{ for } 0 \leq i \leq c - 2.$$

To find $\lim_{\mu \to \infty} \mu \sum_{j=1}^{\infty} p_{c-1,j}$ divide both sides of equation (2.73)
by $j + 1$ and sum with respect to $j \geq 0$:

$$\mu \sum_{j=1}^{\infty} p_{c-1,j} = (\lambda + c)\sum_{j=0}^{\infty} \frac{p_{cj}}{j+1} - \lambda \sum_{j=1}^{\infty} \frac{p_{c,j-1}}{j+1} - \lambda \sum_{j=0}^{\infty} \frac{p_{c-1,j}}{j+1}.$$

Each sum in the right-hand side of this equation as $\mu \to \infty$ has a
limit equal to the sum of limits (by general theorems about weak
convergence of distributions). Therefore,

$$\lim_{\mu \to \infty} \mu \sum_{j=1}^{\infty} p_{c-1,j} = (\lambda + c)\sum_{j=0}^{\infty} \left(\frac{\lambda}{c}\right)^j \frac{\lambda^c}{c!} \frac{p_{00}(\infty)}{j+1}$$

$$- \lambda \sum_{j=1}^{\infty} \left(\frac{\lambda}{c}\right)^{j-1} \frac{\lambda^c}{c!} \frac{p_{00}(\infty)}{j+1}$$

$$- \lambda \frac{\lambda^{c-1}}{(c-1)!} p_{00}(\infty)$$

$$= \frac{\lambda^c}{(c-1)!} \ln\left(1 - \frac{\lambda}{c}\right) p_{00}(\infty).$$

It follows from the above results that:

$$a_{00} = \frac{\frac{\lambda^c}{(c-1)!} \ln\left(1 - \frac{\lambda}{c}\right) p_{00}(\infty)}{\sum\limits_{i=0}^{c-1}(c-i)\frac{\lambda^i}{i!}} = \frac{\lambda^c}{(c-1)!} \ln\left(1 - \frac{\lambda}{c}\right) \frac{p_{00}^2(\infty)}{c-\lambda}.$$

**4.** Now find limit (2.79) for $0 \leq i \leq c-1$, $j = 0$. With this goal use equation (2.76); subtracting from this equation the obvious relation $p_{i0}(\infty) = \frac{\lambda^i}{i!} p_{00}(\infty)$ and multiplying both sides by $\mu$ we get:

$$\mu(p_{i0} - p_{i0}(\infty)) = \frac{\lambda^i}{i!} \mu(p_{00} - p_{00}(\infty))$$

$$- \mu^2 \frac{\lambda^i}{i!} \sum_{l=1}^{i} \frac{(l-1)!}{\lambda^l} \sum_{k=0}^{l-2} p_{k1},$$

$$i = 0, ..., c.$$

Since we have already established that there exist

$$\lim_{\mu \to \infty} \mu(p_{00} - p_{00}(\infty))$$

and

$$\lim_{\mu \to \infty} \mu^2 p_{k1} = 0 \text{ (for } 0 \leq k \leq c-3),$$

we can guarantee that for $i \leq c-1$ there exist

$$a_{i0} \equiv \lim_{\mu \to \infty} \mu(p_{i0} - p_{i0}(\infty))$$

and

$$a_{i0} = \frac{\lambda^i}{i!} a_{00}, \ 0 \leq i \leq c-1.$$

To find $a_{c0}$ we must calculate $\lim_{\mu \to \infty} \mu^2 p_{c-2,1}$. With this goal consider equation (2.82) for $k = c-2$ and $j \geq 1$:

$$\lim_{\mu \to \infty} \mu^2 p_{c-2,j} = \frac{\lambda}{j} \lim_{\mu \to \infty} \mu p_{c-3,j}$$

$$+ \frac{j+1}{j} \lim_{\mu \to \infty} \mu^2 p_{c-3,j+1}$$

$$+ \ \frac{c-1}{j} \lim_{\mu \to \infty} \mu p_{c-1,j}.$$

The first and the second limits in the right-hand side of this equation are equal to zero by (2.74) and (2.83) respectively, and the third was calculated above in part 2 of the present proof (see (2.80)). Thus

$$\lim_{\mu \to \infty} \mu^2 p_{c-2,j} = \frac{c-1}{j^2} \frac{\lambda^c}{c!} \frac{\lambda^j}{c^{j-1}} p_{00}(\infty), \ j \geq 1,$$

so that

$$a_{c0} = \frac{\lambda^c}{c!} a_{00} - \frac{c-1}{c} \frac{\lambda^{c+1}}{c!} p_{00}(\infty). \tag{2.84}$$

**5.** Now let $i = c$, $j = 1$. From equation (2.78) for $j = 0$ we have:

$$\begin{aligned}
\mu(p_{c1} - p_{c1}(\infty)) &= \frac{\lambda + c}{c} \mu(p_{c0} - p_{c0}(\infty)) \\
&+ \frac{\lambda + c - 1}{c} \mu p_{c-1,1} \\
&- \frac{\lambda}{c} \mu(p_{c-1,0} - p_{c-1,0}(\infty)) \\
&- \frac{\lambda}{c} \mu p_{c-2,1} - \frac{2}{c} \mu^2 p_{c-2,2}.
\end{aligned}$$

Here we used the fact that $c p_{c1}(\infty) = (\lambda + c) p_{c0}(\infty) - \lambda p_{c-1,0}(\infty)$. Because we already know that all terms in the right-hand side of this equation have limits as $\mu \to \infty$, we can guarantee that there exists

$$a_{c1} = \frac{\lambda^{c+1}}{c!c} a_{00} - \frac{c-3}{2} \frac{\lambda^{c+2}}{c!c^2} p_{00}(\infty). \tag{2.85}$$

**6.** Now let $i = c$, $j \geq 1$. It is easy to check that the following relations hold:

$$\begin{aligned}
c p_{cj}(\infty) &= \lambda p_{c,j-1}(\infty), \ j \geq 1; \\
\lambda p_{cj}(\infty) &= c p_{c,j+1}(\infty), \ j \geq 1.
\end{aligned}$$

Subtract these equations from (2.78) and multiply both sides of the obtained equation by $\mu$ :

$$(\lambda + c)\mu(p_{cj} - p_{cj}(\infty)) + (\lambda + c - 1)\mu p_{c-1,j+1}$$

$$= \lambda \mu p_{c-1,j} + \lambda \mu p_{c-2,j+1} + (j+2)\mu^2 p_{c-2,j+2}$$

$$+ c\mu(p_{c,j+1} - p_{c,j+1}(\infty)) + \lambda\mu(p_{c,j-1} - p_{c,j-1}(\infty)), \ j \geq 1. \tag{2.86}$$

We know that for $j \geq 1$ there exist

$$\lim_{\mu \to \infty} \mu p_{c-1,j+1},$$
$$\lim_{\mu \to \infty} \mu p_{c-1,j},$$
$$\lim_{\mu \to \infty} \mu p_{c-2,j+1},$$
$$\lim_{\mu \to \infty} \mu^2 p_{c-2,j+2}.$$

Thus by induction with respect to $j$ (using existence of limits (2.84) and (2.85)) one can show that $\lim_{\mu \to \infty} \mu(p_{cj} - p_{cj}(\infty)) = a_{cj}$ exists for all $j \geq 2$. Taking limits of both sides of (2.86) we get a recursive relation for variables $a_{cj}$, from which all these variables can be recursively determined.                                                                □

The above results concerning the asymptotic behaviour of the stationary probabilities of microstates $p_{ij}$ allow us to obtain asymptotic expansions of the macrocharacteristics of the system. We start with the blocking probability. A formal expansion for this probability may be obtained by substitution, into series $B = \sum_{j=0}^{\infty} p_{cj}$, the expansion of $p_{cj}$ which is given by Lemmas 2.1 and 2.2: $p_{cj} = p_{cj}(\infty) + \frac{1}{\mu} a_{cj} + o\left(\frac{1}{\mu}\right)$. However this approach is not rigorous since we do not know any estimate of the remainder $o\left(\frac{1}{\mu}\right)$. Probably the most convenient way to get the expansion for $B$ consists in the following.

Using the normalizing condition

$$\sum_{i=0}^{c} p_i = 1$$

and the relation

$$\sum_{i=0}^{c} i p_i = \lambda,$$

we can express the blocking probability in terms of probabilities $p_0, \ldots, p_{c-2}$ as follows:

$$B = \sum_{i=0}^{c-2} (c - 1 - i)p_i + \lambda - c + 1.$$

Because for the standard $M/M/c/\infty$ queueing system similar equations $\sum_{i=0}^{c} p_i(\infty) = 1$ and $\sum_{i=0}^{c} i p_i(\infty) = \lambda$ hold, we have:

$$B(\infty) = \sum_{i=0}^{c-2} (c - 1 - i)p_i(\infty) + \lambda - c + 1,$$

so that

$$\mu(B - B(\infty)) = \mu \sum_{i=0}^{c-2}(c - 1 - i)(p_i - p_i(\infty))$$

$$= \sum_{i=0}^{c-2}(c - 1 - i)\mu(p_{i0} - p_{i0}(\infty))$$

$$+ \sum_{i=0}^{c-2}(c - 1 - i)\mu \sum_{j=1}^{\infty} p_{ij}.$$

But we know from Lemma 2.2 that for $0 \leq i \leq c - 2$ there exist $\lim_{\mu \to \infty} \mu(p_{i0} - p_{i0}(\infty)) = a_{i0}$, $\lim_{\mu \to \infty} \mu \sum_{j=1}^{\infty} p_{ij} = 0$. Thus, there exists

$$\lim_{\mu \to \infty} \mu(B - B(\infty)) = (c - 1 - \lambda + B(\infty)) \cdot B(\infty) \cdot \ln\left(1 - \frac{\lambda}{c}\right).$$

To prove the formula for $N$ we will use relation (2.25), which implies that we must get an asymptotic expansion only for the second moment of the number of busy servers $\mathrm{E}\,(C(t))^2 = \sum_{i=0}^{c} i^2 p_i$. With this goal we express probabilities $p_c$ and $p_{c-1}$ in terms of probabilities $p_0, ..., p_{c-2}$, which gives

$$\mu \sum_{i=0}^{c} i^2 (p_i - p_i(\infty)) = \sum_{i=0}^{c-2}(c - i)(c - 1 - i)\mu(p_{i0} - p_{i0}(\infty))$$

$$+ \sum_{i=0}^{c-2}(c - i)(c - 1 - i)\mu \sum_{j=1}^{\infty} p_{ij}.$$

The first sum in the right-hand side of this equation has a limit equal to $\sum_{i=0}^{c-2}(c - i)(c - 1 - i)a_{i0}$. The limit of the second term is equal to zero. This yields the desired formula for $N$.    $\square$

As results of numerical calculations show, the above asymptotic formulas for the blocking probability and the mean queue length are sufficiently accurate in the domain of the parameter $\mu$ of practical interest (namely, for $\mu > 10$ and $c$ not too large). As an example in Table 2.6 we give the exact and approximate values of the blocking probability $B$ for the 5-server retrial queue for various values of offered traffic $\lambda$ and retrial intensity $\mu$. The exact values of $B$ were calculated by means of direct solution of the Kolmogorov equations for the stationary micro-probabilities $p_{ij}$ and the approximate ones

Table 2.6 *The exact and approximate values of the blocking probability for the 5-server retrial queue*

|  |  | $\mu = 5$ | $\mu = 10$ | $\mu = 20$ | $\mu = \infty$ |
|---|---|---|---|---|---|
| $\lambda = 1$ | Exact | 0.00355 | 0.00365 | 0.00373 | 0.00383 |
|  | Approx. | 0.00332 | 0.00357 | 0.00370 | 0.00383 |
| $\lambda = 2$ | Exact | 0.05310 | 0.05545 | 0.05719 | 0.05970 |
|  | Approx. | 0.04714 | 0.05342 | 0.05656 | 0.05970 |
| $\lambda = 3$ | Exact | 0.20863 | 0.21838 | 0.22559 | 0.23615 |
|  | Approx. | 0.18266 | 0.20940 | 0.22278 | 0.23615 |
| $\lambda = 4$ | Exact | 0.50348 | 0.52147 | 0.53470 | 0.55411 |
|  | Approx. | 0.45528 | 0.50470 | 0.52940 | 0.55411 |

were obtained from expression (2.69). The relative error is about 3–4% if $\mu = 10$ and about 1% if $\mu = 20$. Thus our formula can be directly used to obtain numerical values.

In order to understand how much the second term of the asymptotic formula for $B = B(\mu)$ improves the approximation quality, we give in the column $\mu = \infty$ the value of the blocking probability in the corresponding standard $M/M/c/\infty$ queueing system. From the table it is seen that the relative error of calculation by the formula $B \approx B(\infty)$ is twice as much and equals 5–9% for $\mu = 10$, and is about four times as large and equals 3–5% if $\mu = 20$.

### 2.7.2 Low rate of retrials

The limit behaviour of retrial queues as $\mu \to 0$ is of interest on account of the weak dependence of the stationary distribution $\{p_n(\mu)\}$ of the number of busy servers upon $\mu$ (a fact illustrated by numerical data; see section 2.6). Because for complex systems $\lim_{\mu \to 0} p_n(\mu)$ can be found more simply than $\lim_{\mu \to \infty} p_n(\mu)$, it is natural to use this limit as an approximate value of $p_n(\mu)$ for all $\mu \in (0, +\infty)$.

The following theorem gives a solution of this problem for the main model in the steady state.

**Theorem 2.6** *Let* $r = r(c; \lambda)$ *be the root of the equation*

$$r \sum_{k=0}^{c-1} \frac{(\lambda + r)^k}{k!} = \lambda \frac{(\lambda + r)^c}{c!} \tag{2.87}$$

*in the interval* $0 < r < +\infty$ *(for* $\lambda < c$ *this root exists and is unique), and* $\Lambda = \lambda + r$. *Also denote*

$$D = r + \frac{r^3}{\lambda - r(c - \lambda)} \cdot \frac{c!}{\Lambda^{c+1}} \cdot \sum_{k=0}^{c-1} \frac{k!}{\Lambda^k} \left( \sum_{n=0}^{k} \frac{\Lambda^n}{n!} \right)^2.$$

*Then as* $\mu \to 0$

$$\mathrm{E} \left\{ \exp \left( it \frac{\mu N(u) - r}{\sqrt{\mu}} \right) ; C(u) = n \right\} = \frac{\frac{\Lambda^n}{n!}}{\sum_{k=0}^{c} \frac{\Lambda^k}{k!}} \cdot \exp \left( -\frac{Dt^2}{2} \right),$$

*i.e. asymptotically*

- *the number of busy servers and the scaled queue length are independent;*
- *the number of busy servers has an Erlang loss distribution with parameter* $\Lambda$;
- *the number of sources of repeated calls is Gaussian with mean* $\frac{r}{\mu}$ *and variance* $\frac{D}{\mu}$.

*Proof.* First of all consider the problem about nonnegative roots of equation (2.87) or equivalently zeros of the function

$$f(z) = (z - \lambda) \sum_{k=0}^{c-1} \frac{z^k}{k!} - \lambda \frac{z^c}{c!}. \tag{2.88}$$

It is easy to see that

$$f(z) = \sum_{k=0}^{c} \frac{z^k}{k!} \cdot (Y(z) - \lambda),$$

where

$$Y(z) = z(1 - E_c(z))$$

and

$$E_c(z) = \frac{z^c}{c!} \Bigg/ \sum_{k=0}^{c} \frac{z^k}{k!}$$

are the mean number of busy servers and the blocking probability respectively in the Erlang loss model with $c$ servers and arrival rate $z$. The number of busy servers in the Erlang loss model is a birth and death process with rate of birth $\lambda_n = z$ and rate of death $\mu_n = n$. Due to general theorems about stochastic monotonicity of birth and death processes, we can guarantee that the function $Y(z)$ is increasing. Besides, $Y(0) = 0, Y(+\infty) = c$. Thus, as $\lambda < c$ equation $Y(z) = \lambda$ has a unique positive root $z_0$.

Since $Y(z) = z(1 - E_c(z)) < z$, this root in fact is greater than $\lambda$. On the other hand, $Y'(z_0) > 0$. But

$$Y'(z) = 1 - E_c(z) - cE_c(z) + zE_c(z)(1 - E_c(z))$$

and

$$E_c(z_0) = 1 - \frac{\lambda}{z_0}.$$

Therefore

$$Y'(z_0) = \frac{\lambda - (c - \lambda)(z_0 - \lambda)}{z_0}$$

and the inequality $Y'(z_0) > 0$ gives the following estimate for the root $z_0$ :

$$z_0 < \lambda + \frac{\lambda}{c - \lambda}.$$

The above implies that equation (2.87) indeed has a unique positive root $r = z_0 - \lambda$ (which is less than $\frac{\lambda}{c-\lambda}$).

Now we are in position to prove the theorem.

Analysis will be based on equations (2.19, 2.20) of section 2.3, where, however, we replace equation (2.20) by an equation which is obtained by summing these equations with respect to $n = 0, 1, ..., c$. Thus as the main set of equations we will use the following set:

$$(\lambda + n)p_n(z) + \mu z p'_n(z) = \lambda p_{n-1}(z) + \mu p'_{n-1}(z) + p_{n+1}(z),$$
$$0 \le n \le c - 1,$$

$$\lambda p_c(z) = \mu \sum_{n=0}^{c-1} p'_n(z).$$

Consider the shifted and scaled queue length $N^*(u) = \frac{\mu N(u) - r}{\sqrt{\mu}}$. The joint distribution of random variables $N^*(u)$ and $C(u)$ can be described by partial characteristic functions

$$\psi_n(t; \mu) = \mathrm{E}\left(e^{itN^*(u)}; C(u) = n\right),$$

where $i = \sqrt{-1}$. Because $\psi_n(t; \mu) = \exp\left\{-\frac{itr}{\sqrt{\mu}}\right\} p_n\left(e^{it\sqrt{\mu}}\right)$, equations for $p_n(z)$ become the following equations for $\psi_n(t; \mu)$ :

$$(\lambda + n + r)\psi_n - i\sqrt{\mu}\psi_n' = (\lambda + re^{-it\sqrt{\mu}})\psi_{n-1}$$
$$- i\sqrt{\mu}e^{-it\sqrt{\mu}}\psi_{n-1}' + (n+1)\psi_{n+1},$$
$$\text{if } 0 \le n \le c-1,$$

$$\lambda\psi_c = re^{-it\sqrt{\mu}}\sum_{n=0}^{c-1}\psi_n$$
$$- i\sqrt{\mu}e^{-it\sqrt{\mu}}\sum_{n=0}^{c-1}\psi_n'.$$

Solving for the derivatives, we get:

$$-i\sqrt{\mu}\psi_n' = \sum_{k=0}^{c-1}\psi_k\left[\lambda e^{-it(n-k)\sqrt{\mu}} - ke^{-it(n-k+1)\sqrt{\mu}}\right]$$
$$\times \left(e^{it\sqrt{\mu}} - 1\right) - \left(\lambda + r + \left(1 - e^{-it\sqrt{\mu}}\right)n\right)\psi_n$$
$$+ (n+1)\psi_{n+1}, \text{ if } 0 \le n \le c-2,$$

$$-i\sqrt{\mu}\psi_{c-1}' = \frac{c + \lambda\left(1 - e^{it\sqrt{\mu}}\right)}{ce^{-it\sqrt{\mu}} - \lambda}$$
$$\times \sum_{k=0}^{c-2}\psi_k\left[\lambda e^{-it(c-1-k)\sqrt{\mu}} - ke^{-it(c-k)\sqrt{\mu}}\right]$$
$$+ \left(\frac{\lambda}{c - \lambda e^{it\sqrt{\mu}}} - \lambda - r - c + 1\right)\psi_{c-1},$$

or in matrix form

$$-i\sqrt{\mu}\begin{pmatrix} \dfrac{d\psi_0(t;\mu)}{dt} \\ \vdots \\ \dfrac{d\psi_{c-1}(t;\mu)}{dt} \end{pmatrix} = \mathbf{A}(t;\mu) \cdot \begin{pmatrix} \psi_0(t;\mu) \\ \vdots \\ \psi_{c-1}(t;\mu) \end{pmatrix}$$

where the matrix $\mathbf{A}(t;\mu)$ is constructed in an obvious manner from the above equations for $\psi_n(t;\mu)$. Since we consider the case $\mu \to 0$, this set of ordinary linear differential equations can be identified as singular perturbed. Now we can apply the general theory of asymptotic expansions for such equations (Wasow, W. (1965) *Asymptotic Expansions for Ordinary Differential Equations*, Interscience, New

York; Eckhaus, W. (1979) *Asymptotic Analysis of Singular Perturbations*, North-Holland Publishing Company, Amsterdam).

First of all we note that nonperturbed matrix $\mathbf{A}(t;0)$ is

$$
\begin{pmatrix}
-\Lambda & 1 & 0 & \cdots & 0 & 0 \\
0 & -\Lambda & 2 & \cdots & 0 & 0 \\
0 & 0 & -\Lambda & \cdots & 0 & 0 \\
\vdots & \vdots & \vdots & \ddots & \vdots & \vdots \\
0 & 0 & 0 & \cdots & -\Lambda & c-1 \\
\frac{c\lambda}{c-\lambda} & \frac{c(\lambda-1)}{c-\lambda} & \frac{c(\lambda-1)}{c-\lambda} & \cdots & \frac{c(\lambda-c+2)}{c-\lambda} & \frac{c(\lambda-c+1)}{c-\lambda} - \Lambda
\end{pmatrix}
$$

Thus the characteristic equation $\det(\mathbf{A}(t;0) - x\mathbf{I}) = 0$ is

$$
(\Lambda - \lambda + x) \sum_{k=0}^{c-1} \frac{(\Lambda + x)^k}{k!} - \lambda \frac{(\Lambda + x)^c}{c!} = 0.
$$

Introduce a new variable $z = \Lambda + x$. Then the characteristic equation becomes $f(z) = 0$, where function $f(z)$ was introduced earlier by equation (2.88). This function does not have multiple zeros. Indeed, consider the set of equations

$$
\begin{cases}
f(z) & = & 0 \\
f'(z) & = & 0,
\end{cases}
$$

or equivalently

$$
\begin{cases}
(z - \lambda) \sum_{k=0}^{c-1} \frac{z^k}{k!} - \lambda \frac{z^c}{c!} & = & 0 \\
\sum_{k=0}^{c-1} \frac{z^k}{k!} + (z - \lambda) \sum_{k=0}^{c-2} \frac{z^k}{k!} - \lambda \frac{z^{c-1}}{(c-1)!} & = & 0.
\end{cases}
$$

Eliminating $\sum_{k=0}^{c-1} z^k/k!$ we find that $z = 0$ or $z = \lambda + \frac{\lambda}{c-\lambda}$. But neither $z = 0$ nor $z = \lambda + \frac{\lambda}{c-\lambda}$ is a root of the equation $f(z) = 0$, since as we have shown, if $\lambda < c$ this equation has a unique positive root $z_0 \in \left(\lambda, \lambda + \frac{\lambda}{c-\lambda}\right)$.

Thus all eigenvalues $x_0, ..., x_{c-1}$ of the matrix $\mathbf{A}(t;0)$ are different and do not depend on $t$. Besides, it is clear that $x_0 = 0$ is an eigenvalue (i.e. the degenerate problem can be solved) if and only if the parameter $r$ is the root $r(c;\lambda)$ of equation (2.87).

Using these facts and boundedness of $|\psi_n(t,\mu)|$ it can be proven that $\psi_n(t;\mu)$ can be represented with the help of a regular asymptotic series (which does not contain a boundary layer):

$$\psi_n(t;\mu) = a_n(t) + \sqrt{\mu}b_n(t) + \ldots.$$

The coefficients of the expansion can be found with the help of substitution into the original set of differential equations.

Equations for $a_n(t)$ are

$$(\Lambda + n)a_n(t) = \Lambda a_{n-1}(t) + (n+1)a_{n+1}(t), \ 0 \le n \le c-1,$$
$$\lambda a_c(t) = (\Lambda - \lambda) \sum_{n=0}^{c-1} a_n(t).$$

The first equation implies that

$$a_n(t) = \frac{\Lambda^n}{n!}a_0(t),$$

and the second reduces to equation (2.87) for the parameter $r = r(c; \lambda)$ and thus is of no interest.

Equations for $b_n(t)$ are:

$$(\Lambda + n)b_n(t) - ia'_n(t) = \Lambda b_{n-1}(t) + (n+1)b_{n+1}(t)$$
$$- it(\Lambda - \lambda)a_{n-1}(t)$$
$$- ia'_{n-1}(t),$$
$$\text{if } 0 \le n \le c-1,$$
$$\lambda b_c(t) = (\Lambda - \lambda) \sum_{n=0}^{c-1} b_n(t)$$
$$- it(\Lambda - \lambda) \sum_{n=0}^{c-1} a_n(t)$$
$$- i \sum_{n=0}^{c-1} a'_n(t).$$

From the first equation we get:

$$b_n(t) = it(\Lambda - \lambda)\frac{\Lambda^n}{n!} \sum_{k=0}^{n-1} \frac{k!}{\Lambda^{k+1}} \sum_{m=0}^{k-1} \frac{\Lambda^n}{n!}a_0(t)$$
$$- i\frac{\Lambda^{n-1}}{(n-1)!}a'_0(t),$$

This allows us to transform the second equation to the following form:

$$a'_0(t) = -Dta_0(t),$$

where variable $D$ was introduced above. This equation gives:

$$a_0(t) = a_0(0) \cdot e^{-\frac{Dt^2}{2}},$$

from which the required result follows. $\qquad\square$

The above result in fact has deeper roots and is connected with convergence of the centered and normalized queue length process to the Ornstein–Uhlenbeck process. The main problem here is that only the second component of the process $(C(t), N(t))$ converges to a diffusion process as $\mu \to 0$. To overcome this difficulty, we apply a method of proving functional limit theorems in queueing theory developed by D.Y.Burman (Burman, D.Y. (1979) *An Analytic Approach to Diffusion Approximation in Queueing*, Ph.D.Thesis, Department of Applied Mathematics, Courant Institute of Mathematics, New York University, New York).

**Theorem 2.7** *If $\lambda < c$, then as $\mu \to 0$, the finite-dimensional distributions of the process*

$$\frac{\mu N(\frac{t}{\mu}) - r}{\sqrt{\mu}},$$

*where $r$ is a positive solution of the equation (2.87), converge to the corresponding distributions of the Ornstein–Uhlenbeck process with a shift coefficient*

$$-\alpha x = -\frac{\lambda - r(c - \lambda)}{\lambda + r} x$$

*and a diffusion coefficient*

$$\sigma^2 = 2 \frac{\lambda - r(c - \lambda)}{\lambda + r} D.$$

*Proof.* The infinitesimal generator of the process

$$\left( C\left(\frac{t}{\mu}\right), \frac{\mu N\left(\frac{t}{\mu}\right) - r}{\sqrt{\mu}} \right)$$

has the form:

$$
\begin{aligned}
A_\mu f(n, x) &= \frac{\lambda}{\mu} [f(n + 1, x) - f(n, x)] \\
&+ \frac{n}{\mu} [f(n - 1, x) - f(n, x)]
\end{aligned}
$$

$$+ \left( \frac{x}{\sqrt{\mu}} + \frac{r}{\mu} \right) [f(n+1, x - \sqrt{\mu}) - f(n, x)],$$

$$\text{if } 0 \le n \le c - 1;$$

$$A_\mu f(c, x) = \frac{\lambda}{\mu} [f(c, x + \sqrt{\mu}) - f(c, x)]$$

$$+ \frac{c}{\mu} [f(c-1, x) - f(c, x)].$$

For an arbitrary twice differentiable function $f(x)$, put

$$f_\mu(n, x) = f(x) + \sqrt{\mu} f'(x) g_n + \mu f''(x) h_n,$$

where the constants $g_n$ and $h_n$, $0 \le n \le c$, will be defined below.

For such functions $f_\mu(n, x)$ we find as $\mu \to 0$ that

$$A_\mu f_\mu(n, x) = \frac{1}{\sqrt{\mu}} \{ \Lambda(g_{n+1} - g_n) - r + n(g_{n-1} - g_n) \} f'(x)$$

$$+ (g_{n+1} - g_n - 1) x f'(x)$$

$$+ \left\{ \Lambda(h_{n+1} - h_n) + \frac{r}{2} - r g_{n+1} \right\} f''(x)$$

$$+ n(h_{n-1} - h_n) f''(x) + o(1), \text{ if } 0 \le n \le c - 1;$$

$$A_\mu f_\mu(c, x) = \frac{1}{\sqrt{\mu}} f'(x) \{ \lambda + c(h_{c-1} - h_c) \}$$

$$+ f''(x) \left\{ \frac{\lambda}{2} + \lambda g_c + c(h_{c-1} - h_c) \right\} + o(1).$$

Thus it is clear that $A_\mu f_\mu(n, x)$ may converge as $\mu \to 0$ to a limit not depending on $n$ if $g_n$ and $h_n$ are chosen so that, for some function $F(x)$,

$$\begin{cases} \Lambda(g_{n+1} - g_n) - r + n(g_{n-1} - g_n) = 0, \\[4pt] \qquad\qquad \text{if } 0 \le n \le c - 1, \qquad\qquad (2.89) \\[4pt] \qquad \lambda + c(g_{c-1} - g_c) = 0, \end{cases}$$

$$\begin{cases} f''(x)\left\{\Lambda(h_{n+1}-h_n)+\frac{r}{2}-rg_{n+1}+n(h_{n-1}-h_n)\right\} \\ \qquad\qquad +xf'(x)(g_{n+1}-g_n-1)=F(x), \\ \qquad\qquad\qquad\qquad 0\le n\le c-1, \\ f''(x)\left\{\frac{\lambda}{2}+\lambda g_c+c(h_{c-1}-h_c)\right\}=F(x). \end{cases} \qquad (2.90)$$

The system (2.89) in the unknowns $g_0, ..., g_c$ has a solution if and only if the centering parameter $r$ satisfies equation (2.87), in which case

$$g_n = g_0 + \frac{\Lambda-\lambda}{\Lambda}\sum_{j=0}^{n-1}\frac{j!}{\Lambda^j}\sum_{k=0}^{j}\frac{\Lambda^k}{k!}, \ 0\le n\le c.$$

Hence the system (2.90) in the unknowns $h_0, ..., h_c$ has a solution if and only if

$$\begin{aligned} F(x) &= \ \left\{(\Lambda-\lambda)f''(x)\sum_{j=0}^{c-1}\frac{\Lambda^j}{j!}\left(\frac{1}{2}-g_{j+1}\right)\right. \\ &+ \ xf'(x)\sum_{j=0}^{c-1}\frac{\Lambda^j}{j!}(g_{j+1}-g_j-1) \\ &+ \ \left.\left(\frac{\lambda}{2}+\lambda g_c\right)f''(x)\frac{\Lambda^c}{c!}\right\}\times\left\{\sum_{j=0}^{c}\frac{\Lambda^j}{j!}\right\}^{-1}. \quad (2.91) \end{aligned}$$

For this choice of $g_n$ and $h_n$, we find that

$$\lim_{\mu\to 0}A_\mu f_\mu(n,x)=F(x).$$

Using (2.87) we can transform expression (2.91) for $F(x)$ into

$$F(x)=-\alpha xf'(x)+\frac{\sigma^2}{2}f''(x).$$

The right-hand side is the operator corresponding to a diffusion process with shift coefficient $-\alpha x$ and diffusion coefficient $\sigma^2$.

To complete the proof, it suffices to refer to the results from the thesis by D.Y.Burman mentioned above, which guarantee that the above reasonings imply the required result.  $\square$

*2.7.3 Transient phenomena under heavy traffic*

**Theorem 2.8** *As $\lambda \to 2 - 0$, the asymptotic expansion of the stationary blocking probability in a two-server retrial queue has the following form (below $\varepsilon = 1 - \frac{\lambda}{2}$):*

$$B = 1 - 2\varepsilon + \begin{cases} \frac{2}{1-\mu}\varepsilon^2 \cdot (1 + o(1)), & \text{if } \mu < 1, \\[2mm] 2\varepsilon^2 \ln \frac{1}{\varepsilon} \cdot (1 + o(1)), & \text{if } \mu = 1, \\[2mm] \varepsilon^{(1+1/\mu)} \frac{\Gamma(4/\mu+1)\cdot\Gamma(1-1/\mu)}{2\Gamma(3/\mu+1)} \cdot (1 + o(1)), & \text{if } \mu > 1. \end{cases}$$

*Proof.* The proof will be based on the explicit formula (2.16) for $B$ in terms of the ratio of two hypergeometric functions:

$$B = \frac{\lambda^2 + (\lambda - 1)g}{2 + \lambda + g},$$

where

$$g = \frac{\lambda^3}{2 + 3\lambda + 2\mu} \cdot \frac{F(a+1, b+1, c+1; \frac{\lambda}{2})}{F(a, b, c; \frac{\lambda}{2})},$$

$F$ is a hypergeometric function and the variables $a, b, c$ were defined in Theorem 2.1 by (2.13).

Thus we must investigate behaviour of the variable $g$ as $\lambda \to 2 - 0$, which clearly reduces to analysis of the behaviour of the hypergeometric functions $F(a, b, c; z)$ and $F(a+1, b+1, c+1; z)$ near the singular point $z = 1$.

It is well known that if

$$a + b - c < 0, \tag{2.92}$$

then

$$F(a, b, c; 1) = \frac{\Gamma(c) \cdot \Gamma(c - a - b)}{\Gamma(c - a) \cdot \Gamma(c - b)} < \infty. \tag{2.93}$$

If $a + b - c > 0$ then applying formula

$$F(a, b, c; z) = (1 - z)^{c-a-b} \cdot F(c - a, c - b, c; z) \tag{2.94}$$

we reduce the problem to analysis of the behaviour of the hypergeometric function $F(c - a, c - b, c; z)$ for which condition (2.92) is already satisfied, so that (2.93) gives:

$$F(c - a, c - b, c; 1) = \frac{\Gamma(c) \cdot \Gamma(a + b - c)}{\Gamma(a) \cdot \Gamma(b)} < \infty.$$

When $\lambda \to 2 - 0$ we have:

$$a \to a^* = \frac{4}{\mu}$$

$$b \to b^* = \frac{1}{\mu}$$

$$c \to c^* = \frac{4}{\mu} + 1.$$

Since $a^* + b^* - c^* = \frac{1}{\mu} - 1$, the asymptotic character of $F(a, b, c; \frac{\lambda}{2})$ as $\lambda \to 2 - 0$ depends on the value of $\mu$.

**Case 1.** $\mu > 1$.

In this case $a^* + b^* - c^* < 0$ and (2.93) implies that

$$F(a, b, c; \frac{\lambda}{2}) = \frac{\Gamma(\frac{4}{\mu} + 1) \cdot \Gamma(1 - \frac{1}{\mu})}{\Gamma(\frac{3}{\mu} + 1)} + o(1).$$

**Case 2.** $\mu < 1$.

In this case equation (2.94) yields that as $\lambda \to 2 - 0$

$$
\begin{aligned}
F(a, b, c; \frac{\lambda}{2}) &= \left(1 - \frac{\lambda}{2}\right)^{1 - \frac{\lambda}{2\mu}} \left\{ \frac{\Gamma(\frac{4}{\mu} + 1) \cdot \Gamma(\frac{1}{\mu} - 1)}{\Gamma(\frac{4}{\mu}) \cdot \Gamma(\frac{1}{\mu})} + o(1) \right\} \\
&= \varepsilon^{1 - \frac{1}{\mu}} \left\{ \frac{4}{1 - \mu} + o(1) \right\}.
\end{aligned}
$$

**Case 3.** $\mu = 1$.

In this case $a^* + b^* - c^* = 0$ and we first apply the following formula

$$
\begin{aligned}
F(a, b, c; z) &= \frac{\Gamma(c) \cdot \Gamma(c - a - b)}{\Gamma(c - a) \cdot \Gamma(c - b)} F(a, b, a + b - c + 1; 1 - z) \\
&\quad + (1 - z)^{c - a - b} \frac{\Gamma(c) \cdot \Gamma(a + b - c)}{\Gamma(a) \cdot \Gamma(b)} \\
&\quad \times F(c - a, c - b, c - a - b + 1; 1 - z). \quad (2.95)
\end{aligned}
$$

This formula holds if $a + b - c$ is not an integer. In our case $a + b - c = \frac{\lambda}{2\mu} - 1 = \frac{\lambda - 2}{2}$ and for $0 < \lambda < 2$ this quantity lies in the interval $(-1, 0)$.

If $z = \frac{\lambda}{2} \to 1$, then the hypergeometric functions in the right-hand side of this relation are (we recall that we consider the case $\mu = 1$):

$$F(a, b, a + b - c + 1; 1 - \frac{\lambda}{2}) = 1 + 4\varepsilon + o(\varepsilon),$$

$$F(c-a, c-b, c-a-b+1; 1-\frac{\lambda}{2}) \quad = \quad 1 + 4\varepsilon + o(\varepsilon).$$

All gamma functions in the right-hand side of (2.95) except for $\Gamma(c-a-b)$ and $\Gamma(a+b-c)$ can be expanded with the help of the formula

$$
\begin{aligned}
\Gamma(z) &= \Gamma(z^*) + \Gamma'(z^*) \cdot (z - z^*) + o(z - z^*) \\
&= \Gamma(z^*) \cdot [1 + \psi(z^*) \cdot (z - z^*) + o(z - z^*)], \quad (2.96)
\end{aligned}
$$

where

$$\psi(z) = \frac{\Gamma'(z)}{\Gamma(z)}$$

is the logarithmic derivative of the gamma function.

Recall that $\psi(1) = -C$, where $C$ is the Euler–Mascheroni constant, and $\psi(z+1) = \psi(z) + \frac{1}{z}$. Therefore,

$$
\begin{aligned}
\psi(2) &= 1 - C, \\
\psi(3) &= \frac{3 - 2C}{2}, \\
\psi(4) &= \frac{11 - 6C}{6}, \\
\psi(5) &= \frac{50 - 24C}{24},
\end{aligned}
$$

so that

$$
\begin{aligned}
\Gamma'(1) &= -C, \\
\Gamma'(2) &= 1 - C, \\
\Gamma'(3) &= 3 - 2C, \\
\Gamma'(4) &= 11 - 6C, \\
\Gamma'(5) &= 50 - 24C.
\end{aligned}
$$

These values of $\Gamma'$ will be used below without special reference.

Now applying (2.96) we get

$$
\begin{aligned}
\Gamma(c) &= 24 - (150 - 72C)\varepsilon + o(\varepsilon), \\
\Gamma(c-a) &= 1 + \frac{C\varepsilon}{3} + o(\varepsilon), \\
\Gamma(c-b) &= 6 - \frac{5}{3}(11 - 6C)\varepsilon + o(\varepsilon), \\
\Gamma(a) &= 6 - \frac{8}{3}(11 - 6C)\varepsilon + o(\varepsilon),
\end{aligned}
$$

$$\Gamma(b) = 1 + \frac{4C\epsilon}{3} + o(\epsilon).$$

In order to get asymptotic expansions for $\Gamma(c - a - b) = \Gamma(\varepsilon)$ and $\Gamma(a + b - c) = \Gamma(-\varepsilon)$, whose arguments tend to zero, we apply the formula $\Gamma(z + 1) = z\Gamma(z)$:

$$\Gamma(\varepsilon) = \frac{\Gamma(\varepsilon + 1)}{\varepsilon} = \frac{\Gamma(1) + \Gamma'(1)\varepsilon + o(\varepsilon)}{\varepsilon} = \frac{1 - C\varepsilon + o(\varepsilon)}{\varepsilon},$$

$$\Gamma(-\varepsilon) = \frac{\Gamma(-\varepsilon + 1)}{-\varepsilon} = \frac{\Gamma(1) + \Gamma'(1)(-\varepsilon) + o(\varepsilon)}{-\varepsilon} = \frac{1 + C\varepsilon + o(\varepsilon)}{-\varepsilon}.$$

Substituting these asymptotic expansions into (2.95) and taking into account that $\mu = 1$, we have:

$$F\left(a, b, c; \frac{\lambda}{2}\right) = 4\frac{1 - \varepsilon^\varepsilon}{\varepsilon} + \frac{96C - 264}{36} + o(1).$$

Since $\varepsilon^\varepsilon - 1 \sim \varepsilon \ln \varepsilon$, this formula can be rewritten as

$$F\left(a, b, c; \frac{\lambda}{2}\right) = \ln \frac{1}{\varepsilon} \cdot (4 + o(1)).$$

Now consider asymptotic behaviour as $\lambda \to 2 - 0$ of the function $F(a + 1, b + 1, c + 1; \frac{\lambda}{2})$. Because for this function condition (2.92) does not hold (indeed, $(a + 1) + (b + 1) - (c + 1) = a + b - c + 1 = \frac{\lambda}{2\mu} \to \frac{1}{\mu} > 0$), applying transformation (2.94) we get:

$$F(a + 1, b + 1, c + 1; \frac{\lambda}{2}) = \varepsilon^{-\frac{1}{\mu}} \cdot (4 + \mu + o(1)).$$

Thus

$$\frac{1}{g} = \left(\frac{1}{4} + o(1)\right) \cdot \varepsilon^{\frac{1}{\mu}} \cdot F(a, b, c; \frac{\lambda}{2})$$

and so applying the above results about asymptotic behaviour of the function $F(a, b, c; \frac{\lambda}{2})$ we get:

$$\frac{1}{g} \sim \begin{cases} \varepsilon^{1/\mu} \cdot \frac{\Gamma(4/\mu + 1) \cdot \Gamma(1 - 1/\mu)}{4 \cdot \Gamma(3/\mu + 1)}, & \text{if } \mu > 1, \\[2mm] \frac{1}{1 - \mu}\varepsilon, & \text{if } \mu < 1, \\[2mm] \varepsilon \ln \frac{1}{\varepsilon}, & \text{if } \mu = 1. \end{cases}$$

which immediately implies the desired statement. $\qquad\square$

## 2.8 Approximations

### 2.8.1 Approximation with the help of a loss model

In a general retrial queueing model, a full description of the system state includes the state of the group of servers and the state of the pool of subscribers served by these servers. Such a model of a real system can be thought of as a model from the point of view of an omniscient observer.

Consider now the functioning of a real system with repeated calls from the point of view of an observer which has only information about the state of the trunk group. This observer sees that from time to time calls arrive into the system. If there is a free server at the time of arrival, the call starts to be served immediately and then leaves the system after service. If all servers are occupied at the time of arrival then (from the point of view of our observer) the call disappears.

Thus, from the point of view of an observer which has only information about the state of the trunk group any retrial queue functions as a loss system. To put this in other words, both the retrial queue and the corresponding loss queue are equally good (or bad) models of a real system where blocked calls can be reinitiated. However, it is important to realize that if the real system is modelled with the help of a loss queue then the input flow is composed of all calls (both primary and repeated), i.e. is a composite of the initial flow of primary calls, which reflects real needs in service, and the flow of repeated calls, which is connected with delays in service. Thus, this joint flow has a more complex structure than the flow of primary calls (as a matter of fact, we cannot even describe this structure). We can, however, make some simplified assumptions about the structure of the joint flow, which allow us to transform the loss model into a tractable model. For example, if we assume that the flow of repeated calls is Poisson and does not depend on the (Poisson) flow of primary calls, then the joint flow is Poisson with rate $\Lambda = \lambda + r$, where $\lambda$ is the rate of flow of primary calls and $r$ is the rate of flow of repeated calls. Such a loss model is the classical Erlang model and allows explicit calculation of the main performance characteristics. Say, the distribution of the number of busy servers is a truncated Poisson distribution (or, in other terms, Erlang loss distribution):

$$p_i = \frac{\Lambda^i}{i!} \Big/ \sum_{k=0}^{c} \frac{\Lambda^k}{k!}, \quad 0 \le i \le c, \qquad (2.97)$$

with mean

$$Y = \Lambda \cdot [1 - E_c(\Lambda)]$$

and variance

$$V = Y - \Lambda \cdot E_c(\Lambda) \cdot [c - Y],$$

where

$$E_c(\Lambda) \equiv p_c = \frac{\Lambda^c}{c!} \Big/ \sum_{k=0}^{c} \frac{\Lambda^k}{k!}$$

is the loss probability (which for a loss model equals the blocking probability).

The unknown rate $\Lambda$ of the joint flow (or equivalently, the unknown rate of the flow of repeated calls $r$) can be found from the condition

$$Y = \lambda. \qquad (2.98)$$

This condition is intuitively clear and expresses equality of arrival and carried traffic, and in fact is a consequence of the general Little formula. With its help the above formulas for $E_c(\Lambda)$ and $V$ can be rewritten in a simpler form:

$$
\begin{aligned}
E_c(\Lambda) &= 1 - \frac{\lambda}{\Lambda} \equiv \frac{r}{\lambda + r}, \\
V &= \lambda - (\Lambda - \lambda)(c - \lambda).
\end{aligned}
$$

Since the loss model described above is a model of the same real system as the retrial queue, one may expect that both models should have approximately identical distributions of the number of busy servers (which is included in both models).

In the Table 2.7 we give values of the blocking probability $B$ in the main retrial queue with $c = 10$ servers, the blocking probability $B_{\mathrm{appr}} \equiv E_c(\Lambda)$ in the corresponding (in the above described sense) Erlang loss model and relative error $\frac{B - B_{\mathrm{appr}}}{B_{\mathrm{appr}}}$.

In Table 2.8 we give values of the variance $V$ of the number of busy servers (as we saw, the mean queue length in a retrial queue can be expressed in terms of this characteristic) in the main retrial queue with $c = 10$ servers, the variance of the number of busy servers $V_{\mathrm{appr}}$ in the corresponding (in the above described sense) Erlang loss model and relative error $\frac{V - V_{\mathrm{appr}}}{V_{\mathrm{appr}}}$.

Table 2.7 *Exact value of the blocking probability for the main model with* $c = 10$ *servers and its approximation with the help of the corresponding Erlang loss model*

|  |  | $\mu = 0.1$ | $\mu = 1$ | $\mu = 10$ |
|---|---|---|---|---|
| | $B$ | 0.0056 | 0.0061 | 0.0076 |
| $\lambda/c = 0.4$ | $B_{\mathrm{appr}}$ | 0.0055 | 0.0055 | 0.0055 |
| | error | 1.53% | 11.61% | 39.17% |
| | $B$ | 0.0558 | 0.0647 | 0.0862 |
| $\lambda/c = 0.6$ | $B_{\mathrm{appr}}$ | 0.0543 | 0.0543 | 0.0543 |
| | error | 2.64% | 19.09% | 58.55% |
| | $B$ | 0.2424 | 0.2792 | 0.3589 |
| $\lambda/c = 0.8$ | $B_{\mathrm{appr}}$ | 0.2362 | 0.2362 | 0.2362 |
| | error | 2.64% | 18.23% | 51.95% |

Table 2.8 *Exact value of* $V \equiv \mathrm{Var}C(t)$ *for the main model with* $c = 10$ *servers and its approximation with the help of the corresponding Erlang loss model*

|  |  | $\mu = 0.1$ | $\mu = 1$ | $\mu = 10$ |
|---|---|---|---|---|
| | $V$ | 3.8755 | 3.9118 | 3.9529 |
| $\lambda/c = 0.4$ | $V_{\mathrm{appr}}$ | 3.8676 | 3.8676 | 3.8676 |
| | error | 0.20% | 1.14% | 2.21% |
| | $V$ | 4.6746 | 4.9338 | 5.2756 |
| $\lambda/c = 0.6$ | $V_{\mathrm{appr}}$ | 4.6210 | 4.6210 | 4.6210 |
| | error | 1.16% | 6.77% | 14.17% |
| | $V$ | 3.1474 | 3.6462 | 4.4211 |
| $\lambda/c = 0.8$ | $V_{\mathrm{appr}}$ | 3.0527 | 3.0527 | 3.0527 |
| | error | 3.10% | 19.44% | 44.83% |

It is easy to see that for low rate of retrial the accuracy of the approximation with the help of the corresponding Erlang loss model is quite high. This is a consequence of the limit theorem about behaviour of the performance characteristics of the retrial queue as $\mu \to 0$.

The approximation under consideration is obtained as a consequence of the assumption that the flow of repeated calls is Poisson and does not depend on the flow of primary calls. Several other simplified assumptions lead to the same approximation.

1. Let $r_i = \mu N_i/p_i$ be the rate of flow of repeated calls given that the number of busy servers equals $i$. Assume that $r_i$ does not depend on $i : r_i \equiv r$. Then from equations (2.27), (2.28), (2.29) of Section 2.3 we get that the stationary distribution of the number of busy servers is given by equation (2.97), where $\Lambda = \lambda + r$ and for parameter $\Lambda$ equation (2.98) holds. These equations describe our approximation.

2. Let $R_i \equiv \mu N_i/(\mu N)$ be the fraction of repeated calls which find $i$ busy servers at the time of their arrival. For primary calls the corresponding characteristic equals the steady state probability $p_i$ that in the steady state exactly $i$ servers are occupied. Assume that repeated calls experience the same state of the trunk group as primary calls, i.e. $R_i = p_i$, $0 \leq i \leq c$. This relation is equivalent to the relation $N_i = Np_i$, $0 \leq i \leq c$. Thus for $r_i$, the rate of flow of repeated calls given that the number of busy servers equals $i$, we get: $r_i \equiv \mu N$, $0 \leq i \leq c$, which, as we have seen, implies the approximation under consideration.

### 2.8.2 Interpolation between low and high rates of retrials

For large values of $\mu$ the approximation using the Erlang loss model does not provide satisfactory accuracy. The situation is even worse for a still larger number of servers. For example, in the case $c = 200$, $\lambda = 180$, $\mu = 5$ the exact blocking probability $B = 0.0311$, whereas the approximate value is $B_{\text{appr}} = 0.0138$, so that relative error equals 125%.

This is not too surprising, since the approximation with the help of the Erlang loss model corresponds to low retrial rate. For a high retrial rate the retrial queue should be approximated with the help of the standard $M/M/c/\infty$ queueing system, so that as an approximate value of the blocking probability we should take

$$B(\infty) = \frac{\lambda^c}{(c-1)!} \left/ \left( (c-\lambda) \sum_{k=0}^{c-1} \frac{\lambda^k}{k!} + \frac{\lambda^c}{(c-1)!} \right) \right.$$

and as an approximate value of the variance of the number of busy servers we should take

$$V(\infty) = \lambda \left( 1 - B(\infty) \right).$$

For intermediate values of $\mu$, as an approximation of the blocking probability $B(\mu)$ (variance of the number of busy servers $V(\mu)$) we may take some intermediate value between $\lim_{\mu \to 0} B(\mu) = B_{\text{appr}}$ and $\lim_{\mu \to \infty} B(\mu) \equiv B(\infty)$ (correspondingly, some intermediate value between $\lim_{\mu \to 0} V(\mu) \equiv V_{\text{appr}}$ and $\lim_{\mu \to \infty} V(\mu) \equiv V(\infty)$).

The simplest and the most natural interpolations are:

$$B(\mu) \approx \frac{1}{1+\mu} B_{\text{appr}} + \frac{\mu}{1+\mu} B(\infty), \qquad (2.99)$$

$$V(\mu) \approx \frac{V_{\text{appr}} + \mu V(\infty)}{1+\mu}. \qquad (2.100)$$

Equations (2.100) and (2.24) yield the following approximation for the mean queue length:

$$\mu N \approx N_{\text{appr}} + \mu N(\infty)$$

$$\equiv r + \frac{\lambda \mu}{c - \lambda} B(\infty).$$

Since for $Q = \text{E}\left(N(t) | C(t) = c\right) = N_c / B$ we have:

$$\lim_{\mu \to 0} \mu Q = r, \quad \lim_{\mu \to \infty} Q = \frac{\lambda}{c - \lambda},$$

a parallel equation for $Q$ is:

$$\mu Q \approx r + \frac{\lambda \mu}{c - \lambda}.$$

Using equation (2.22) we have:

$$B = \frac{\mu N}{\lambda + \mu Q},$$

so that these approximations for $N$ and $Q$ imply:

$$B \approx \frac{r(c - \lambda) + \lambda \mu B(\infty)}{(\lambda + r)(c - \lambda) + \lambda \mu}.$$

Since $r = \frac{\lambda B_{\text{appr}}}{1 - B_{\text{appr}}}$, we get:

$$B(\mu) \approx \frac{(c - \lambda) B_{\text{appr}} + \mu \left(1 - B_{\text{appr}}\right) B(\infty)}{(c - \lambda) + \mu \left(1 - B_{\text{appr}}\right)}. \qquad (2.101)$$

The simple approximation (2.99) for $B(\mu)$ (which is similar to the interpolation formula (2.100) for $V(\mu)$) is less accurate than that given by equation (2.101). Sometimes equation (2.99) is even less accurate than the simplest approximation $B(\mu) \approx B_{\text{appr}}$. For

Table 2.9 *Exact value of the blocking probability for the main model with $c = 10$ servers and its approximation with the help of interpolation between corresponding $M/M/c/0$ and $M/M/c/\infty$ models*

|                  |                     | $\mu = 0.1$ | $\mu = 1$ | $\mu = 10$ |
|------------------|---------------------|-------------|-----------|------------|
|                  | $B(\mu)$            | .0056       | .0061     | .0076      |
| $\lambda/c = 0.4$ | $B_{\text{appr}}(\mu)$ | .0055       | .0060     | .0076      |
|                  | error               | .53%        | 2.75%     | .96%       |
|                  | $B(\mu)$            | .0558       | .0647     | .0862      |
| $\lambda/c = 0.6$ | $B_{\text{appr}}(\mu)$ | .0554       | .0633     | .0873      |
|                  | error               | .63%        | 2.20%     | −1.36%     |
|                  | $B(\mu)$            | .2424       | .2792     | .3589      |
| $\lambda/c = 0.8$ | $B_{\text{appr}}(\mu)$ | .2425       | .2840     | .3733      |
|                  | error               | −0.05%      | −1.67%    | −3.86%     |

example, if $c = 10$, $\lambda = 6$, $\mu = 0.1$ then $B = 0.0558$, $B_{\text{appr}} = 0.0543$ (relative error equals 2.64%), and equation (2.101) gives $B_{(\mu)} \approx 0.554$ (relative error equals 0.63%), whereas approximation (2.99) gives for the blocking probability value 0.0586 (relative error equals $-4.84\%$). Thus we will use relations (2.101) and (2.100) to approximate the blocking probability and the variance of the number of customers in orbit respectively.

In Table 2.9 we give values of the blocking probability $B(\mu)$ in the main retrial queue with $c = 10$ servers, the approximate blocking probability $B_{\text{appr}}(\mu)$ (calculated with the help of the right-hand side of (2.101)) and the relative error $\frac{B(\mu) - B_{\text{appr}}(\mu)}{B_{\text{appr}}(\mu)}$.

In Table 2.10 we give values of the variance of the number of busy servers $V(\mu)$ in the main retrial queue with $c = 10$ servers, the approximate variance of the number of busy servers $V_{\text{appr}}(\mu)$ (calculated with the help of the right-hand side of equation (2.100)) and the relative error $\frac{V(\mu) - V_{\text{appr}}(\mu)}{V_{\text{appr}}(\mu)}$. As one can see from these tables the interpolation between the corresponding $M/M/c/0$ and $M/M/c/\infty$ models provides extremely high accuracy for all values of the system parameters.

Table 2.10 *Exact value of* Var$C(t)$ *for the main model with* $c = 10$ *servers and its approximation with the help of interpolation between corresponding* $M/M/c/0$ *and* $M/M/c/\infty$ *models*

|              |                       | $\mu = 0.1$ | $\mu = 1$ | $\mu = 10$ |
|--------------|-----------------------|-------------|-----------|------------|
|              | $V(\mu)$              | 3.8755      | 3.9118    | 3.9529     |
| $\lambda/c = 0.4$ | $V_{\mathrm{appr}}(\mu)$ | 3.8764      | 3.9162    | 3.9559     |
|              | error                 | −0.025%     | −0.11%    | −0.075%    |
|              | $V(\mu)$              | 4.6746      | 4.9338    | 5.2756     |
| $\lambda/c = 0.6$ | $V_{\mathrm{appr}}(\mu)$ | 4.6911      | 5.0066    | 5.3221     |
|              | error                 | −0.35%      | −1.45%    | −0.87%     |
|              | $V(\mu)$              | 3.1474      | 3.6462    | 4.4211     |
| $\lambda/c = 0.8$ | $V_{\mathrm{appr}}(\mu)$ | 3.2049      | 3.8896    | 4.5744     |
|              | error                 | −1.79%      | −6.26%    | −3.35%     |

# CHAPTER 3

# Advanced single-server models

## 3.1 A single-server batch arrival retrial queue

### 3.1.1 Model description

In the batch arrival retrial queue it is assumed that at every arrival epoch a batch of $k$ primary calls arrives with probability $c_k$. If the server is busy at the arrival epoch, then all these calls join the orbit, whereas if the server is free, then one of the arriving customers begins his service and the others form sources of repeated calls.

Behaviour of customers in orbit is the same as in the main model, i.e. every such customer produces a Poisson flow of repeated calls with rate $\mu$. If an incoming repeated call finds the server free it is served and leaves the system after service, while the customer which produced this repeated call disappears. Otherwise, if the server is occupied at the time of the repeated call arrival then the system state does not change.

We assume that the input flow of primary calls (i.e. arrival epochs and sizes of batches), intervals between repeated trials, and service times are mutually independent.

Denote by $c(z) = \sum_{k=1}^{\infty} z^k c_k$ the generating function of the batch size distribution, $\bar{c} = c'(1)$ the mean batch size, $\rho = \lambda \beta_1 \bar{c}$. All other notations will be the same as in Chapter 1 unless otherwise indicated.

### 3.1.2 Joint distribution of the server state and the queue length in the steady state

In this section we carry out the simplest and simultaneously the most important (from an applied point of view) analysis of the system. Namely, we investigate the joint distribution of the server state and the queue length in the steady state. As we will show later on, the stationary regime exists if and only if $\rho < 1$, so the condition $\rho < 1$ is assumed to hold from now on.

**Theorem 3.1** *For an $M/G/1$ batch arrival retrial queue in the steady state the joint distribution of the server state and queue length*

$$
\begin{aligned}
p_{0n} &= \mathrm{P}\{C(t) = 0, N(t) = n\}, \\
p_{1n}(x) &= \frac{d}{dx}\mathrm{P}\{C(t) = 1, \xi(t) < x, N(t) = n\}
\end{aligned}
$$

*has partial generating functions*

$$
p_0(z) \equiv \sum_{n=0}^{\infty} z^n p_{0n}
$$

$$
= (1-\rho)\exp\left\{\frac{\lambda}{\mu}\int_1^z \frac{1 - \beta(\lambda - \lambda c(u)) \cdot \frac{c(u)}{u}}{\beta(\lambda - \lambda c(u)) - u}\,du\right\},
$$

$$
p_1(z, x) \equiv \sum_{n=0}^{\infty} z^n p_{1n}(x)
$$

$$
= \lambda\frac{1 - c(z)}{\beta(\lambda - \lambda c(z)) - z}p_0(z)[1 - B(x)]e^{-(\lambda - \lambda c(z))x}.
$$

*If in the case $C(t) = 1$ we neglect the elapsed service time $\xi(t)$, then for the probabilities $p_{1n} = \mathrm{P}\{C(t) = 1, N(t) = n\}$ we have*

$$
p_1(z) \equiv \sum_{n=0}^{\infty} z^n p_{1n} = \frac{1 - \beta(\lambda - \lambda c(z))}{\beta(\lambda - \lambda c(z)) - z}p_0(z).
$$

*Proof.* In a general way we obtain the equations of statistical equilibrium:

$$
(\lambda + n\mu)p_{0n} = \int_0^{\infty} p_{1n}(x)b(x)\,dx,
$$

$$
p'_{1n}(x) = -(\lambda + b(x))p_{1n}(x) + \lambda\sum_{k=1}^{n} c_k p_{1,n-k}(x),
$$

$$
p_{1n}(0) = \lambda\sum_{k=1}^{n+1} c_k p_{0,n-k+1} + (n+1)\mu p_{0,n+1}.
$$

For generating functions $p_0(z)$ and $p_1(z, x)$ these equations give:

$$
\lambda p_0(z) + \mu z\frac{dp_0(z)}{dz} = \int_0^{+\infty} p_1(z, x)b(x)\,dx, \qquad (3.1)
$$

$$
\frac{\partial p_1(z, x)}{\partial x} = -(\lambda - \lambda c(z) + b(x))p_1(z, x), \qquad (3.2)
$$

$$p_1(z,0) = \lambda \frac{c(z)}{z} p_0(z) + \mu \frac{dp_0(z)}{dz}. \tag{3.3}$$

From (3.2) we find that $p_1(z,x)$ depends upon $x$ as follows:

$$p_1(z,x) = p_1(z,0)[1 - B(x)]e^{-(\lambda - \lambda c(z))x}. \tag{3.4}$$

With the help of (3.4), equation (3.1) can be rewritten as

$$\lambda p_0(z) + \mu z \frac{dp_0(z)}{dz} = \beta(\lambda - \lambda c(z))p_1(z,0). \tag{3.5}$$

Eliminating $p_1(z,0)$ from (3.5) and (3.3) we get:

$$\mu[\beta(\lambda - \lambda c(z)) - z]\frac{dp_0(z)}{dz} = \lambda \left[1 - \beta(\lambda - \lambda c(z)) \cdot \frac{c(z)}{z}\right]p_0(z). \tag{3.6}$$

Consider the coefficient $f(z) = \beta(\lambda - \lambda c(z)) - z$. Note that:

1. $f(1) = \beta(0) - 1 = 1 - 1 = 0$;
2. $f'(z) = -\lambda c'(z)\beta'(\lambda - \lambda c(z)) - 1$ and thus $f'(1) = \rho - 1 < 0$;
3. $f''(z) = -\lambda c''(z)\beta'(\lambda - \lambda c(z)) + \lambda^2(c'(z))^2\beta''(\lambda - \lambda c(z)) \geq 0$.

Therefore the function $f(z)$ is decreasing on the interval $[0,1]$, $z = 1$ is the only zero there and for $z \in [0,1)$ the function is positive, i.e. (as $\rho < 1$) for $z \in [0,1)$ we have:

$$z < \beta(\lambda - \lambda c(z)) \leq 1.$$

Besides,

$$\lim_{z \to 1-0} \frac{1 - \beta(\lambda - \lambda c(z)) \cdot \frac{c(z)}{z}}{\beta(\lambda - \lambda c(z)) - z} = \frac{\rho + \bar{c} - 1}{1 - \rho} < \infty,$$

i.e. the function $\frac{1 - \beta(\lambda - \lambda c(z)) \cdot \frac{c(z)}{z}}{\beta(\lambda - \lambda c(z)) - z}$ can be defined at the point $z = 1$ as $\frac{\rho + \bar{c} - 1}{1 - \rho}$. This means that for $z \in [0;1]$ we can rewrite the equation (3.6) as

$$\frac{dp_0(z)}{dz} = \frac{\lambda}{\mu} \frac{1 - \beta(\lambda - \lambda c(z)) \cdot \frac{c(z)}{z}}{\beta(\lambda - \lambda c(z)) - z} p_0(z),$$

which implies that

$$p_0(z) = p_0(1) \exp\left\{\frac{\lambda}{\mu} \int_1^z \frac{1 - \beta(\lambda - \lambda c(u)) \cdot \frac{c(u)}{u}}{\beta(\lambda - \lambda c(u)) - u} du\right\}.$$

Now from (3.3)

$$p_1(z,0) = \lambda \frac{1 - c(z)}{\beta(\lambda - \lambda c(z)) - z} p_0(z),$$

and so from (3.4)

$$p_1(z, x) = \lambda \frac{1 - c(z)}{\beta(\lambda - \lambda c(z)) - z} p_0(z)[1 - B(x)]e^{-(\lambda - \lambda c(z))x}. \quad (3.7)$$

From (3.7) we have that

$$p_1(z) = \int_0^{+\infty} p_1(z, x)dx = \frac{1 - \beta(\lambda - \lambda c(z))}{\beta(\lambda - \lambda c(z)) - z} p_0(z). \quad (3.8)$$

The unknown constant $p_0(1)$ can be found from the normalizing condition $p_0(1) + p_1(1) = 1$. Using (3.8) we have:

$$p_1(1) = \frac{\rho}{1 - \rho} p_0(1).$$

Thus,

$$1 = p_0(1) + p_1(1) = p_0(1) + \frac{\rho}{1 - \rho} p_0(1) = \frac{1}{1 - \rho} p_0(1),$$

i.e.

$$p_0(1) = 1 - \rho.$$

which completes the proof.                                            □

With the help of generating functions $p_0(z)$, $p_1(z)$ we can find various performance characteristics of the system. For example, the distribution of the number of customers in orbit has generating function

$$p(z) = (1 - \rho)\frac{1 - z}{\beta(\lambda - \lambda c(z)) - z}$$

$$\times \exp\left\{\frac{\lambda}{\mu} \int_1^z \frac{1 - \beta(\lambda - \lambda c(u)) \cdot \frac{c(u)}{u}}{\beta(\lambda - \lambda c(u)) - u} du\right\}.$$

In particular, the mean queue length $EN(t) = p'(1)$ is given by:

$$EN(t) = \frac{\lambda^2 [c'(1)]^2 \beta_2 + \rho c''(1)/c'(1)}{2(1 - \rho)} + \frac{\lambda}{\mu} \cdot \frac{\rho + c'(1) - 1}{1 - \rho}. \quad (3.9)$$

### 3.1.3 Embedded Markov chain

*The structure of the embedded Markov chain*

Let $N_i = N(\eta_i)$ be the number of sources of repeated calls at the time $\eta_i$ of the $i$th departure. It is easy to see that

$$N_i = N_{i-1} - B_i + \nu_i, \quad (3.10)$$

where $B_i$ is the number of sources which enter service at time $\xi_i$ (i.e. $B_i = 1$ if the $i$th call is a repeated call and $B_i = 0$ if the $i$th call is a primary call) and $\nu_i$ is the number of primary calls which arrive in the system during the service time $S_i$ of the $i$th call.

The Bernoulli random variable $B_i$ depends on the history of the system before time $\eta_{i-1}$ only through $N_{i-1}$; its conditional distribution is given by

$$P\{B_i = 1 \mid N_{i-1} = n\} = \frac{n\mu}{\lambda + n\mu},$$

$$P\{B_i = 0 \mid N_{i-1} = n\} = \frac{\lambda}{\lambda + n\mu}.$$

The random variable $\nu_i$ does not depend on events which have occurred before epoch $\xi_i$ and has distribution

$$k_n \equiv P\{\nu_i = n\} = \int_0^\infty \sum_j \frac{(\lambda x)^j}{j!} e^{-\lambda x} c_n^{(j)} dB(x),$$

where $c_n^{(j)}$ is the $j$-fold convolution of the sequence $c_n$. Thus

$$k(z) \equiv \sum_{n=0}^\infty k_n z^n = \beta(\lambda - \lambda c(z))$$

and

$$E\nu_i = \sum_{n=0}^\infty n k_n = \rho.$$

The above remarks mean that the sequence of random variables $N_i$ forms a Markov chain, which is the embedded chain for our queueing system.

Its one-step transition probabilities $r_{mn} = P\{N_i = n \mid N_{i-1} = m\}$ are given by the formula

$$r_{mn} = \frac{\lambda}{\lambda + m\mu} \sum_{i=1}^{n-m+1} c_i k_{n-m+1-i} + \frac{m\mu}{\lambda + m\mu} k_{n-m+1}.$$

*Ergodicity*

As usual, the first question to be investigated is the ergodicity of the chain.

Because of the recursive structure of the equation (3.10) we will use criteria based on mean drift. For the Markov chain under consideration we have:

$$
\begin{aligned}
x_n &\equiv \mathrm{E}(N_{i+1} - N_i \mid N_i = n) \\
&= \mathrm{E}(-B_{i+1} + \nu_{i+1} \mid N_i = n) \\
&= -\mathrm{E}(B_{i+1} \mid N_i = n) + \mathrm{E}(\nu_{i+1} \mid N_i = n) \\
&= -\mathrm{P}(B_{i+1} = 1 \mid N_i = n) + \mathrm{E}(\nu_{i+1}) \\
&= -\frac{n\mu}{\lambda + n\mu} + \rho.
\end{aligned}
$$

As $n \to \infty$, there exists $\lim x_n = -1 + \rho$. This limit is negative iff $\rho < 1$. Applying Foster's criteria (Statement 1, section 1.3) we can guarantee that for $\rho < 1$ the embedded Markov chain is ergodic. For $\rho \geq 1$

$$
x_n = -\frac{n\mu}{\lambda + n\mu} + \rho \geq -\frac{n\mu}{\lambda + n\mu} + 1 = \frac{\lambda}{\lambda + n\mu} > 0,
$$

and since down drifts $N_{i+1} - N_i$ are bounded from below we can guarantee nonergodicity in the case $\rho \geq 1$.

Thus, the embedded Markov chain is ergodic iff $\rho < 1$.

*Stationary distribution*

Our second goal is to find the stationary distribution $\pi_n$ of the embedded Markov chain $\{N_i\}$.

Kolmogorov equations for the distribution $\pi_n$ are

$$
\begin{aligned}
\pi_n &= \sum_{m=0}^{n} \pi_m \frac{\lambda}{\lambda + m\mu} \sum_{i=1}^{n-m+1} c_i k_{n-m+1-i} \\
&+ \sum_{m=1}^{n+1} \pi_m \frac{m\mu}{\lambda + m\mu} k_{n-m+1}, \quad n = 0, 1, \ldots
\end{aligned}
$$

Because of the presence of convolutions, these equations can be transformed with the help of the generating functions

$$
\varphi(z) = \sum_{n=0}^{\infty} z^n \pi_n,
$$

$$
\psi(z) = \sum_{n=0}^{\infty} z^n \frac{\pi_n}{\lambda + n\mu},
$$

to

$$
\varphi(z) = \beta(\lambda - \lambda c(z)) \cdot \left( \lambda \frac{c(z)}{z} \psi(z) + \mu \psi'(z) \right).
$$

Since the ordinary generating function $\varphi(z)$ can be expressed in terms of $\psi(z)$:

$$\varphi(z) = \lambda\psi(z) + \mu z\psi'(z),$$

we get the following equation for the generating function $\psi(z)$:

$$\mu[\beta(\lambda - \lambda c(z)) - z]\psi'(z) = \lambda\left[1 - \beta(\lambda - \lambda c(z)) \cdot \frac{c(z)}{z}\right]\psi(z).$$

To solve this equation we note that it is identical to the equation (3.6) and thus in the case $\rho < 1$ we have

$$\psi(z) = \psi(1) \cdot \exp\left\{\frac{\lambda}{\mu} \int\limits_1^z \frac{1 - \beta(\lambda - \lambda c(u)) \cdot \frac{c(u)}{u}}{\beta(\lambda - \lambda c(u)) - u}du\right\}.$$

From this,

$$\begin{aligned}
\varphi(z) &= \lambda\psi(z) + \mu z\psi'(z) \\
&= \lambda\beta(\lambda - \lambda c(z)) \cdot \frac{1 - c(z)}{\beta(\lambda - \lambda c(z)) - z}\psi(z).
\end{aligned}$$

Since $\varphi(1) = 1$, we have:

$$\psi(1) = \sum_{n=0}^{\infty} \frac{\pi_n}{\lambda + n\mu} = \frac{1 - \rho}{\lambda\overline{c}}.$$

Finally we get the following formula for the generating function $\varphi(z) = \sum_{n=0}^{\infty} z^n \pi_n$ of the stationary distribution of the embedded Markov chain $\{N_i\}$:

$$\begin{aligned}
\varphi(z) &= \frac{1 - \rho}{\overline{c}} \frac{1 - c(z)}{\beta(\lambda - \lambda c(z)) - z}\beta(\lambda - \lambda c(z)) \\
&\times \exp\left\{\frac{\lambda}{\mu} \int\limits_1^z \frac{1 - \beta(\lambda - \lambda c(u)) \cdot \frac{c(u)}{u}}{\beta(\lambda - \lambda c(u)) - u}du\right\}.
\end{aligned}$$

### 3.1.4 Functioning of the system in the nonstationary regime

As in Chapter 1 we consider the queueing process as alternating between busy periods and idle periods. To investigate the process during the busy periods we will use the Kolmogorov differential equations for transient probabilities.

All definitions and notations of Chapter 1 hold, however we will not take into account the number of served customers $I(t)$ and correspondingly will omit in these notations index $i$ which indicates that $I(t) = i$ and argument $y$ in generating functions.

For the batch arrival retrial queue a $k$-busy period can be defined as the period which starts when a batch of $k$ primary calls arrives into an empty system and ends at the next departure epoch when the system is empty.

Assume that a $k$-busy period started at time $t = 0$. Let $L^{(k)}$ be the length of the $k$-busy period, $\Pi^{(k)}(t) = \mathrm{P}(L^{(k)} < t)$, $\pi^{(k)}(s) = \mathrm{E}e^{-sL^{(k)}}$. Besides, let

$$P_{0n}^{(k)}(t) = \mathrm{P}\left\{L^{(k)} > t, C(t) = 0, N(t) = n\right\}$$

be the probability that at time $t$ the $k$-busy period does not expire, the server is free, the number of customers in orbit is equal to $n$, $n \geq 1$, and

$$\begin{aligned}
P_{1n}^{(k)}&(t, x)dx \\
&= \mathrm{P}\left\{L^{(k)} > t, C(t) = 1, x < \xi(t) < x + dx, N(t) = n\right\}
\end{aligned}$$

be the probability that at time $t$ the $k$-busy period does not expire, the server has been busy for the time $\xi(t) \in (x, x+dx)$, the number of customers in orbit is equal to $n$, $n \geq 0$.

The Kolmogorov equations for these probabilities are

$$\frac{dP_{0n}^{(k)}(t)}{dt} = -(\lambda + n\mu)P_{0n}^{(k)}(t) + \int_0^\infty P_{1n}^{(k)}(t, x)b(x)dx, \ n \geq 1,$$

$$\frac{d\Pi^{(k)}(t)}{dt} = \int_0^\infty P_{10}^{(k)}(t, x)b(x)dx,$$

$$\begin{aligned}
\frac{\partial P_{1n}^{(k)}(t, x)}{\partial t} &= -\left(\lambda + b(x) + \frac{\partial}{\partial x}\right)P_{1n}^{(k)}(t, x) \\
&+ \lambda \sum_{k=1}^n c_k P_{1,n-k}^{(k)}(t, x),
\end{aligned}$$

$$P_{1n}^{(k)}(t, 0) = \lambda \sum_{k=1}^{n+1} c_k P_{0,n-k+1}^{(k)}(t) + (n+1)\mu P_{0,n+1}^{(k)}(t).$$

The initial conditions are

$$P_{0n}^{(k)}(0) = 0, P_{1n}^{(k)}(0, x) = \delta(x)\delta_{n,k-1}.$$

For generating functions of Laplace transforms

$$\varphi_0^{(k)}(s,z) = \sum_{n=1}^{\infty} z^n \int_0^{\infty} e^{-st} P_{0n}^{(k)}(t)dt,$$

$$\varphi_1^{(k)}(s,z,x) = \sum_{n=0}^{\infty} z^n \int_0^{\infty} e^{-st} P_{1n}^{(k)}(t,x)dt$$

these equations become:

$$(s+\lambda)\varphi_0^{(k)}(s,z) + \pi^{(k)}(s) + \mu z \frac{\partial \varphi_0^{(k)}(s,z)}{\partial z} = \int_0^{\infty} \varphi_1^{(k)}(s,z,x)b(x)dx,$$

$$(3.11)$$

$$-\delta(x)z^{k-1} + \frac{\partial \varphi_1^{(k)}(s,z,x)}{\partial x} = -(s+\lambda-\lambda c(z)+b(x))\varphi_1^{(k)}(s,z,x),$$

$$(3.12)$$

$$\varphi_1^{(k)}(s,z,0) = \lambda \frac{c(z)}{z}\varphi_0^{(k)}(s,z) + \mu \frac{\partial \varphi_0^{(k)}(s,z)}{\partial z}. \qquad (3.13)$$

From equation (3.12) we find the form of dependence of $\varphi_1^{(k)}(s,z,x)$ upon the variable $x$:

$$\varphi_1^{(k)}(s,z,x) = (1-B(x))e^{-(s+\lambda-\lambda c(z))x}\left\{\varphi_1^{(k)}(s,z,0) + z^{k-1}\right\}.$$

This allows us to rewrite equation (3.11) as

$$(s+\lambda)\varphi_0^{(k)}(s,z) + \pi^{(k)}(s) + \mu z \frac{\partial \varphi_0^{(k)}(s,z)}{\partial z}$$
$$= \beta(s+\lambda-\lambda c(z))\left\{\varphi_1^{(k)}(s,z,0) + z^{k-1}\right\}.$$

Eliminating $\varphi_1^{(k)}(s,z,0)$ with the help of equation (3.13) we get the following differential equation for $\varphi_0^{(k)}(s,z)$ :

$$\mu[\beta(s+\lambda-\lambda c(z)) - z]\frac{\partial \varphi_0^{(k)}(s,z)}{\partial z}$$
$$= \left[s+\lambda-\lambda\frac{c(z)}{z}\beta(s+\lambda-\lambda c(z))\right]\varphi_0^{(k)}(s,z)$$
$$+\pi^{(k)}(s) - z^{k-1}\beta(s+\lambda-\lambda c(z)). \qquad (3.14)$$

The initial condition is $\varphi_0^{(k)}(s,0) = 0$.

Below we will use the following fact. Let $\pi_{\infty}(s)$ be the Laplace transform of the length of busy period in the standard $M/G/1/\infty$ batch arrival queueing system. Then:

1. for $s > 0$ the function $\pi_\infty(s)$ is the unique solution of the equation $\beta(s + \lambda - \lambda c(z)) = z$ on the interval $0 \leq z \leq 1$;

2. if $\rho > 1$ then $\pi_\infty(0) < 1$; if $\rho \leq 1$ then $\pi_\infty(0) = 1$ and $\pi'_\infty(0) = \beta'(0)/(1 - \rho)$.

Thus if $s > 0$ and $0 \leq z < \pi_\infty(s)$ then the coefficient in the left-hand side of equation (3.14) is nonzero. Therefore in this interval the solution of this equation is

$$\varphi_0^{(k)}(s, z) = e(s, z) \int_0^z \frac{\pi^{(k)}(s) - u^{k-1}\beta(s + \lambda - \lambda c(u))}{\mu\left[\beta(s + \lambda - \lambda c(u)) - u\right] e(s, u)} du, \quad (3.15)$$

where the function $e(s, z)$ is given by

$$e(s, z) = \exp\left\{\frac{1}{\mu} \int_0^z \frac{s + \lambda - \lambda\frac{c(u)}{u}\beta(s + \lambda - \lambda c(u))}{\beta(s + \lambda - \lambda c(u)) - u} du\right\}.$$

If $s > 0$ then $\lim_{z \to \pi_\infty(s) - 0} e(s, z) = +\infty$. On the other hand

$$\varphi_0^{(k)}(s, \pi_\infty(s)) < \infty.$$

Thus the integral in the right-hand side of equation (3.15) must tend to zero, i.e.

$$\int_0^{\pi_\infty(s)} \frac{\pi^{(k)}(s) - u^{k-1}\beta(s + \lambda - \lambda c(u))}{\mu\left[\beta(s + \lambda - \lambda c(u)) - u\right] e(s, u)} du = 0, \quad s > 0, \quad (3.16)$$

Using (3.16) we can rewrite equation (3.15) as

$$\varphi_0^{(k)}(s, z) = \int_{\pi_\infty(s)}^z \frac{\pi^{(k)}(s) - u^{k-1}\beta(s + \lambda - \lambda c(u))}{\mu\left[\beta(s + \lambda - \lambda c(u)) - u\right]}$$

$$\times \exp\left\{\int_u^z \frac{s + \lambda - \lambda\frac{c(v)}{v}\beta(s + \lambda - \lambda c(v))}{\mu[\beta(s + \lambda - \lambda c(v)) - v]} dv\right\} du. \quad (3.17)$$

Consider now the interval $\pi_\infty(s) < z \leq 1$. For this $z$ the coefficient $\beta(s + \lambda - \lambda c(z)) - z \neq 0$ (in fact it is negative) and thus,

$$
\varphi_0^{(k)}(s, z) = \left\{ C + \int\limits_1^z \frac{\pi^{(k)}(s) - u^{k-1}\beta(s + \lambda - \lambda c(u))}{\mu\left[\beta(s + \lambda - \lambda c(u)) - u\right] e_1(s, u)} du \right\}
$$
$$
\times \quad e_1(s, z), \tag{3.18}
$$

where for $\pi_\infty(s) < z \leq 1$

$$
e_1(s, z) = \exp\left\{ \frac{1}{\mu} \int\limits_1^z \frac{s + \lambda - \lambda\frac{c(u)}{u}\beta(s + \lambda - \lambda c(u))}{\beta(s + \lambda - \lambda c(u)) - u} du \right\}.
$$

As $z \to \pi_\infty(s) + 0$ the function $e_1(s, z) \to +\infty$. On the other hand $\varphi_0^{(k)}(s, \pi_\infty(s)) < \infty$. Thus,

$$
C = \int\limits_{\pi_\infty(s)}^1 \frac{\pi^{(k)}(s) - u^{k-1}\beta(s + \lambda - \lambda c(u))}{\mu\left[\beta(s + \lambda - \lambda c(u)) - u\right] e_1(s, u)} du,
$$

which allows us to transform (3.18) to the same form as (3.17). Thus, we can guarantee that (3.17) holds for all $z \neq \pi_\infty(s)$.

For $z = \pi_\infty(s)$ we have directly from (3.14) :

$$
\varphi_0^{(k)}(s, z) = \frac{(\pi_\infty(s))^k - \pi^{(k)}(s)}{s + \lambda - \lambda c\left(\pi_\infty(s)\right)}.
$$

In order to find the Laplace transform of the $k$-busy period we consider equation (3.16). Since for $s > 0$ integrals

$$
\int\limits_0^{\pi_\infty(s)} \frac{u^{k-1}\beta(s + \lambda - \lambda c(u))}{\left[\beta(s + \lambda - \lambda c(u)) - u\right] e(s, u)} du
$$

and

$$
\int\limits_0^{\pi_\infty(s)} \frac{1}{\left[\beta(s + \lambda - \lambda c(u)) - u\right] e(s, u)} du
$$

are finite, we have:

$$
\pi^{(k)}(s) = \frac{\int\limits_0^{\pi_\infty(s)} \frac{u^{k-1}\beta(s+\lambda-\lambda c(u))}{[\beta(s+\lambda-\lambda c(u))-u]e(s,u)} du}{\int\limits_0^{\pi_\infty(s)} \frac{1}{[\beta(s+\lambda-\lambda c(u))-u]e(s,u)} du}, \quad s > 0. \tag{3.19}
$$

For the batch arrival retrial queue, define an ordinary busy period as a period which starts when a batch of primary calls arrives into an empty system and ends at the next departure epoch when the system is empty. Thus with probability $c_k$ the ordinary busy period has the same distribution as the $k$-busy period. It should be noted that for the batch arrival queue, the ordinary busy period and 1-busy period are not identical. Let $L$ be the length of the ordinary busy period and $\pi(s) = \mathrm{E}e^{-sL}$ its Laplace transform. Then from (3.19) we have:

$$\frac{1}{s + \lambda - \lambda \mathrm{E}e^{-sL}} = \int_0^{\pi_\infty(s)} \frac{1}{\mu \left[ \beta \left( s + \lambda - \lambda c(u) \right) - u \right]}$$

$$\times \exp \left\{ -\int_0^u \frac{s + \lambda - \lambda \frac{c(v)}{v} \beta \left( s + \lambda - \lambda c(v) \right)}{\mu \left[ \beta \left( s + \lambda - \lambda c(v) \right) - v \right]} dv \right\} du.$$

Now let $\rho < 1$. In order to get $\mathrm{E}L^{(k)}$ we put $z = \pi_\infty(s)$ in equation (3.14):

$$\pi^{(k)}(s) = [\pi_\infty(s)]^k - [s + \lambda - \lambda c(\pi_\infty(s))] \cdot \varphi_0^{(k)}(s, \pi_\infty(s)).$$

Since $\pi_\infty(0) = 1$, $\pi_\infty'(0) = -\beta_1/(1 - \rho)$, we have:

$$
\begin{aligned}
\mathrm{E}L^{(k)} &= \lim_{s \to 0} \frac{1 - \pi^{(k)}(s)}{s} \\
&= \lim_{s \to 0} \frac{1 - [\pi_\infty(s)]^k}{s} \\
&+ \lim_{s \to 0} \frac{s + \lambda - \lambda c(\pi_\infty(s))}{s} \cdot \varphi_0^{(k)}(0, 1) \\
&= \frac{k\beta_1}{1 - \rho} + \frac{1}{1 - \rho} \cdot \varphi_0^{(k)}(0, 1).
\end{aligned}
$$

On the other hand putting $s = 0$ in equation (3.14), we get the following equation for $\varphi_0^{(k)}(0, z)$ :

$$\mu[\beta(\lambda - \lambda c(z)) - z] \frac{\partial \varphi_0^{(k)}(0, z)}{\partial z}$$

$$= \lambda \left[ 1 - \frac{c(z)}{z} \beta(\lambda - \lambda c(z)) \right] \varphi_0^{(k)}(0, z)$$

$$+ 1 - z^{k-1} \beta(\lambda - \lambda c(z)),$$

which implies that

$$
\varphi_0^{(k)}(0, z) = \int_0^z \frac{1 - u^{k-1}\beta(\lambda - \lambda c(u))}{\mu(\beta(\lambda - \lambda c(u)) - u)}
$$
$$
\times \quad \exp\left\{\frac{\lambda}{\mu} \int_u^z \frac{1 - \frac{c(v)}{v}\beta(\lambda - \lambda c(v))}{\beta(\lambda - \lambda c(v)) - v} dv\right\} du.
$$

Thus,

$$
EL^{(k)} = \frac{k\beta_1}{1 - \rho}
$$
$$
+ \quad \frac{1}{1 - \rho} \int_0^1 \frac{1 - u^{k-1}\beta(\lambda - \lambda c(u))}{\mu(\beta(\lambda - \lambda c(u)) - u)}
$$
$$
\times \quad \exp\left\{\frac{\lambda}{\mu} \int_u^1 \frac{1 - \frac{c(v)}{v}\beta(\lambda - \lambda c(v))}{\beta(\lambda - \lambda c(v)) - v} dv\right\} du.
$$

For the ordinary busy period this formula can be simplified to

$$
EL = -\frac{1}{\lambda} + \frac{1}{\lambda(1 - \rho)} \exp\left\{\frac{\lambda}{\mu} \int_0^1 \frac{1 - \frac{c(u)}{u}\beta(\lambda - \lambda c(u))}{\beta(\lambda - \lambda c(u)) - u} du\right\}.
$$

Now consider the main process at an arbitrary time. To avoid unnecessary complication with minor details we will assume that the system was empty at the initial epoch $t_0 = 0$, i.e. $C(0) = 0$, $N(0) = 0$. Let

$$
p_{0n}(t) = P\{C(t) = 0, N(t) = n\},
$$
$$
p_{1n}(t, x) = \frac{d}{dx}P\{C(t) = 1, N(t) = n, \xi(t) < x\}
$$

be the transient distribution of the process $(C(t), \xi(t), N(t))$ in the nonstationary regime and

$$
p_0^*(s, z) = \int_0^\infty e^{-st} \sum_{n=0}^\infty z^n p_{0n}(t)dt,
$$
$$
p_1^*(s, z, x) = \int_0^\infty e^{-st} \sum_{n=0}^\infty z^n p_{1n}(t, x)dt,
$$

be corresponding Laplace transforms of generating functions.

Verbatim repetition of the proof of Theorem 1.19 of Section 1.7 gives the following result:

$$p_1^*(s, z, x) = \frac{(s + \lambda - \lambda c(z))p_0^*(s, z) - 1}{\beta\left(s + \lambda - \lambda c(z)\right) - z} \\ \times \left[1 - B(x)\right] e^{-(s+\lambda-\lambda c(z))x};$$

$$p_0^*(s, z) = \int\limits_z^{\pi_\infty(s)} \frac{1}{\mu[\beta\left(s + \lambda - \lambda c(u)\right) - u]} \\ \times \exp\left\{\frac{1}{\mu} \int\limits_u^z \frac{s + \lambda - \lambda\frac{c(v)}{v}\beta\left(s + \lambda - \lambda c(v)\right)}{\beta\left(s + \lambda - \lambda c(v)\right) - v} dv\right\} du,$$

if $z \neq \pi_\infty(s)$. Also $p_0^*(s, \pi_\infty(s)) = 1/(s + \lambda - \lambda c(\pi_\infty(s)))$.

## 3.2 A single-server model with priority subscribers

### 3.2.1 Model description

Consider a single-server queueing system in which two different types of primary customers arrive according to independent Poisson flows with rates $\lambda_1$ and $\lambda_2$ respectively. Service times for customers from the first (second) flow are independent and identically distributed positive random variables with a common distribution function $B_1(x)$ (respectively, $B_2(x)$). If the server is free at the time of any primary call arrival, this call begins to be served immediately and leaves the system after service completion.

However, behaviour of a blocked customer, i.e. one who finds the server occupied at the time of arrival, depends on its type. Customers from the first flow are queued after blocking and then are served in some discipline such as FCFS or random order. Customers from the second flow who find the server busy upon arrival cannot be queued and leave the service area, but after some random delay repeat an attempt to get service. We assume that intervals between retrials are exponentially distributed with parameter $\mu$.

It is easy to see that a second type customer can be admitted for service only if there is no queue of first type customers. Thus the first type customers have a priority which is usually based on the fact that they have a direct access to the server and therefore can detect the epoch of the server release and immediately enter service. This type of priority is similar to the standard head-of-the-line priority discipline. Moreover, as $\mu$ tends to infinity, the model under consideration can be thought of as the standard queueing system with the head-of-the-line priority discipline. Taking this into account we shall refer to the first type customers as priority customers and to the second type customers as nonpriority (or low priority) customers.

As usual, the input flows of primary calls, intervals between repetitions, and service times are mutually independent.

At time $t$ let $N_1(t)$ and $N_2(t)$ be the number of customers in priority and nonpriority queue respectively, $C(t)$ is 0, 1 or 2 according as the server is free or occupied by a first or a second type customer. The process $(C(t), N_1(t), N_2(t))$ which describes the number of customers in the system is the simplest and simultaneously the most important process associated with the above queueing system. Clearly that if $C(t) = 0$ then $N_1(t) = 0$. If the service time distribution is not exponential then the process $(C(t), N_1(t), N_2(t))$ is not Markov. In this case we introduce a supplementary variable: if $C(t) = 1$ or 2, we define $\xi(t)$ as the elapsed service time of the call being served.

The queueing process evolves in the following manner. Suppose that the $(i-1)$th call completes its service at epoch $\eta_{i-1}$ (the calls are numbered in the order of service) and the server becomes free. If the priority queue is not empty, i.e. $N_{1,i-1} \equiv N_1(\eta_{i-1} - 0) > 0$, then one of the waiting priority customers according to the queueing discipline immediately enters service, i.e. the $i$th call's service starts at epoch $\xi_i = \eta_{i-1}$. Otherwise, if the priority queue is empty, then the next, $i$th, call enters service only after some time interval $R_i$ during which the server is free while there may be waiting low priority customers. If the number of sources of repeated calls at the time $\eta_{i-1}$, $N_{2,i-1} \equiv N_2(\eta_{i-1} - 0)$, is equal to $n$, then the random variable $R_i$ has an exponential distribution with parameter $\lambda_1 + \lambda_2 + n\mu$. The $i$th call is a primary priority call with probability $\frac{\lambda_1}{\lambda_1 + \lambda_2 + n\mu}$, it is a primary low priority call with probability $\frac{\lambda_2}{\lambda_1 + \lambda_2 + n\mu}$, and it is a repeated low priority call with probability $\frac{n\mu}{\lambda_1 + \lambda_2 + n\mu}$. At epoch $\xi_i = \eta_{i-1} + R_i$ the $i$th call's service starts. All primary calls arriving during the service time join the corresponding (priority or low priority) queue. Repeated low priority calls which arrive during this time interval do not influence the process. Then, at epoch $\eta_i = \xi_i + S_i$ (where $S_i$ is the service time of the $i$th call) the $i$th call completes service and the server becomes free again.

Let $\beta_i(s) = \int_0^\infty e^{-sx} dB_i(x)$ be the Laplace–Stieltjes transform of the service time distribution function $B_i(x)$, $i = 1, 2$, $\beta_{i,k} = (-1)^k \beta_i^{(k)}(0)$ be the $k$th moment of the $i$th type customer service time about the origin, $\rho_i = \lambda_i \beta_{i,1}$ the system load due to $i$th type primary calls, $\rho = \rho_1 + \rho_2$, $\lambda = \lambda_1 + \lambda_2$, $b_i(x) = B_i'(x)/(1 - B_i(x))$ be the instantaneous service intensity for $i$th type customers given that the elapsed service time is equal to $x$,

$$k_i(z_1, z_2) = \beta_i(\lambda_1 - \lambda_1 z_1 + \lambda_2 - \lambda_2 z_2).$$

It is easy to see that

$$k_i(z_1, z_2) = \sum_{m=0}^\infty \sum_{n=0}^\infty k_{i,m,n} z_1^n z_2^m,$$

where

$$k_{i,m,n} = \int_0^\infty \frac{(\lambda_1 x)^m}{m!} e^{-\lambda_1 x} \frac{(\lambda_2 x)^n}{n!} e^{-\lambda_2 x} dB_i(x)$$

is the joint distribution of the number of primary calls of both types which arrive during the service time of a $i$th type call.

### 3.2.2 Joint distribution of the server state and the queue length in the steady state

In this section we carry out the simplest and simultaneously the most important (from an applied point of view) analysis of the system. Namely, we investigate the joint distribution of the server state and the queue length in the steady state. As we will show later on the stationary regime exists if and only if $\rho < 1$, so the condition $\rho < 1$ is assumed to hold from now on.

First introduce the following functions which describe the joint distribution of the server state and queue length in the steady state:

$$
\begin{aligned}
p_{0ij} &= \mathrm{P}\{C(t) = 0, N_1(t) = i, N_2(t) = j\}, \\
p_{1ij}(x) &= \frac{d}{dx} \mathrm{P}\{C(t) = 1, \xi(t) < x, N_1(t) = i, N_2(t) = j\}, \\
p_{2ij}(x) &= \frac{d}{dx} \mathrm{P}\{C(t) = 2, \xi(t) < x, N_1(t) = i, N_2(t) = j\},
\end{aligned}
$$

and corresponding partial generating functions

$$p_0(z_1, z_2) \equiv \sum_{i=0}^{\infty} \sum_{j=0}^{\infty} z_1^i z_2^j p_{0ij},$$

$$p_1(z_1, z_2, x) \equiv \sum_{i=0}^{\infty} \sum_{j=0}^{\infty} z_1^i z_2^j p_{1ij}(x),$$

$$p_2(z_1, z_2, x) \equiv \sum_{i=0}^{\infty} \sum_{j=0}^{\infty} z_1^i z_2^j p_{2ij}(x).$$

Since $N_1(t) = 0$ if $C(t) = 0$, then $p_{0,i,j} = 0$ for $i \geq 1$. Thus the generating function $p_0(z_1, z_2) \equiv p_0(0, z_2)$, i.e. does not depend on variable $z_1$, and so later on we shall denote it as $p_0(z_2)$.

**Theorem 3.2** *For an $M/G/1$ retrial queue with priority subscribers*

$$p_0(z_2) = (1 - \rho_1 - \rho_2)$$
$$\times \exp \left\{ \frac{1}{\mu} \int\limits_{1}^{z_2} \frac{\lambda_1 - \lambda_1 h(u) + \lambda_2 - \lambda_2 k_2(h(u), u)}{k_2(h(u), u) - u} du \right\},$$

$$p_1(z_1, z_2, x)$$
$$= \{(\lambda_1 - \lambda_1 h(z_2) + \lambda_2 - \lambda_2 k_2(h(z_2), z_2))\,(k_2(z_1, z_2) - z_2)$$
$$- (\lambda_1 - \lambda_1 z_1 + \lambda_2 - \lambda_2 k_2(z_1, z_2))\,(k_2(h(z_2), z_2) - z_2)\}$$
$$\times \{(k_2(h(z_2), z_2) - z_2)\,(z_1 - k_1(z_1, z_2))\}^{-1}$$
$$\times p_0(z_2)\,[1 - B_1(x)]\,e^{-(\lambda_1 - \lambda_1 z_1 + \lambda_2 - \lambda_2 z_2)x}, \qquad (3.20)$$

$$p_2(z_1, z_2, x) = \frac{\lambda_1 - \lambda_1 h(z_2) + \lambda_2 - \lambda_2 z_2}{k_2(h(z_2), z_2) - z_2}$$
$$\times p_0(z_2)\,[1 - B_2(x)]\,e^{-(\lambda_1 - \lambda_1 z_1 + \lambda_2 - \lambda_2 z_2)x}. \qquad (3.21)$$

*If in the cases $C(t) = 1$, $C(t) = 2$ we neglect the elapsed service time $\xi(t)$, then for the probabilities $p_{1ij} = \mathrm{P}\{C(t) = 1, N_1(t) = i, N_2(t) = j\}$, $p_{2ij} = \mathrm{P}\{C(t) = 2, N_1(t) = i, N_2(t) = j\}$ we have*

$$p_1(z_1, z_2) \equiv \sum_{i=0}^{\infty} \sum_{j=0}^{\infty} z_1^i z_2^j p_{1ij}$$
$$= \{(\lambda_1 - \lambda_1 h(z_2) + \lambda_2 - \lambda_2 k_2(h(z_2), z_2))\,(k_2(z_1, z_2) - z_2)$$
$$- (\lambda_1 - \lambda_1 z_1 + \lambda_2 - \lambda_2 k_2(z_1, z_2))\,(k_2(h(z_2), z_2) - z_2)\}$$

$$\times \left[ (k_2(h(z_2), z_2) - z_2)(z_1 - k_1(z_1, z_2)) \right]^{-1}$$

$$\times \frac{1 - k_1(z_1, z_2)}{[\lambda_1 - \lambda_1 z_1 + \lambda_2 - \lambda_2 z_2]} p_0(z_2),$$

$$p_2(z_1, z_2) \equiv \sum_{i=0}^{\infty} \sum_{j=0}^{\infty} z_1^i z_2^j p_{2ij}$$

$$= \frac{\lambda_1 - \lambda_1 h(z_2) + \lambda_2 - \lambda_2 z_2}{k_2(h(z_2), z_2) - z_2}$$

$$\times \frac{1 - k_2(z_1, z_2)}{\lambda_1 - \lambda_1 z_1 + \lambda_2 - \lambda_2 z_2} p_0(z_2).$$

*Proof.* In a general way we obtain the equations of statistical equilibrium:

$$(\lambda_1 + \lambda_2 + j\mu)p_{0,0,j} = \int_0^{\infty} (p_{1,0,j}(x)b_1(x) + p_{2,0,j}(x)b_2(x))dx,$$

$$p_{0,i,j} = 0 \text{ if } i \geq 1,$$

$$p'_{1ij}(x) = -(\lambda_1 + \lambda_2 + b_1(x))p_{1ij}(x) + \lambda_1 p_{1,i-1,j} + \lambda_2 p_{1,i,j-1},$$

$$p'_{2ij}(x) = -(\lambda_1 + \lambda_2 + b_2(x))p_{2ij}(x) + \lambda_1 p_{2,i-1,j} + \lambda_2 p_{2,i,j-1},$$

$$p_{1ij}(0) = \lambda_1 p_{0,0,j}\delta_{i,0} + \int_0^{\infty} p_{1,i+1,j}(x)b_1(x)dx$$

$$+ \int_0^{\infty} p_{2,i+1,j}(x)b_2(x)dx,$$

$$p_{2ij}(0) = \begin{cases} 0, & \text{if } i \geq 1, \\ \\ \lambda_2 p_{00j} + (j+1)\mu p_{00j+1}, & \text{if } i = 0. \end{cases}$$

For the generating functions $p_0(z_2)$, $p_1(z_1, z_2, x)$ and $p_2(z_1, z_2, x)$ these equations give:

$$\mu z_2 \frac{dp_0(z_2)}{dz_2} = -(\lambda_1 + \lambda_2)p_0(z_2)$$

$$+ \int_0^{+\infty} p_1(0, z_2, x)b_1(x)dx$$

$$+ \int_0^{+\infty} p_2(0, z_2, x)b_2(x)dx, \qquad (3.22)$$

$$\frac{\partial p_1(z_1, z_2, x)}{\partial x} = -(\lambda_1 - \lambda_1 z_1 + \lambda_2 - \lambda_2 z_2 + b_1(x))$$

$$\times p_1(z_1, z_2, x), \qquad (3.23)$$

$$\frac{\partial p_2(z_1, z_2, x)}{\partial x} = -(\lambda_1 - \lambda_1 z_1 + \lambda_2 - \lambda_2 z_2 + b_2(x))$$
$$\times \quad p_2(z_1, z_2, x), \tag{3.24}$$

$$z_1 p_1(z_1, z_2, 0) = \int_0^{+\infty} (p_1(z_1, z_2, x) - p_1(0, z_2, x)) \, b_1(x) dx$$
$$+ \int_0^{+\infty} (p_2(z_1, z_2, x) - p_2(0, z_2, x)) \, b_2(x) dx,$$
$$+ \quad \lambda_1 z_1 p_0(z_2) \tag{3.25}$$

$$p_2(z_1, z_2, 0) = \lambda_2 p_0(z_2) + \mu \frac{dp_0(z_2)}{dz_2}. \tag{3.26}$$

From (3.23) and (3.24) we find that $p_1(z_1, z_2, x)$ and $p_2(z_1, z_2, x)$ depend upon $x$ as follows:

$$p_1(z_1, z_2, x) = p_1(z_1, z_2, 0)$$
$$\times \quad [1 - B_1(x)] e^{-(\lambda_1 - \lambda_1 z_1 + \lambda_2 - \lambda_2 z_2)x}, \tag{3.27}$$

$$p_2(z_1, z_2, x) = p_2(z_1, z_2, 0)$$
$$\times \quad [1 - B_2(x)] e^{-(\lambda_1 - \lambda_1 z_1 + \lambda_2 - \lambda_2 z_2)x}. \tag{3.28}$$

With the help of (3.27), (3.28), from equations (3.22) and (3.25) we have:

$$k_2(z_1, z_2) \cdot p_2(z_1, z_2, 0) = [z_1 - k_1(z_1, z_2)] \cdot p_1(z_1, z_2, 0)$$
$$+ \quad (\lambda_1 - \lambda_1 z_1 + \lambda_2) p_0(z_2)$$
$$+ \quad \mu z_2 \frac{dp_0(z_2)}{dz_2}. \tag{3.29}$$

Eliminating $p_2(z_1, z_2, 0)$ from (3.29) and (3.26) we get:

$$\mu [k_2(z_1, z_2) - z_2] \frac{dp_0(z_2)}{dz_2}$$
$$= [\lambda_1 - \lambda_1 z_1 + \lambda_2 - \lambda_2 k_2(z_1, z_2)] p_0(z_2)$$
$$+ [z_1 - k_1(z_1, z_2)] p_1(z_1, z_2, 0). \tag{3.30}$$

Consider equation
$$z_1 - k_1(z_1, z_2) = 0. \tag{3.31}$$

It can be rewritten as

$$z_1 - \beta_1(s + \lambda_1 - \lambda_1 z_1) = 0, \tag{3.32}$$

where $s = \lambda_2 - \lambda_2 z_2$. As we have noted in section 1.6, if $\rho_1 < 1$ then for $\text{Re}\, s \geq 0$ equation (3.32) has a unique root $z_1(s)$ in the unit disk $|z_1| \leq 1$. The function $z_1(s)$ can be thought of as the Laplace transform of the length of the busy period in an $M/G/1/\infty$ queue which serves only the priority customers. Since $\mathrm{Re}(\lambda_2 - \lambda_2 z_2) \geq 0$ when $|z_2| \leq 1$, we can guarantee that for $|z_2| \leq 1$ equation (3.31) has a unique root $z_1 = h(z_2)$ in the unit disk $|z_1| \leq 1$. The function $h(z_2)$ can be thought of as the generating function of the number of low priority calls which arrive during the busy period formed only by priority calls. It is easy to show that:

1. $h'(1) = \lambda_2 \beta_{1,1}/(1 - \rho_1)$;
2. $h''(1) = \lambda_2 \beta_{1,2}/(1 - \rho_1)^3$;
3. if $\rho_1 + \rho_2 < 1$ then $k_2(h(z), z) = z$ if and only if $z = 1$.

Now we are in position to solve equation (3.30). Replacing $z_1 = h(z_2)$ we get:

$$\mu\left[k_2(h(z_2), z_2) - z_2\right] \frac{dp_0(z_2)}{dz_2}$$
$$= \left[\lambda_1 - \lambda_1 h(z_2) + \lambda_2 - \lambda_2 k_2(h(z_2), z_2)\right] p_0(z_2). \quad (3.33)$$

As we have noted the coefficient $k_2(h(z_2), z_2) - z_2$ of the derivative in the left-hand side never vanishes for $z_2 \in [0, 1)$. Besides,

$$\lim_{z \to 1} \frac{\lambda_1 - \lambda_1 h(z) + \lambda_2 - \lambda_2 k_2(h(z), z)}{k_2(h(z), z) - z} = \frac{\lambda_2 (\rho_1 + \rho_2)}{1 - \rho_1 - \rho_2} < \infty.$$

Thus the function

$$\frac{\lambda_1 - \lambda_1 h(z) + \lambda_2 - \lambda_2 k_2(h(z), z)}{k_2(h(z), z) - z}$$

is analytical in the open disk $|z| < 1$ and is continuous in the closed disk $|z| \leq 1$. Therefore for $|z| \leq 1$ the solution of equation (3.33) is

$$p_0(z) = p_0(1) \exp\left\{\frac{1}{\mu} \int_1^z \frac{\lambda_1 - \lambda_1 h(u) + \lambda_2 - \lambda_2 k_2(h(u), u)}{k_2(h(u), u) - u} du\right\}.$$

Now from (3.30) we can find $p_1(z_1, z_2, 0)$ and thus from (3.27) $p_1(z_1, z_2, x)$. This implies formula (3.20). Similarly, from (3.26) we can find $p_2(z_1, z_2, 0)$ and thus from (3.28) $p_2(z_1, z_2, x)$. This implies formula (3.21). The normalizing constant $p_0(1)$ now can be determined from the normalizing condition $p_0(1) + p_1(1, 1) + p_2(1, 1) = 1$:

$$p_0(1) = 1 - \rho_1 - \rho_2,$$

which completes the proof. $\qquad\qquad\square$

With the help of generating functions $p_0(z_2), p_1(z_1, z_2), p_2(z_1, z_2)$ we can get various performance characteristics of the system. Say,

- probability that the server is occupied by a priority customer (carried priority traffic):

$$p_1 = \rho_1;$$

- probability that the server is occupied by a low priority customer (carried low priority traffic):

$$p_2 = \rho_2;$$

- the mean number of subscribers in the priority queue

$$EN_1(t) = \frac{\lambda_1 (\lambda_1 \beta_{1,2} + \lambda_2 \beta_{2,2})}{2 (1 - \rho_1)}; \qquad (3.34)$$

- the mean number of subscribers in the low priority queue

$$EN_2(t) = \frac{\lambda_2 (\lambda_1 \beta_{1,2} + \lambda_2 \beta_{2,2})}{2 (1 - \rho_1) (1 - \rho_1 - \rho_2)} + \frac{\lambda_2 (\rho_1 + \rho_2)}{\mu (1 - \rho_1 - \rho_2)}. \qquad (3.35)$$

The mean waiting time for each type of subscriber can be obtained with the help of Little's formula.

### 3.2.3 Embedded Markov chain

#### The structure of the embedded Markov chain

Let $\eta_d$ be the time of the $d$th departure, $C_d \equiv C(\eta_d - 0)$ be the type of the $d$th served call, $N_{1,d} \equiv N_1(\eta_d - 0)$ ($N_{2,d} \equiv N_2(\eta_d - 0)$) the number of customers in priority (respectively, low priority) queue just before the time $\eta_d$. Obviously,

$$\begin{cases} N_{1,d} &= (N_{1,d-1} - 1)^+ + \nu_{1,d} \\ N_{2.d} &= N_{2,d-1} - B_d + \nu_{2,d}, \end{cases} \qquad (3.36)$$

where $B_d$ is the number of customers from the low priority queue which enter service at time $\xi_d$ (i.e. $B_d = 1$ if the $d$th call is a repeated call and $B_d = 0$ if the $d$th call is a primary call), $\nu_{1,d}$ ($\nu_{2,d}$) is the number of the first (respectively, second) type primary calls which arrive in the system during the service time $S_d$ of the $d$th call.

The random vector $(C_d, B_d)$ depends on the history of the system before time $\eta_{d-1}$ only through the vector $(N_{1,d-1}, N_{2,d-1})$; its conditional distribution is given by the following formulas:

- if $i \geq 1$ then

$$P\{C_d = 1, B_d = 0 | (N_{1,d-1}, N_{2,d-1}) = (i,j)\} = 1,$$

$$P\{C_d = 1, B_d = 1 | (N_{1,d-1}, N_{2,d-1}) = (i,j)\} = 0,$$
$$P\{C_d = 2, B_d = 0 | (N_{1,d-1}, N_{2,d-1}) = (i,j)\} = 0,$$
$$P\{C_d = 2, B_d = 1 | (N_{1,d-1}, N_{2,d-1}) = (i,j)\} = 0;$$

- if $i = 0$ then

$$P\{C_d = 1, B_d = 0 | (N_{1,d-1}, N_{2,d-1}) = (i,j)\} = \frac{\lambda_1}{\lambda + j\mu},$$

$$P\{C_d = 1, B_d = 1 | (N_{1,d-1}, N_{2,d-1}) = (i,j)\} = 0,$$

$$P\{C_d = 2, B_d = 0 | (N_{1,d-1}, N_{2,d-1}) = (i,j)\} = \frac{\lambda_2}{\lambda + j\mu},$$

$$P\{C_d = 2, B_d = 1 | (N_{1,d-1}, N_{2,d-1}) = (i,j)\} = \frac{j\mu}{\lambda + j\mu}.$$

The random vector $(\nu_{1,d}, \nu_{2,d})$ depends on events which have occurred before epoch $\xi_d$ only through $C_d$; its conditional distribution is given by

$$P\{(\nu_{1,d}, \nu_{2,d}) = (m,n) | C_d = l\}$$
$$= \int_0^\infty \frac{(\lambda_1 x)^m}{m!} e^{-\lambda_1 x} \frac{(\lambda_2 x)^n}{n!} e^{-\lambda_2 x} dB_l(x)$$
$$\equiv k_{l,n,m}, \ l = 1, 2,$$

and thus has the following generating function

$$E\left(z_1^{\nu_{1,d}} z_2^{\nu_{2,d}} | C_d = l\right) = \beta_l(\lambda_1 - \lambda_1 z_1 + \lambda_2 - \lambda_2 z_2)$$
$$\equiv k_l(z_1, z_2), \ l = 1, 2.$$

The above remarks yield that the sequence of random vectors $X_d \equiv (C_d, N_{1,d}, N_{2,d})$ forms a Markov chain with $\{1,2\} \times Z_+^2$ as the state space, which is the embedded chain for our queueing system.

Its one-step transition probabilities

$$r_{(l,m,n)(k,i,j)} = P\{X_d = (k,i,j) \mid X_{d-1} = (l,m,n)\}$$

are given by the formulas:

- if $m \geq 1$ then

$$r_{(l,m,n)(1,i,j)} = k_{1,i-m+1,j-n},$$
$$r_{(l,m,n)(2,i,j)} = 0,$$

- if $m = 0$ then

$$r_{(l,m,n)(1,i,j)} = \frac{\lambda_1}{\lambda_1 + \lambda_2 + m\mu} k_{1,i,j-n},$$

$$r_{(l,m,n)(2,i,j)} = \frac{\lambda_2}{\lambda_1 + \lambda_2 + n\mu} k_{2,i,j-n}$$
$$+ \frac{n\mu}{\lambda_1 + \lambda_2 + n\mu} k_{2,i,j-n+1}.$$

*Ergodicity*

As usual, the first question to be investigated is the ergodicity of the chain.

Because of the recursive structure of the equations (3.36) which describe the stochastic dynamics of the chain we will use Foster's criterion.

Consider the following Lyapunov function on the state space:

$$f(l, m, n) = (\lambda_2\beta_{1,1} + 1 - \rho_2)\, m + (\lambda_1\beta_{2,1} + 1 - \rho_1)\, n.$$

Its mean drift

$$x_{l,m,n} \equiv \mathrm{E}\left(f(X_d) - f(X_{d-1})\,|\,X_{d-1} = (l, m, n)\right)$$

is given by

$$x_{l,m,n} = \begin{cases} \rho_1 + \rho_2 - 1, & \text{if } m \geq 1, \\ \rho_1 + \rho_2 - 1 + \dfrac{\lambda_1 + \lambda_2}{\lambda_1 + \lambda_2 + n\mu}, & \text{if } m = 0. \end{cases}$$

Indeed, let $m \geq 1$. Then $C_d = 1$ and the main recursive equations (3.36) become:

$$N_{1,d} = N_{1,d-1} - 1 + \nu_{1,d},$$
$$N_{2,d} = N_{2,d-1} + \nu_{2,d}.$$

Thus,

$$\begin{aligned} x_{l,m,n} &= (\lambda_2\beta_{1,1} + 1 - \rho_2)\,(-1 + \mathrm{E}\,(\nu_{1,d}|C_d = 1)) \\ &+ (\lambda_1\beta_{2,1} + 1 - \rho_1)\,\mathrm{E}\,(\nu_{2,d}|C_d = 1) \\ &= (\lambda_2\beta_{1,1} + 1 - \rho_2)\,(-1 + \lambda_1\beta_{1,1}) \\ &+ (\lambda_1\beta_{2,1} + 1 - \rho_1)\,\lambda_2\beta_{1,1} \\ &= \rho_1 + \rho_2 - 1. \end{aligned}$$

Now let $m = 0$. Then the main recursive equations (3.36) become:

$$N_{1,d} = \nu_{1,d},$$
$$N_{2,d} = N_{2,d-1} - B_d + \nu_{2,d}.$$

Thus

$$x_{l,m,n} = (\lambda_2\beta_{1,1} + 1 - \rho_2)\,\mathrm{E}\,(\nu_{1,d}\,|\,N_{1,d-1} = 0, N_{2,d-1} = n)$$

$$+ \quad (\lambda_1\beta_{2,1} + 1 - \rho_1)\left\{-E\left(B_d \,|\, N_{1,d-1} = 0, N_{2,d-1} = n\right)\right.$$
$$+ \quad \left.E\left(\nu_{2,d} \,|\, N_{1,d-1} = 0, N_{2,d-1} = n\right)\right\}.$$

But,

$$E\left(\nu_{1,d} \,|\, N_{1,d-1} = 0, N_{2,d-1} = n\right) = \frac{\lambda_1}{\lambda_1 + \lambda_2 + n\mu}\lambda_1\beta_{1,1}$$
$$+ \quad \frac{\lambda_2 + n\mu}{\lambda_1 + \lambda_2 + n\mu}\lambda_1\beta_{2,1};$$

$$E\left(\nu_{2,d} \,|\, N_{1,d-1} = 0, N_{2,d-1} = n\right) = \frac{\lambda_1}{\lambda_1 + \lambda_2 + n\mu}\lambda_2\beta_{1,1}$$
$$+ \quad \frac{\lambda_2 + n\mu}{\lambda_1 + \lambda_2 + n\mu}\lambda_2\beta_{2,1};$$

$$E\left(B_d \,|\, N_{1,d-1} = 0, N_{2,d-1} = n\right) = \frac{n\mu}{\lambda_1 + \lambda_2 + n\mu},$$

which implies that

$$x_{l,0,n} = \rho_1 + \rho_2 - 1 + \frac{\lambda_1 + \lambda_2}{\lambda_1 + \lambda_2 + n\mu}.$$

Let $\rho_1 + \rho_2 < 1$. Then $\varepsilon = \frac{1 - \rho_1 - \rho_2}{2}$ is positive and the set

$$A = \left\{(l, m, n) \,\bigg|\, l = 1, 2; m = 0; n < \frac{(\lambda_1 + \lambda_2)(1 - \varepsilon)}{\mu\varepsilon}\right\}$$

is finite. For all states $(l, m, n)$ with $m \geq 1$ we have:

$$x_{l,m,n} = -2\varepsilon < -\varepsilon,$$

and for all states $(l, m, n)$ with $m = 0, n \geq \frac{(\lambda_1 + \lambda_2)(1 - \varepsilon)}{\mu\varepsilon}$ we have:

$$x_{l,m,n} = -2\varepsilon + \frac{\lambda_1 + \lambda_2}{\lambda_1 + \lambda_2 + n\mu} \leq -\varepsilon.$$

Thus, $x_{l,m,n} \leq -\varepsilon$ for all states except for a finite number, and so the chain is ergodic.

On the other hand, the condition $\rho_1 + \rho_2 < 1$ is necessary for ergodicity. Indeed, since subscribers cannot be lost, in the steady state carried traffic is equal to offered traffic. But offered traffic is $\rho_1 + \rho_2$ and carried traffic is equal to the mean number of busy servers, i.e. to the probability that the server is busy. This probability is obviously less than 1 and thus $\rho_1 + \rho_2 < 1$.

*Stationary distribution*

Our second goal is to find the stationary distribution

$$\pi_{l,i,j} = \lim_{d \to \infty} P\left(C_d = l, N_{1,d} = i, N_{2,d} = j\right)$$

of the embedded Markov chain $\{C_d = l, N_{1,d} = i, N_{2,d} = j\}$.

Kolmogorov equations for the distribution $\pi_{l,i,j}$ are

$$\pi_{1,i,j} = \sum_{m=1}^{i+1} \sum_{n=0}^{j} \pi_{mn} k_{1,i+1-m,j-n}$$

$$+ \sum_{n=0}^{j} \pi_{0n} \frac{\lambda_1}{\lambda_1 + \lambda_2 + n\mu} k_{1,i,j-n},$$

$$\pi_{2,i,j} = \sum_{n=0}^{j} \pi_{0n} \frac{\lambda_2}{\lambda_1 + \lambda_2 + n\mu} k_{2,i,j-n}$$

$$+ \sum_{n=1}^{j+1} \pi_{0n} \frac{n\mu}{\lambda_1 + \lambda_2 + n\mu} k_{2,i,j+1-n},$$

where $\pi_{i,j} = \pi_{1,i,j} + \pi_{2,i,j}$. For generating functions

$$\varphi_1(z_1, z_2) = \sum_{i=0}^{\infty} \sum_{j=0}^{\infty} z_1^i z_2^j \pi_{1,i,j},$$

$$\varphi_2(z_1, z_2) = \sum_{i=0}^{\infty} \sum_{j=0}^{\infty} z_1^i z_2^j \pi_{2,i,j},$$

$$\varphi(z_1, z_2) = \sum_{i=0}^{\infty} \sum_{j=0}^{\infty} z_1^i z_2^j \pi_{i,j} = \varphi_1(z_1, z_2) + \varphi_2(z_1, z_2),$$

$$\psi(z) = \sum_{j=0}^{\infty} z^j \frac{\pi_{0,j}}{\lambda_1 + \lambda_2 + j\mu}$$

these equations become

$$z_1 \varphi_1(z_1, z_2) = k_1(z_1, z_2) \cdot [\lambda_1 z_1 \psi(z_2) + \varphi(z_1, z_2) - \varphi(0, z_2)],$$
$$\varphi_2(z_1, z_2) = k_2(z_1, z_2) \cdot [\lambda_2 \psi(z_2) + \mu \psi'(z_2)]. \qquad (3.37)$$

Eliminating $\varphi_2(z_1, z_2)$ and taking into account that

$$\varphi(0, z) = (\lambda_1 + \lambda_2) \psi(z) + \mu z \psi'(z),$$

we get:

$$\left[\frac{z_1}{k_1(z_1, z_2)} - 1\right] \varphi_1(z_1, z_2) = \mu \left[k_2(z_1, z_2) - z_2\right] \psi'(z_2)$$
$$- (\lambda_1 - \lambda_1 z_1 + \lambda_2 - \lambda_2 k_2(z_1, z_2)) \psi(z_2). \qquad (3.38)$$

This equation has the same structure as equation (3.30) and so using the same approach we get the following equation for $\psi(z)$ :

$$\mu \left[k_2(h(z_2), z_2) - z_2\right] \psi'(z_2)$$
$$= (\lambda_1 - \lambda_1 h(z_2) + \lambda_2 - \lambda_2 k_2(h(z_2), z_2)) \psi(z_2), \quad (3.39)$$

which is identical to equation (3.33) for $p_0(z_2)$. Thus

$$\psi(z_2) = \text{Const} \cdot p_0(z_2). \qquad (3.40)$$

With the help of (3.39) we can express from (3.38) and (3.37) functions $\varphi_1(z_1, z_2)$ and $\varphi_2(z_1, z_2)$ in terms of $\psi(z_2)$ :

$$\varphi_1(z_1, z_2)$$
$$= \{(\lambda_1 - \lambda_1 h(z_2) + \lambda_2 - \lambda_2 k_2(h(z_2), z_2)) (k_2(z_1, z_2) - z_2)$$
$$- (\lambda_1 - \lambda_1 z_1 + \lambda_2 - \lambda_2 k_2(z_1, z_2)) (k_2(h(z_2), z_2) - z_2)\}$$
$$\times \{(k_2(h(z_2), z_2) - z_2) (z_1 - k_1(z_1, z_2))\}^{-1}$$
$$\times k_1(z_1, z_2) \psi(z_2), \qquad (3.41)$$

$$\varphi_2(z_1, z_2) = \frac{\lambda_1 - \lambda_1 h(z_2) + \lambda_2 - \lambda_2 z_2}{k_2(h(z_2), z_2) - z_2} k_2(z_1, z_2) \psi(z_2). \quad (3.42)$$

Now from normalizing condition $\varphi(1,1) = 1$ we can find the still unknown constant in equation (3.40):

$$\text{Const} = \frac{1}{\lambda_1 + \lambda_2},$$

so that

$$\psi(z_2) = \frac{1 - \rho_1 - \rho_2}{\lambda_1 + \lambda_2} e^{\frac{\lambda_2}{\mu}(1 - z_2)}$$
$$\times \exp\left\{\frac{1}{\mu} \int_1^{z_2} \frac{\lambda_1 - \lambda_1 h(u) + \lambda_2 - \lambda_2 u}{k_2(h(u), u) - u} du\right\}. \quad (3.43)$$

Equations (3.41), (3.42), (3.43) give closed form formulas for generating functions $\varphi_1(z_1, z_2)$ and $\varphi_2(z_1, z_2)$.

### 3.2.4 Waiting time process

*Waiting time for priority subscribers*

The waiting time distribution for priority customers depends on the queueing discipline and can usually be easily found. For example, when these subscribers are served on an FIFO basis this distribution can be easily found with the help of an embedded Markov chain.

Let us number priority customers in order of arrival and denote by $W_1^{(n)}$ the waiting time of the $n$th priority customer, $S_1^{(n)}$ its service time and $N_1^{(n)}$ the number of priority customers in the system at the time of its departure. Then

$$\mathrm{E}z^{N_1^{(n)}} = \mathrm{E}e^{-(\lambda_1 - \lambda_1 z)W_1^{(n)}} \cdot \mathrm{E}e^{-(\lambda_1 - \lambda_1 z)S_1^{(n)}}. \qquad (3.44)$$

To prove this relation paint all priority calls red with probability $z$ and green with probability $1 - z$; colours for different priority calls are chosen independently. Besides, name the arrival of a green priority customer as a 'catastrophe'. The flow of these 'catastrophes' is Poisson with rate $s = \lambda_1 - \lambda_1 z$. Then $\mathrm{E}z^{N_1^{(n)}}$ can be thought of as the probability that all customers in the priority queue at the time of the $n$th priority call departure are red. This event can occur only if the following independent events occur:

- no green priority customer arrived during the waiting time $W_1^{(n)}$, or equivalently, no 'catastrophe' occurred during the waiting time $W_1^{(n)}$ (probability of this event is $\mathrm{E}e^{-(\lambda_1 - \lambda_1 z)W_1^{(n)}}$);

- no green priority customer arrived during the service time $S_1^{(n)}$, or equivalently, no 'catastrophe' occurred during the service time $S_1^{(n)}$ (probability of this event is $\mathrm{E}e^{-(\lambda_1 - \lambda_1 z)S_1^{(n)}}$).

In the steady state, relation (3.44) becomes:

$$\frac{\varphi_1(z,1)}{\varphi_1(1,1)} = \omega_1(\lambda_1 - \lambda_1 z) \cdot \beta_1(\lambda_1 - \lambda_1 z),$$

where

$$\omega_1(s) = \lim_{n \to \infty} \mathrm{E}e^{-sW_1^{(n)}}.$$

Taking into account formula (3.41) we get:

$$\omega_1(s) = \frac{(1 - \rho_1 - \rho_2)\,s + \lambda_2\,(1 - \beta_2(s))}{s - \lambda_1\,(1 - \beta_1(s))}.$$

*Waiting time for low priority subscribers*

Suppose that at the moment of departure of some customer (considered as the 0th customer) there are $m$ priority and $n \geq 1$ low priority customers in the system. Tag one of the low priority subscribers and denote by $T_{m,n}$ its waiting time.

To find the Laplace transform of the random variable $T_{m,n}$ we will use the method of collective marks. Fix some $s > 0$ and introduce an additional Poisson process with intensity $s$, which is independent of the functioning of the system. The events of this Poisson process will be called 'catastrophes'. Denote by $\tau_s$ the time of the first 'catastrophe' and introduce for $i \geq 0$, $j \geq 1$ the following events:

$$
\begin{aligned}
E_{1,i,j}^{(d)}(s) &= \{C_d = 1, N_{1,d} = i, N_{2,d} = j, T_{m,n} > \eta_d, \tau_s > \eta_d\}, \\
E_{2,i,j}^{(d)}(s) &= \{C_d = 2, N_{1,d} = i, N_{2,d} = j, T_{m,n} > \eta_d, \tau_s > \eta_d\}, \\
E_{i,j}^{(d)}(s) &= \{N_{1,d} = i, N_{2,d} = j, T_{m,n} > \eta_d, \tau_s > \eta_d\} \\
&= E_{1,i,j}^{(d)}(s) \cup E_{2,i,j}^{(d)}(s).
\end{aligned}
$$

Intuitively event $E_{l,i,j}^{(d)}(s)$ means that

- the $d$th served call has type $l$;
- at the time of the $d$th departure there are $i$ priority and $j \geq 1$ low priority subscribers in the system including the tagged one;
- until the time of the $d$th departure no 'catastrophe' occurred.

Let

$$
\begin{aligned}
p_{1,i,j}^{(d)}(s) &= P\left\{E_{1,i,j}^{(d)}(s)\right\}, \\
p_{2,i,j}^{(d)}(s) &= P\left\{E_{2,i,j}^{(d)}(s)\right\}, \\
p_{i,j}^{(d)}(s) &= P\left\{E_{i,j}^{(d)}(s)\right\} = p_{1,i,j}^{(d)}(s) + p_{2,i,j}^{(d)}(s).
\end{aligned}
$$

Using the same reasonings as in Section 1.8 we get the following main equations:

$$
\begin{aligned}
p_{1,i,j}^{(d)}(s) &= \sum_{m=1}^{i+1} \sum_{n=1}^{j} p_{mn}^{(d-1)}(s) k_{1,i+1-m,j-n}(s) \\
&+ \sum_{n=1}^{j} p_{0n}^{(d-1)} \frac{\lambda_1}{s + \lambda_1 + \lambda_2 + n\mu} k_{1,i,j-n}(s),
\end{aligned}
$$

$$p_{2,i,j}^{(d)}(s) = \sum_{n=1}^{j} p_{0n}^{(d-1)}(s)\frac{\lambda_2}{s+\lambda_1+\lambda_2+n\mu}k_{2,i,j-n}(s)$$

$$+ \sum_{n=1}^{j+1} p_{0n}^{(d-1)}(s)\frac{(n-1)\mu}{s+\lambda_1+\lambda_2+n\mu}k_{2,i,j+1-n}(s),$$

where

$$k_{l,m,n}(s) = \int_0^{\infty} e^{-sx}\frac{(\lambda_1 x)^m}{m!}e^{-\lambda_1 x}\frac{(\lambda_2 x)^n}{n!}e^{-\lambda_2 x}dB_l(x)$$

is the probability that during the service time of an $l$th type customer no 'catastrophe' occurred and exactly $m$ priority and $n$ primary low priority customers arrived into the system. Note that

$$k_l(s,z_1,z_2) \equiv \sum_{m=0}^{\infty}\sum_{n=0}^{\infty} z_1^m z_2^n k_{l,m,n}(s)$$

$$= \beta_l(s+\lambda_1-\lambda_1 z_1+\lambda_2-\lambda_2 z_2).$$

For the initial case $d=0$ we obviously have:

$$p_{i,j}^{(0)}(s) = \delta_{i,m}\cdot\delta_{j,n}.$$

For generating functions

$$p_1^{(d)}(s,z_1,z_2) = \sum_{i=0}^{\infty}\sum_{j=1}^{\infty} z_1^i z_2^j p_{1,i,j}^{(d)}(s),$$

$$p_2^{(d)}(s,z_1,z_2) = \sum_{i=0}^{\infty}\sum_{j=1}^{\infty} z_1^i z_2^j p_{2,i,j}^{(d)}(s),$$

$$p^{(d)}(s,z_1,z_2) = \sum_{i=0}^{\infty}\sum_{j=1}^{\infty} z_1^i z_2^j p_{i,j}^{(d)}(s)$$

$$= p_1^{(d)}(s,z_1,z_2) + p_2^{(d)}(s,z_1,z_2),$$

$$\psi^{(d)}(s,z_2) = \sum_{j=1}^{\infty} z_2^j \frac{p_{0,j}^{(d)}}{s+\lambda_1+\lambda_2+j\mu}$$

these equations become

$$p_1^{(d)}(s,z_1,z_2) = k_1(s,z_1,z_2)\left[\frac{p^{(d-1)}(s,z_1,z_2)-p^{(d-1)}(s,0,z_2)}{z_1}\right.$$

$$+ \lambda_1\psi^{(d-1)}(s,z_2)\Big],$$

$$p_2^{(d)}(s, z_1, z_2) = k_2(s, z_1, z_2)\left[\lambda_2\psi^{(d-1)}(s, z_2) + \mu\frac{\partial\psi^{(d-1)}(s, z_2)}{\partial z_2}\right.$$

$$\left. - \frac{\mu}{z_2}\psi^{(d-1)}(s, z_2)\right],$$

$$p^{(0)}(s, z_1, z_2) = z_1^m z_2^n.$$

For generating functions

$$p_1(s, z_1, z_2) = \sum_{d=0}^{\infty} p_1^{(d)}(s, z_1, z_2),$$

$$p_2(s, z_1, z_2) = \sum_{d=0}^{\infty} p_2^{(d)}(s, z_1, z_2),$$

$$p(s, z_1, z_2) = \sum_{d=0}^{\infty} p^{(d)}(s, z_1, z_2),$$

$$\psi(s, z_2) = \sum_{d=0}^{\infty} \psi^{(d)}(s, z_2)$$

we have:

$$p_1(s, z_1, z_2) = p_1^{(0)}(s, z_1, z_2) + k_1(s, z_1, z_2)$$
$$\times \left[\frac{p(s, z_1, z_2) - p(s, 0, z_2)}{z_1} + \lambda_1\psi(s, z_2)\right],$$

$$p_2(s, z_1, z_2) = p_2^{(0)}(s, z_1, z_2) + k_2(s, z_1, z_2)$$
$$\times \left[\lambda_2\psi(s, z_2) + \mu\frac{\partial\psi(s, z_2)}{\partial z_2} - \frac{\mu}{z_2}\psi(s, z_2)\right],$$

$$p^{(0)}(s, z_1, z_2) = z_1^m z_2^n.$$

Summing all three equations we get:

$$p(s, z_1, z_2)$$
$$= k_1(s, z_1, z_2) \cdot \left[\frac{p(s, z_1, z_2) - p(s, 0, z_2)}{z_1} + \lambda_1\psi(s, z_2)\right]$$
$$+ k_2(s, z_1, z_2) \cdot \left[\lambda_2\psi(s, z_2) + \mu\frac{\partial\psi(s, z_2)}{\partial z_2} - \frac{\mu}{z_2}\psi(s, z_2)\right]$$
$$+ z_1^m z_2^n.$$

Since

$$p(s, 0, z_2) = (s + \lambda_1 + \lambda_2)\,\psi(s, z_2) + \mu z_2\frac{\partial\psi(s, z_2)}{\partial z_2},$$

we have:

$$[z_1 - k_1]\, p(s, z_1, z_2)$$
$$= \left[ -(s + \lambda_1 + \lambda_2)\, k_1 + \lambda_1 z_1 k_1 + \lambda_2 z_1 k_2 - \frac{\mu z_1}{z_2} k_2 \right] \psi(s, z_2)$$
$$+ \mu\, [z_1 k_2 - z_2 k_1] \frac{\partial \psi(s, z_2)}{\partial z_2} + z_1^{m+1} z_2^n, \qquad (3.45)$$

where

$$k_1 \equiv k_1(s, z_1, z_2),$$
$$k_2 \equiv k_2(s, z_1, z_2).$$

Consider an $M/G/1/\infty$ queueing system which serves only priority calls and denote by $\pi_\infty(s)$ the Laplace transform of the length $L_\infty$ of a busy period. As we have noted, the function $\pi_\infty(s)$ is a solution of equation $z_1 - \beta_1 (s + \lambda_1 - \lambda_1 z_1) = 0$ in the unit disk $|z_1| \leq 1$. Thus the function $h(s, z_2) = \pi_\infty(s + \lambda_2 - \lambda_2 z_2)$ (which can be thought of as $E e^{-s L_\infty} z_2^\nu$, where $\nu$ is the number of primary low priority calls which arrive during the busy period formed only by priority calls) is a solution of the equation

$$z_1 - \beta_1 (s + \lambda_1 - \lambda_1 z_1 + \lambda_2 - \lambda_2 z_2) \equiv z_1 - k_1(s, z_1, z_2) = 0$$

in the unit disk $|z_1| \leq 1$.

Now we are in position to solve equation (3.45). Replacing in this equation $z_1 = h(s, z_2)$ we get:

$$\mu \left[ k_2^h - z_2 \right] \frac{\partial \psi}{\partial z_2}$$
$$= \left[ s + \lambda_1 + \lambda_2 - \lambda_1 h - \lambda_2 k_2^h + \frac{\mu k_2^h}{z_2} \right] \psi - h^m z_2^n, \quad (3.46)$$

where

$$\psi \equiv \psi(s, z_2),$$
$$h \equiv h(s, z_2),$$
$$k_2^h \equiv k_2(s, h(s, z_2), z_2).$$

Consider the coefficient of the derivative in the left-hand side of this equation as a function of variable $z_2$. For each fixed $s$ with $\text{Re}\, s > 0$ it has a unique zero $z_2 = g(s)$ in the unit disk $|z_2| \leq 1$. The function $g(s)$ is well known in the theory of the $(M_1, M_2)/(G_1, G_2)/1/(\infty, \infty)$ priority queue with head-of-the-line discipline. It is the Laplace transform of the length of a busy period in this system which starts with service of a low priority customer.

Now return to equation (3.46). Applying standard arguments we can write the solution of this equation as follows:

$$
\psi(s, z) \;=\; \int_{z}^{g(s)} \frac{z\,(h(s, u))^{m}\,u^{n-1}}{\mu\,[k_2(s, h(s, u), u) - u]}\,e^{\frac{\lambda_2}{\mu}(u-z)}
$$

$$
\times\;\exp\left\{ \int_{u}^{z} \frac{s + \mu + \lambda - \lambda_1 h(s, v) - \lambda_2 v}{\mu\,[k_2(s, h(s, v), v) - v]}\,dv \right\}\,du.
$$

Since the Laplace transform $Ee^{-sT_{m,n}}$ of the waiting time of the tagged customer equals

$$
\sum_{d=0}^{\infty}\sum_{j=1}^{\infty} p_{0j}^{(d)}(s)\,\frac{\mu}{s + \lambda_1 + \lambda_2 + j\mu} = \mu\psi(s, 1),
$$

we have:

$$
Ee^{-sT_{m,n}}
$$

$$
= \int_{1}^{g(s)} \frac{(h(s, u))^{m}\,u^{n-1}}{\mu\,[k_2(s, h(s, u), u) - u]}\,e^{\frac{\lambda_2}{\mu}(u-1)}
$$

$$
\times\;\exp\left\{ \int_{u}^{1} \frac{\mu + s + \lambda_1 + \lambda_2 - \lambda_1 h(s, v) - \lambda_2 v}{\mu\,[k_2(s, h(s, v), v) - v]}\,dv \right\}\,du.
$$

This equation is a key to further analysis of the waiting time process. It allows us to obtain the Laplace transform of the virtual waiting time, investigate its qualitative properties, etc. As in the preceding analysis this investigation can be carried out with the help of methods developed in section 1.8 and so we omit it.

## 3.3 A single-server model with impatient subscribers

### 3.3.1 Model description

Suppose that a calling subscriber after some unsuccessful retrials gives up further repetitions and abandons the system. Let $H_j$ be the probability that after the $j$th attempt fails, a subscriber will make the $(j+1)$th one. The set of probabilities $\{H_j, j \geq 1\}$ is called the persistence function. We assume that the probability of a call reinitiating after failure of a repeated attempt does not depend on the number of previous attempts (i.e. $H_2 = H_3 = ...$). Statistical measurements in telephone networks show that this is a realistic assumption in applications to such networks.

Thus consider a single server queueing system in which customers arrive in a Poisson process with rate $\lambda$. These customers are identified as primary calls. If the server is free at the time of a primary call arrival, the arriving call begins to be served immediately and leaves the system after service completion. However if the server is busy at the time of arrival of a primary call then with probability $1 - H_1$ the call leaves the system without service and with probability $H_1 > 0$ forms a source of repeated calls. Every source produces a Poisson process of repeated calls with intensity $\mu$. If an incoming repeated call finds the line free, it is served and leaves the system after service, while the source which produced this repeated call disappears. Otherwise, i.e. if the server is occupied at time of a repeated call arrival, with probability $1 - H_2$ the source leaves the system without service and with probability $H_2$ retries for service again.

The input flow of primary calls, intervals between repetitions, service times and decisions whether or not to retry for service are mutually independent. All notations introduced in Chapter 1 (unless otherwise stated) hold in this section as well.

An important feature of the model under consideration is that for many problems the cases $H_2 < 1$ and $H_2 = 1$ yield essentially different solutions. In the case $H_2 = 1$ the models with impatient subscribers can be be analysed in full detail in spirit of Chapter 1. The case $H_1 < 1$ is far more complicated and closed form solution is available only in the case of exponential service time distribution. In the general case a complete closed form solution seems impossible. However, we can get some useful information about steady state performance characteristics.

### 3.3.2 The case $H_2 = 1$

*Joint distribution of the server state and the queue length in the steady state*

In this section we carry out the simplest and simultaneously the most important (from an applied point of view) analysis of the system. Namely, we investigate the joint distribution of the server state and the queue length in the steady state. As we will show later on, the stationary regime exists if and only if $\rho H_1 \equiv \lambda \beta_1 H_1 < 1$, so the condition $\rho H_1 < 1$ is assumed to hold from now on.

**Theorem 3.3** *If $\rho H_1 < 1$ and the system is in the steady state then the joint distribution of the server state and queue length*

$$
\begin{aligned}
p_{0n} &= \mathrm{P}\{C(t) = 0, N(t) = n\}, \\
p_{1n}(x) &= \frac{d}{dx}\mathrm{P}\{C(t) = 1, \xi(t) < x, N(t) = n\}
\end{aligned}
$$

*has partial generating functions*

$$
\begin{aligned}
p_0(z) &\equiv \sum_{n=0}^{\infty} z^n p_{0n} \\
&= \frac{1 - \rho H_1}{1 + \rho(1 - H_1)} \exp\left\{ \frac{\lambda}{\mu} \int_1^z \frac{1 - k(u)}{k(u) - u} du \right\}, \\
p_1(z, x) &\equiv \sum_{n=0}^{\infty} z^n p_{1n}(x) \\
&= \lambda \frac{1 - z}{k(z) - z} p_0(z)[1 - B(x)]e^{-\lambda H_1(1-z)x},
\end{aligned}
$$

*where, as distinguished from Chapter 1, $k(z) = \beta(\lambda H_1 - \lambda H_1 z)$. If in the case $C(t) = 1$ we neglect the elapsed service time $\xi(t)$, then for the probabilities $p_{1n} = \mathrm{P}\{C(t) = 1, N(t) = n\}$ we have*

$$
\begin{aligned}
p_1(z) &\equiv \sum_{n=0}^{\infty} z^n p_{1n} \\
&= \frac{1}{H_1} \frac{1 - k(z)}{k(z) - z} p_0(z).
\end{aligned}
$$

*Proof.* In a general way we obtain the equations of statistical equilibrium:

$$
\begin{aligned}
(\lambda + n\mu)p_{0n} &= \int_0^{\infty} p_{1n}(x)b(x)dx, \\
p'_{1n}(x) &= -(\lambda H_1 + b(x))p_{1n}(x) + \lambda H_1 p_{1n-1}(x), \\
p_{1n}(0) &= \lambda p_{0n} + (n+1)\mu p_{0,n+1}.
\end{aligned}
$$

For generating functions $p_0(z)$ and $p_1(z, x)$ these equations give:

$$
\lambda p_0(z) + \mu z p'_0(z) = \int_0^{+\infty} p_1(z, x)b(x)dx, \qquad (3.47)
$$

$$\frac{\partial p_1(z,x)}{\partial x} = -(\lambda H_1(1-z) + b(x))p_1(z,x), \quad (3.48)$$

$$p_1(z,0) = \lambda p_0(z) + \mu p_0'(z). \quad (3.49)$$

From (3.48) we find that $p_1(z,x)$ depends upon $x$ as follows:

$$p_1(z,x) = p_1(z,0)[1 - B(x)]e^{-\lambda H_1(1-z)x}. \quad (3.50)$$

With the help of (3.50) equation (3.47) can be rewritten as follows:

$$\lambda p_0(z) + \mu z \frac{dp_0(z)}{dz} = k(z)p_1(z,0). \quad (3.51)$$

Eliminating $p_1(z,0)$ from (3.49) and (3.51) we get:

$$\mu[k(z) - z]\frac{dp_0(z)}{dz} = \lambda[1 - k(z)]p_0(z). \quad (3.52)$$

Consider the coefficient $f(z) = k(z) - z \equiv \beta(\lambda H_1(1-z)) - z$. Note that:

1. $f(1) = \beta(0) - 1 = 1 - 1 = 0$;

2. $f'(z) = -\lambda H_1 \beta'(\lambda H_1 - \lambda H_1 z) - 1$, so that $f'(1) = \rho H_1 - 1 < 0$;

3. $f''(z) = (\lambda H_1)^2 \beta''(\lambda H_1 - \lambda H_1 z) \geq 0$.

Therefore the function $f(z)$ is decreasing on the interval $[0,1]$, $z = 1$ is the unique zero there and for $z \in [0,1)$ the function is positive, i.e. (as $\rho H_1 < 1$) for $z \in [0,1)$ we have:

$$z < k(z) \leq 1.$$

Besides,

$$\lim_{z \to 1-0} \frac{1 - k(z)}{k(z) - z} = \lim_{z \to 1-0} \frac{\lambda H_1 \beta'(\lambda H_1(1-z))}{-\lambda H_1 \beta'(\lambda H_1(1-z)) - 1}$$

$$= \frac{\lambda H_1 \beta'(0)}{-\lambda H_1 \beta'(0) - 1}$$

$$= \frac{\rho H_1}{1 - \rho H_1} < \infty,$$

i.e. the function $(1 - k(z))/(k(z) - z)$ can be defined at the point $z = 1$ as $\rho H_1/(1 - \rho H_1)$. This means that for $z \in [0;1]$ we can rewrite equation (3.52) as

$$\frac{dp_0(z)}{dz} = \frac{\lambda}{\mu}\frac{1 - k(z)}{k(z) - z}p_0(z),$$

which implies that:

$$p_0(z) = p_0(1) \exp\left\{ \frac{\lambda}{\mu} \int_1^z \frac{1 - k(u)}{k(u) - u} du \right\}.$$

Now from (3.49)

$$p_1(z, 0) = \lambda \frac{1 - z}{k(z) - z} p_0(z),$$

and so from (3.50)

$$p_1(z, x) = \lambda \frac{1 - z}{k(z) - z} p_0(z)[1 - B(x)]e^{-\lambda H_1(1-z)x}. \qquad (3.53)$$

From (3.53) we have that

$$p_1(z) = \int_0^{+\infty} p_1(z, x)dx = \frac{1}{H_1} \frac{1 - k(z)}{k(z) - z} p_0(z). \qquad (3.54)$$

The unknown constant $p_0(1)$ can be found from the normalizing condition $p_0(1) + p_1(1) = 1$. Using (3.54) we have:

$$p_1(1) = \frac{\rho}{1 - \rho H_1} p_0(1).$$

Thus,

$$1 = p_0(1) + p_1(1) = \frac{1 + \rho(1 - H_1)}{1 - \rho H_1} p_0(1),$$

i.e.

$$p_0(1) = \frac{1 - \rho H_1}{1 + \rho(1 - H_1)}.$$

which completes the proof.                                                    □

With the help of generating functions $p_0(z)$, $p_1(z)$ we can get various performance characteristics of the system:

(a) The distribution of the number of customers in orbit has generating function:

$$\begin{aligned} p(z) &= p_0(z) + p_1(z) \\ &= \frac{1 - \rho H_1}{1 + \rho(1 - H_1)} \frac{1 - k(z) + H_1(k(z) - z)}{H_1(k(z) - z)} \\ &\quad \times \exp\left\{ \frac{\lambda}{\mu} \int_1^z \frac{1 - k(u)}{k(u) - u} du \right\}. \end{aligned}$$

In particular, the mean queue length is given by:

$$\mathrm{E}N(t) = p'(1) = \frac{\lambda^2 H_1}{1 - \rho H_1} \left( \frac{\beta_1}{\mu} + \frac{\beta_2}{2(1 + \rho(1 - H_1))} \right).$$

(b) The blocking probability $p_1$ is given by

$$p_1 = p_1(1) = \frac{\rho}{1 + \rho(1 - H_1)}.$$

*Embedded Markov chain*

*The structure of the embedded Markov chain*   Let $N_i = N(\eta_i)$ be the number of sources at the time $\eta_i$ of the $i$th departure. It is easy to see that

$$N_i = N_{i-1} - B_i + \nu_i, \tag{3.55}$$

where $B_i$ is the number of sources which enter service at time $\xi_i$ (i.e. $B_i = 1$ if the $i$th call is a repeated call and $B_i = 0$ if the $i$th call is a primary call) and $\nu_i$ is the number of primary calls which arrive in the system during the service time $S_i$ of the $i$th call and do not leave the system after subsequent blocking.

The Bernoulli random variable $B_i$ depends on the history of the system before time $\eta_{i-1}$ only through $N_{i-1}$; its conditional distribution is given by

$$\mathrm{P}\{B_i = 1 \mid N_{i-1} = n\} = \frac{n\mu}{\lambda + n\mu},$$

$$\mathrm{P}\{B_i = 0 \mid N_{i-1} = n\} = \frac{\lambda}{\lambda + n\mu}.$$

The random variable $\nu_i$ does not depend on events which have occurred before epoch $\xi_i$. The flow of primary calls which arrived during the service time and are not lost can be thought of as a thinning of the original Poisson flow of primary calls. Thus it is Poisson with the rate $\lambda H_1$. This implies that $\nu_i$ has distribution

$$k_n = \mathrm{P}\{\nu_i = n\} = \int_0^\infty \frac{(\lambda H_1 x)^n}{n!} e^{-\lambda H_1 x} dB(x)$$

with generating function

$$\sum_{n=0}^\infty k_n z^n = \beta(\lambda H_1(1 - z)) \equiv k(z).$$

and mean value

$$\mathrm{E}\nu_i = \sum_{n=0}^{\infty} nk_n = \rho H_1.$$

The above remarks mean that the sequence of random variables $N_i$ forms a Markov chain, which is the embedded chain for our queueing system.

Its one-step transition probabilities $r_{mn} = \mathrm{P}\{N_i = n \mid N_{i-1} = m\}$ are given by the formula

$$r_{mn} = \frac{\lambda}{\lambda + m\mu} k_{n-m} + \frac{m\mu}{\lambda + m\mu} k_{n-m+1}.$$

*Ergodicity* As usual, the first question to be investigated is the ergodicity of the chain.

Because of the recursive structure of equation (3.55), we will use criteria based on mean drift. For the Markov chain under consideration we have:

$$
\begin{aligned}
x_n &\equiv \mathrm{E}(N_{i+1} - N_i \mid N_i = n) \\
&= \mathrm{E}(-B_{i+1} + \nu_{i+1} \mid N_i = n) \\
&= -\mathrm{E}(B_{i+1} \mid N_i = n) + \mathrm{E}(\nu_{i+1} \mid N_i = n) \\
&= -\mathrm{P}(B_{i+1} = 1 \mid N_i = n) + \mathrm{E}(\nu_{i+1}) \\
&= -\frac{n\mu}{\lambda + n\mu} + \rho H_1.
\end{aligned}
$$

As $n \to \infty$ there exists $\lim x_n = -1 + \rho H_1$. This limit is negative iff $\rho H_1 < 1$. Thus we can guarantee that for $\rho H_1 < 1$ the embedded Markov chain is ergodic.

For $\rho H_1 \geq 1$ we have:

$$x_n = -\frac{n\mu}{\lambda + n\mu} + \rho H_1 \geq -\frac{n\mu}{\lambda + n\mu} + 1 = \frac{\lambda}{\lambda + n\mu} > 0.$$

Since down drifts $N_{i+1} - N_i$ are bounded from below, we can guarantee nonergodicity in the case $\rho H_1 \geq 1$.

Thus, we have proved that the embedded Markov chain is ergodic iff $\rho H_1 < 1$.

*Stationary distribution* Our second goal is to find the stationary distribution $\pi_n$ of the embedded Markov chain $\{N_i\}$. This can be done without difficulty if we note that the structure of the one-step transition probabilities for the embedded Markov chain under consideration are identical to the one-step transition probabilities for the embedded Markov chain for the main retrial queue. Thus from the results of section 1.3 we get:

$$
\psi(z) \;\equiv\; \sum_{n=0}^{\infty} z^n \frac{\pi_n}{\lambda + n\mu} = \psi(1) \cdot \exp\left\{ \frac{\lambda}{\mu} \int_{1}^{z} \frac{1 - k(u)}{k(u) - u}\,du \right\},
$$

$$
\varphi(z) \;\equiv\; \sum_{n=0}^{\infty} z^n \pi_n = \lambda k(z) \cdot \frac{1 - z}{k(z) - z}\psi(z).
$$

Since $\varphi(1) = 1$, we have:

$$
\psi(1) = \sum_{n=0}^{\infty} \frac{\pi_n}{\lambda + n\mu} = \frac{1 - \rho H_1}{\lambda}.
$$

Finally we get the following formula for the generating function $\varphi(z)$ of the stationary distribution of the embedded Markov chain $\{N_i\}$:

$$
\varphi(z) = (1 - \rho H_1)k(z) \cdot \frac{1 - z}{k(z) - z} \exp\left\{ \frac{\lambda}{\mu} \int_{1}^{z} \frac{1 - k(u)}{k(u) - u}\,du \right\}.
$$

*Functioning of the system in the nonstationary regime*

As in Chapter 1 we consider the queueing process as alternating between busy periods and idle periods. To investigate the process during the busy periods we will use the Kolmogorov differential equations for transient probabilities.

All definitions and notations of Chapter 1 hold, however we will not take into account the number of served customers $I(t)$ and correspondingly will omit in these notations the index $i$ which indicates that $I(t) = i$, and also the argument $y$ in generating functions.

Assume that a $k$-busy period starts at time $t = 0$. Let $L^{(k)}$ be the length of the $k$-busy period, $\Pi^{(k)}(t) = \mathrm{P}(L^{(k)} < t)$, $\pi^{(k)} = \mathrm{E}e^{-sL^{(k)}}$. Besides, let

$$
P_{0n}^{(k)}(t) \;=\; \mathrm{P}\left\{ L^{(k)} > t, C(t) = 0, N(t) = n \right\},
$$

$$
P_{1n}^{(k)}(t, x) \;=\; \frac{d}{dx}\mathrm{P}\left\{ L^{(k)} > t, C(t) = 1, \xi(t) < x, N(t) = n \right\}.
$$

The Kolmogorov equations for probabilities $P_{0n}^{(k)}(t)$, $P_{1n}^{(k)}(t, x)$

are

$$\frac{dP_{0n}^{(k)}(t)}{dt} = -(\lambda + n\mu)P_{0n}^{(k)}(t) + \int_0^\infty P_{1n}^{(k)}(t,x)b(x)dx, \ n \geq 1,$$

$$\frac{d\Pi^{(k)}(t)}{dt} = \int_0^\infty P_{1,0}^{(k)}(t,x)b(x)dx,$$

$$\frac{\partial P_{1n}^{(k)}(t,x)}{\partial t} + \frac{\partial P_{1n}^{(k)}(t,x)}{\partial x} = -(\lambda H_1 + b(x))P_{1n}^{(k)}(t,x)$$
$$+ \ \lambda H_1 P_{1,n-1}^{(k)}(t,x), \ n \geq 0,$$

$$P_{1n}^{(k)}(t,0) = \lambda P_{0n}^{(k)}(t) + (n+1)\mu P_{0,n+1}^{(k)}(t), \ n \geq 0.$$

The initial conditions are:

$$P_{0n}^{(k)}(0) = 0, \ P_{1n}^{(k)}(0,x) = \delta(x)\delta_{n,k-1},$$

where $\delta(x)$ is the Dirac delta function and $\delta_{i,j}$ is Kronecker's delta.

For Laplace transforms

$$\varphi_{0n}^{(k)}(s) = \int_0^\infty e^{-st}P_{0n}^{(k)}(t)dt,$$

$$\varphi_{1n}^{(k)}(s,x) = \int_0^\infty e^{-st}P_{1n}^{(k)}(t,x)dt,$$

these equations become:

$$(s + \lambda + n\mu)\varphi_{0n}^{(k)}(s) = \int_0^\infty \varphi_{1,n}^{(k)}(s,x)b(x)dx, \ n \geq 1,$$

$$\pi^{(k)}(s) = \int_0^\infty \varphi_{1,0}^{(k)}(s,x)b(x)dx,$$

$$-\delta(x)\delta_{n,k-1} + \frac{\partial \varphi_{1n}^{(k)}(s,x)}{\partial x} = -(s + \lambda H_1 + b(x))\varphi_{1n}^{(k)}(s,x)$$
$$+ \ \lambda H_1 \varphi_{1,n-1}^{(k)}(s,x), \ n \geq 0,$$

$$\varphi_{1n}^{(k)}(s,0) = \lambda \varphi_{0n}^{(k)}(s) + (n+1)\mu \varphi_{0,n+1}^{(k)}(s),$$
$$n \geq 0,$$

and for generating functions

$$\varphi_0^{(k)}(s,z) = \sum_{n=1}^{\infty} z^n \varphi_{0n}^{(k)}(s), \quad \varphi_1^{(k)}(s,z,x) = \sum_{n=0}^{\infty} z^n \varphi_{1n}^{(k)}(s,x)$$

we have:

$$(s+\lambda)\varphi_0^{(k)}(s,z) + \pi^{(k)}(s) + \mu z \frac{\partial \varphi_0^{(k)}(s,z)}{\partial z} = \int_0^{\infty} \varphi_1^{(k)}(s,z,x)b(x)dx,$$

$$(3.56)$$

$$-\delta(x)z^{k-1} + \frac{\partial \varphi_1^{(k)}(s,z,x)}{\partial x} = -(s+\lambda H_1 - \lambda H_1 z + b(x))\varphi_1^{(k)}(s,z,x),$$

$$(3.57)$$

$$\varphi_1^{(k)}(s,z,0) = \lambda \varphi_0^{(k)}(s,z) + \mu \frac{\partial \varphi_0^{(k)}(s,z)}{\partial z}. \qquad (3.58)$$

From equation (3.57) we find the form of dependence of $\varphi_1^{(k)}(s,z,x)$ upon the variable $x$:

$$\varphi_1^{(k)}(s,z,x) = (1 - B(x))e^{-(s+\lambda H_1 - \lambda H_1 z)x}\left\{\varphi_1^{(k)}(s,z,0) + z^{k-1}\right\}.$$

$$(3.59)$$

This allows us to rewrite equation (3.56) as

$$(s+\lambda)\varphi_0^{(k)}(s,z) + \pi^{(k)}(s) + \mu z \frac{\partial \varphi_0^{(k)}(s,z)}{\partial z}$$
$$= \beta(s+\lambda H_1 - \lambda H_1 z) \cdot \left\{\varphi_1^{(k)}(s,z,0) + z^{k-1}\right\}. \quad (3.60)$$

Eliminating $\varphi_1^{(k)}(s,z,0)$ with the help of equations (3.58), (3.60) we get the following differential equation for $\varphi_0^{(k)}(s,z,y)$:

$$\mu[\beta(s+\lambda H_1 - \lambda H_1 z) - z]\frac{\partial \varphi_0^{(k)}(s,z)}{\partial z}$$
$$= [s + \lambda - \lambda\beta(s+\lambda H_1 - \lambda H_1 z)]\varphi_0^{(k)}(s,z)$$
$$+ \pi^{(k)}(s) - z^{k-1}\beta(s+\lambda H_1 - \lambda H_1 z). \qquad (3.61)$$

The initial condition is $\varphi_0^{(k)}(s,0) = 0$.

Let $\pi_\infty(s|\lambda)$ be the Laplace transform of the length of a busy period in the standard $M/G/1/\infty$ queueing system with arrival rate $\lambda$. As we noted in section 1.6,

1. for $s > 0$, function $\pi_\infty(s|\lambda)$ is the unique solution of the equation $\beta(s + \lambda - \lambda z) = z$ on the interval $0 \le z \le 1$;

2. if $\lambda\beta_1 > 1$ then $\pi_\infty(0|\lambda) < 1$; if $\lambda\beta_1 \leq 1$ then $\pi_\infty(0|\lambda) = 1$ and $\pi'_\infty(0|\lambda) = \beta'(0)/(1 - \lambda\beta_1)$.

Thus if $s > 0$ and $0 \leq z < \pi_\infty(s|\lambda H_1)$ the coefficient in the left-hand side of equation (3.61) is nonzero. Therefore on this interval the solution of this equation is

$$\varphi_0^{(k)}(s, z) = e(s, z) \int_0^z \frac{\pi^{(k)}(s) - u^{k-1}\beta(s + \lambda H_1 - \lambda H_1 u)}{\mu\left[\beta(s + \lambda H_1 - \lambda H_1 u) - u\right] e(s, u)} du, \tag{3.62}$$

where the function $e(s, z)$ is given by

$$e(s, z) = \exp\left\{\frac{1}{\mu}\int_0^z \frac{s + \lambda - \lambda\beta(s + \lambda H_1 - \lambda H_1 u)}{\beta(s + \lambda H_1 - \lambda H_1 u) - u} du\right\}.$$

If $s > 0$ then

$$\lim_{z \to \pi_\infty(s|\lambda H_1) - 0} e(s, z) = +\infty.$$

On the other hand

$$\varphi_0^{(k)}(s, \pi_\infty(s|\lambda H_1)) < \infty.$$

Thus the integral in the right-hand side of equation (3.62) must tend to zero, i.e.

$$\int_0^{\pi_\infty(s|\lambda H_1)} \frac{\pi^{(k)}(s) - u^{k-1}\beta(s + \lambda H_1 - \lambda H_1 u)}{\mu\left[\beta(s + \lambda H_1 - \lambda H_1 u) - u\right] e(s, u)} du = 0, \quad s > 0, \tag{3.63}$$

Using (3.63) we can rewrite (3.62) as follows:

$$\varphi_0^{(k)}(s, z) = \int_{\pi_\infty(s|\lambda H_1)}^z \frac{\pi^{(k)}(s) - u^{k-1}\beta(s + \lambda H_1 - \lambda H_1 u)}{\mu\left[\beta(s + \lambda H_1 - \lambda H_1 u) - u\right]}$$

$$\times \exp\left\{\frac{1}{\mu}\int_u^z \frac{s + \lambda - \lambda\beta(s + \lambda H_1 - \lambda H_1 v)}{\beta(s + \lambda H_1 - \lambda H_1 v) - v} dv\right\} du. \tag{3.64}$$

Consider now the interval $\pi_\infty(s|\lambda H_1) < z \leq 1$. On this interval the coefficient $\beta(s + \lambda H_1 - \lambda H_1 z) - z \neq 0$ (in fact it is negative) and thus,

$$\varphi_0^{(k)}(s, z) = e_1(s, z)$$

$$\times \quad \left\{ C + \int\limits_1^z \frac{\pi^{(k)}(s) - u^{k-1}\beta(s + \lambda H_1 - \lambda H_1 u)}{\mu[\beta(s + \lambda H_1 - \lambda H_1 u) - u] e_1(s, u)} du \right\} \quad (3.65)$$

where for $\pi_\infty(s|\lambda H_1) < z \leq 1$

$$e_1(s, z) \quad = \quad \exp\left\{ \frac{1}{\mu} \int\limits_1^z \frac{s + \lambda - \lambda\beta(s + \lambda H_1 - \lambda H_1 u)}{\beta(s + \lambda H_1 - \lambda H_1 u) - u} du \right\}.$$

If $s > 0$ then

$$\lim_{z \to \pi_\infty(s|\lambda H_1) + 0} e_1(s, z) = +\infty.$$

On the other hand

$$\varphi_0^{(k)}(s, \pi_\infty(s|\lambda H_1)) < \infty.$$

Thus,

$$C = \int\limits_{\pi_\infty(s|\lambda H_1)}^1 \frac{\pi^{(k)}(s) - u^{k-1}\beta(s + \lambda H_1 - \lambda H_1 u)}{\mu\left[\beta(s + \lambda H_1 - \lambda H_1 u) - u\right] e_1(s, u)} du,$$

which allows us to transform formula (3.65) to the same form as (3.64). Thus, we can guarantee that (3.64) holds for all $z \neq \pi_\infty(s)$.

For $z = \pi_\infty(s)$ we have directly from (3.61) :

$$\varphi_0^{(k)}(s, z) = \frac{(\pi_\infty(s|\lambda H_1))^k - \pi^{(k)}(s)}{s + \lambda - \lambda\pi_\infty(s|\lambda H_1)}.$$

In order to find the Laplace transform of the $k$-busy period we consider equation (3.63). Since for $s > 0$ integrals

$$\int\limits_0^{\pi_\infty(s|\lambda H_1)} \frac{u^{k-1}\beta(s + \lambda H_1 - \lambda H_1 u)}{[\beta(s + \lambda H_1 - \lambda H_1 u) - u] e(s, u)} du$$

and

$$\int\limits_0^{\pi_\infty(s|\lambda H_1)} \frac{1}{[\beta(s + \lambda H_1 - \lambda H_1 u) - u] e(s, u)} du$$

are finite, we have:

$$\pi^{(k)}(s) = \frac{\int\limits_0^{\pi_\infty(s|\lambda H_1)} \frac{u^{k-1}\beta(s+\lambda H_1-\lambda H_1 u)}{[\beta(s+\lambda H_1-\lambda H_1 u)-u]e(s,u)} du}{\int\limits_0^{\pi_\infty(s|\lambda H_1)} \frac{1}{[\beta(s+\lambda H_1-\lambda H_1 u)-u]e(s,u)} du}. \quad (3.66)$$

Let $L$ be the length of the ordinary busy period (which corresponds to the case $k = 1$) and $\pi(s) = \mathrm{E}e^{-sL}$ its Laplace transform. Then from (3.66) we have:

$$\mathrm{E}e^{-sL} = \frac{s+\lambda}{\lambda}$$

$$- \frac{\mu}{\lambda} \left[ \int_0^{\pi_\infty(s|\lambda H_1)} \frac{1}{[\beta(s + \lambda H_1(1-u)) - u]\,e(s,u)} du \right]^{-1}.$$

Now let $\rho H_1 < 1$. In order to get $\mathrm{E}L^{(k)}$ we put $z = \pi_\infty(s|\lambda H_1)$ in equation (3.61):

$$\pi^{(k)}(s) = [\pi_\infty(s|\lambda H_1)]^k - [s + \lambda - \lambda\pi_\infty(s|\lambda H_1)]\cdot\varphi_0^{(k)}(s, \pi_\infty(s|\lambda H_1)).$$

Since $\pi_\infty(0|\lambda H_1) = 1$ and $\pi'_\infty(0|\lambda H_1) = -\beta_1/(1 - \rho H_1)$, we have:

$$\begin{aligned}
\mathrm{E}L^{(k)} &= \lim_{s\to 0} \frac{1 - \pi^{(k)}(s)}{s} \\
&= \lim_{s\to 0} \frac{1 - [\pi_\infty(s|\lambda H_1)]^k}{s} \\
&\quad + \lim_{s\to 0} \frac{s + \lambda - \lambda\pi_\infty(s|\lambda H_1)}{s} \cdot \varphi_0^{(k)}(0, 1) \\
&= \frac{k\beta_1}{1 - \rho H_1} + \frac{1 + \rho - \rho H_1}{1 - \rho H_1} \cdot \varphi_0^{(k)}(0, 1).
\end{aligned}$$

On the other hand, putting $s = 0$ in equation (3.61) we get the following equation for $\varphi_0^{(k)}(0, z)$ :

$$\mu\left[\beta(\lambda H_1 - \lambda H_1 z) - z\right] \frac{\partial \varphi_0^{(k)}(0, z)}{\partial z}$$
$$= \lambda\left[1 - \beta(\lambda H_1 - \lambda H_1 z)\right] \varphi_0^{(k)}(0, z)$$
$$+ 1 - z^{k-1}\beta(\lambda H_1 - \lambda H_1 z),$$

which implies that

$$\varphi_0^{(k)}(0, z) = \int_0^z \frac{1 - u^{k-1}\beta(\lambda H_1 - \lambda H_1 u)}{\mu\left[\beta(\lambda H_1 - \lambda H_1 u) - u\right]}$$

$$\times \quad \exp\left\{ \frac{\lambda}{\mu} \int_u^z \frac{1 - \beta(\lambda H_1 - \lambda H_1 v)}{\beta(\lambda H_1 - \lambda H_1 v) - v} dv \right\} du.$$

Thus

$$
\begin{aligned}
EL^{(k)} &= \frac{k\beta_1}{1 - \rho H_1} + \frac{1 + \rho - \rho H_1}{1 - \rho H_1} \\
&\times \int_0^1 \frac{1 - u^{k-1}\beta(\lambda H_1 - \lambda H_1 u)}{\mu\left[\beta(\lambda H_1 - \lambda H_1 u) - u\right]} \\
&\times \exp\left\{\frac{\lambda}{\mu}\int_u^1 \frac{1 - \beta(\lambda H_1 - \lambda H_1 v)}{\beta(\lambda H_1 - \lambda H_1 v) - v}\,dv\right\}\,du.
\end{aligned}
$$

For the ordinary busy period this formula can be simplified to

$$
\begin{aligned}
EL &= -\frac{1}{\lambda} + \frac{1 + \rho - \rho H_1}{\lambda(1 - \rho H_1)} \\
&\times \exp\left\{\frac{\lambda}{\mu}\int_0^1 \frac{1 - \beta(\lambda H_1 - \lambda H_1 u)}{\beta(\lambda H_1 - \lambda H_1 u) - u}\,du\right\}.
\end{aligned}
$$

Now consider the main process at an arbitrary time. To avoid unnecessary complication with minor details we will assume that the system was empty at the initial epoch $t_0 = 0$, i.e. $C(0) = 0$, $N(0) = 0$. Let

$$
\begin{aligned}
p_{0n}(t) &= P\left\{C(t) = 0, N(t) = n\right\}, \\
p_{1n}(t, x) &= \frac{d}{dx}P\left\{C(t) = 1, N(t) = n, \xi(t) < x\right\}
\end{aligned}
$$

be the transient distribution of the process $(C(t), \xi(t), N(t))$ in the nonstationary regime and

$$
\begin{aligned}
p_0^*(s, z) &= \int_0^\infty e^{-st}\sum_{n=0}^\infty z^n p_{0n}(t)\,dt, \\
p_1^*(s, z, x) &= \int_0^\infty e^{-st}\sum_{n=0}^\infty z^n p_{1n}(t, x)\,dt,
\end{aligned}
$$

be the corresponding Laplace transforms of generating functions.

Verbatim repetition of the proof of Theorem 1.19 of section 1.7 gives the following result:

$$
\begin{aligned}
p_1^*(s, z, x) &= \frac{(s + \lambda - \lambda z)p_0^*(s, z) - 1}{\beta\left(s + \lambda H_1 - \lambda H_1 z\right) - z} \\
&\times [1 - B(x)]\,e^{-(s + \lambda H_1 - \lambda H_1 z)x},
\end{aligned}
$$

$$p_0^*(s, z) = \int\limits_{z}^{\pi_\infty(s|\lambda H_1)} \frac{1}{\mu[\beta(s + \lambda H_1 - \lambda H_1 u) - u]} e^{\frac{\lambda}{\mu}(u-z)}$$

$$\times \quad \exp\left\{\frac{1}{\mu}\int\limits_{u}^{z} \frac{s + \lambda - \lambda v}{\beta(s + \lambda H_1 - \lambda H_1 v) - v} dv\right\} du,$$

if $z \neq \pi_\infty(s|\lambda H_1)$. Also

$$p_0^*(s, \pi_\infty(s|\lambda H_1)) = 1/(s + \lambda - \lambda\pi_\infty(s|\lambda H_1)).$$

### 3.3.3 The case $H_2 < 1$

*Joint distribution of the server state and the queue length in the steady state*

First consider the important particular case of exponential service time distribution when $B(x) = 1 - e^{-\nu x}$.

**Theorem 3.4** *For an $M/M/1$ retrial queue with impatient subscribers in the steady state, the joint distribution of the server state $C(t)$ and queue length $N(t)$ is given by*

$$p_{0n} = \frac{\gamma^n}{n!} \prod_{i=0}^{n-1} \frac{a+i}{c+i} \cdot \frac{1}{\Phi(a, c, \gamma) + \rho\Phi(a+1, c, \gamma)},$$

$$p_{1n} = \rho\frac{\gamma^n}{n!} \prod_{i=0}^{n-1} \frac{a+1+i}{c+i} \cdot \frac{1}{\Phi(a, c, \gamma) + \rho\Phi(a+1, c, \gamma)},$$

*where*

$$a = \frac{\lambda}{\mu}, \quad c = \frac{\nu + (1 - H_2)(\lambda + \mu)}{\mu(1 - H_2)}, \quad \gamma = \frac{\lambda H_1}{\mu(1 - H_2)}, \qquad (3.67)$$

*and*

$$\Phi(a, c, x) = \sum_{n=0}^{\infty} \frac{x^n}{n!} \prod_{i=0}^{n-1} \frac{a+i}{c+i}$$

*is the Kummer confluent function.*

 *Corresponding partial generating functions $p_0(z)$, $p_1(z)$ are given by*

$$p_0(z) = \frac{\Phi(a, c, \gamma z)}{\Phi(a, c, \gamma) + \rho\Phi(a+1, c, \gamma)},$$

$$p_1(z) = \frac{\rho\Phi(a+1, c, \gamma z)}{\Phi(a, c, \gamma) + \rho\Phi(a+1, c, \gamma)}.$$

*Proof.* In the case of exponentially distributed service times the process $(C(t), N(t))$ is a Markov process with $\{0; 1\} \times Z_+$ as the state space.

From a state $(0, n)$ only transitions into the following states are possible:

1. $(1, n)$ with rate $\lambda$;

2. $(1, n - 1)$ with rate $n\mu$.

The first transition is due to arrival of a primary call and the second is due to arrival of a repeated call. Since the state $(0, n)$ means that the server is free, there is no transition corresponding to a service completion.

Reaching state $(0, n)$ is possible only from state $(1, n)$ with rate $\nu$.

From a state $(1, n)$ only transitions into the following states are possible:

1. $(1, n + 1)$ with rate $\lambda H_1$;

2. $(0, n)$ with rate $\nu$;

3. $(1, n - 1)$ with rate $n\mu(1 - H_2)$.

The first transition is due to arrival of a primary call which is blocked and decides to try again, the second is due to a service completion and the third is due to arrival of a repeated call from a source which was blocked again and then decided to leave the system without service.

Reaching state $(1, n)$ is possible only from the states:

1. $(0, n)$ with rate $\lambda$;

2. $(0, n + 1)$ with rate $(n + 1)\mu$;

3. $(1, n - 1)$ with rate $\lambda H_1$;

4. $(1, n + 1)$ with rate $(n + 1)\mu(1 - H_2)$.

Thus the set of statistical equilibrium equations for the probabilities $p_{0n}$, $p_{1n}$ is

$$(\lambda + n\mu)p_{0n} = \nu p_{1n}, \tag{3.68}$$

$$(\lambda H_1 + \nu + n\mu(1 - H_2))p_{1n} = \lambda p_{0n} + (n + 1)\mu p_{0,n+1}$$
$$+(n + 1)\mu(1 - H_2)p_{1,n+1} + \lambda H_1 p_{1,n-1}.$$

Eliminate the probabilities $p_{1n}$ with the help of equation (3.68) and rewrite the resulting equation as

$$(n + 1)\mu[\nu + (1 - H_2)(\lambda + (n + 1)\mu)]p_{0n+1}$$

$$-\lambda H_1(\lambda + n\mu)p_{0n}$$
$$= n\mu[\nu + (1 - H_2)(\lambda + n\mu)]p_{0n}$$
$$-\lambda H_1(\lambda + (n-1)\mu)p_{0,n-1}.$$

This implies that

$$n\mu[\nu + (1 - H_2)(\lambda + n\mu)]p_{0n} - \lambda H_1(\lambda + (n-1)\mu)p_{0,n-1} = 0,$$

i.e.

$$
\begin{aligned}
p_{0n} &= \frac{\lambda H_1(\lambda + (n-1)\mu)}{n\mu[\nu + (1 - H_2)(\lambda + n\mu)]}p_{0,n-1} \\
&= \left(\frac{\lambda H_1}{\mu}\right)^n \frac{1}{n!} \frac{\prod\limits_{i=0}^{n-1}(\lambda + i\mu)}{\prod\limits_{i=1}^{n}[\nu + (1 - H_2)(\lambda + i\mu)]}p_{00} \\
&= \frac{\gamma^n}{n!} \prod_{i=0}^{n-1}\frac{a+i}{c+i}p_{00},
\end{aligned}
\tag{3.69}
$$

where variables $a, c, \gamma$ are given by (3.67). Now from equation (3.68) we have:

$$p_{1n} = \rho\frac{\gamma^n}{n!}\prod_{i=0}^{n-1}\frac{a+1+i}{c+i}p_{00}. \tag{3.70}$$

From (3.69), (3.70) we get the following formulas for generating functions $p_0(z)$ and $p_1(z)$:

$$
\begin{aligned}
p_0(z) &= \Phi(a, c, \gamma z)p_{00}, \\
p_1(z) &= \rho\Phi(a + 1, c, \gamma z)p_{00}.
\end{aligned}
$$

Using the normalizing condition $p_0(1) + p_1(1) = 1$ we can find the probability $p_{00}$ :

$$p_{00} = \frac{1}{\Phi(a, c, \gamma) + \rho\Phi(a + 1, c, \gamma)},$$

which yields the desired formulas.                              □

Now we can get various performance characteristics of the system in the steady state:

1. the distribution of the server state is given by:

$$p_0 = \mathrm{P}(C(t) = 0) = \frac{1}{1 + \Lambda},$$

$$p_1 \;=\; P(C(t) = 1) = \frac{\Lambda}{1 + \Lambda}, \tag{3.71}$$

where

$$\Lambda = \rho \frac{\Phi(a + 1, c, \gamma)}{\Phi(a, c, \gamma)}.$$

Thus if we are interested only in the state of the server in the stationary regime then our system can be thought of as an Erlang loss model $M/M/1/0$ with offered traffic $\Lambda$.

2. the mean number of customers in orbit is given by:

$$EN(t) = \frac{\lambda H_2 + (\lambda H_1 - \nu H_2)\Lambda}{\mu(1 - H_2)(1 + \Lambda)}.$$

To get this relation we used the following well-known relations for the Kummer function:

$$\Phi'(a, c, x) \;=\; \frac{a}{c}\Phi(a + 1, c + 1, x),$$

$$(a + 1)\Phi(a + 2, c + 1, x) \;=\; (x + 2a - c + 1)\Phi(a + 1, c + 1, x)$$
$$+ (c - a)\Phi(a, c + 1, x),$$

$$a\Phi(a + 1, c + 1, x) \;=\; (a - c)\Phi(a, c + 1, x) + c\Phi(a, c, x),$$

$$x\Phi(a + 1, c + 1, x) \;=\; c\Phi(a + 1, c, x) - c\Phi(a, c, x).$$

Consider now the case of a general distribution $B(x)$. In a general way we obtain the equations of statistical equilibrum:

$$(\lambda + n\mu)p_{0n} \;=\; \int_0^\infty p_{1n}(x)b(x)dx,$$

$$p'_{1n}(x) \;=\; -(\lambda H_1 + \mu(1 - H_2)n + b(x))p_{1n}(x)$$
$$+ \; \lambda H_1 p_{1,n-1}(x) + \mu(1 - H_2)(n + 1)p_{1,n+1}(x),$$

$$p_{1n}(0) \;=\; \lambda p_{0n} + (n + 1)\mu p_{0,n+1}.$$

For the generating functions $p_0(z)$ and $p_1(z, x)$ these equations give:

$$\lambda p_0(z) + \mu z \frac{dp_0(z)}{dz} \;=\; \int_0^{+\infty} p_1(z, x)b(x)dx, \tag{3.72}$$

$$\frac{\partial p_1(z, x)}{\partial x} \;=\; \mu(1 - H_2)(1 - z)\frac{\partial p_1(z, x)}{\partial z}$$
$$- \; (\lambda H_1(1 - z) + b(x))p_1(z, x), \tag{3.73}$$

$$p_1(z, 0) \;=\; \lambda p_0(z) + \mu \frac{dp_0(z)}{dz}. \tag{3.74}$$

The solution of partial differential equation (3.73) is

$$p_1(z,x) = p_1\left(1-(1-z)e^{-(1-H_2)\mu x},0\right)[1-B(x)]$$
$$\times \exp\left\{-\frac{\lambda H_1(1-z)}{\mu(1-H_2)}\left(1-e^{-(1-H_2)\mu x}\right)\right\}. (3.75)$$

Now (3.72) becomes:

$$\mu z\frac{dp_0(z)}{dz} = -\lambda p_0(z)$$
$$+ \int_0^\infty p_1\left(1-(1-z)e^{-(1-H_2)\mu x},0\right)$$
$$\times \exp\left\{-\frac{\lambda H_1(1-z)}{\mu(1-H_2)}\left(1-e^{-(1-H_2)\mu x}\right)\right\} dB(x).$$

Although we can eliminate $p_1(z,0)$ with the help of (3.74) and get an equation for $p_0(z)$, this equation is such that complete closed form solution seems impossible. However, we can get some useful information about steady state performance characteristics. First we can express moments of the queue length in terms of the server utilization $p_1 = P(C(t) = 1)$.

Denote:

$$p_1(x)dx = P(C(t) = 1, \xi(t) \in (x, x + dx)),$$
$$p_1 = P(C(t) = 1) = \int_0^\infty p_1(x)dx,$$
$$p_0 = P(C(t) = 0) = 1 - p_1,$$
$$N_0 = E(N(t); C(t) = 0),$$
$$N_1(x)dx = E(N(t); C(t) = 1, \xi(t) \in (x, x + dx)),$$
$$N_1 = E(N(t); C(t) = 1) = \int_0^\infty N_1(x)dx.$$

Putting $z = 1$ in equation (3.73) we get:

$$\frac{d}{dx}p_1(x) = -b(x)p_1(x),$$

which implies that

$$p_1(x) = (1 - B(x))p_1(0).$$

Thus

$$p_1 = \beta_1 p_1(0),$$

i.e.

$$p_1(0) = \frac{1}{\beta_1}p_1, \; p_1(x) = \frac{1 - B(x)}{\beta_1}p_1. \tag{3.76}$$

Also, from (3.74) we have:

$$\mu N_0 = \frac{(1+\rho)p_1 - \rho}{\beta_1}. \tag{3.77}$$

Now differentiate equations (3.72), (3.73) and (3.74) with respect to $z$ at the point $z = 1$:

$$(\lambda + \mu)N_0 + \mu p_0''(1) = \int_0^\infty N_1(x)b(x)dx, \tag{3.78}$$

$$\frac{dN_1(x)}{dx} = -(\mu(1 - H_2) + b(x))N_1(x)$$
$$+ \lambda H_1 p_1(x), \tag{3.79}$$

$$N_1(0) = \lambda N_0 + \mu p_0''(1). \tag{3.80}$$

Using (3.76) we can solve differential equation (3.79):

$$N_1(x) = \frac{1 - B(x)}{\beta_1} \left\{ \frac{\lambda H_1}{\mu(1 - H_2)} \left(1 - e^{-\mu(1-H_2)x}\right) p_1 \right.$$
$$+ \left. N_1(0)\beta_1 e^{-\mu(1-H_2)x} \right\}. \tag{3.81}$$

Eliminating $\mu p_0''(1)$ from (3.78) and (3.80) and using (3.81) we can find $N_1(0)$:

$$\beta_1 N_1(0) = \frac{\lambda H_1}{\mu(1 - H_2)}p_1 - \frac{(1+\rho)p_1 - \rho}{1 - \beta(\mu(1 - H_2))}.$$

Thus

$$N_1(x) = \frac{1 - B(x)}{\beta_1}$$
$$\times \left\{ \frac{\lambda H_1}{\mu(1 - H_2)}p_1 - \frac{(1+\rho)p_1 - \rho}{1 - \beta(\mu(1 - H_2))}e^{-\mu(1-H_2)x} \right\},$$

$$N_1 = \int_0^\infty N_1(x)dx = \frac{\rho - (1 + \rho - \rho H_1)p_1}{\mu(1 - H_2)\beta_1},$$

and so

$$N = N_0 + N_1 = \frac{\rho H_2 + (\rho H_1 - \rho H_2 - H_2)p_1}{\mu(1 - H_2)\beta_1}.$$

The above analysis can be generalized in order to show that all partial factorial moments of the queue length, $\Phi_n = E\left((N(t))_n\right)$, are linear functions of the server utilization $p_1$:

$$\Phi_n = E_n + F_n \cdot p_1,$$

where coefficients $E_n, F_n$ can be calculated with the help of a recursive procedure in terms of system parameters $\lambda, \mu, H_1, H_2$ and $B(x)$.

Consider partial factorial moments

$$\Phi_{0n} \equiv \mathrm{E}\left((N(t))_n ; C(t) = 0\right) = \frac{d^n}{dz^n} p_0(z) \Big|_{z=1},$$

$$\Phi_{1n} \equiv \mathrm{E}\left((N(t))_n ; C(t) = 1\right) = \frac{d^n}{dz^n} p_1(z) \Big|_{z=1},$$

$$\Phi_{1n}(x)dx \equiv \mathrm{E}\left((N(t))_n ; C(t) = 1, \xi(t) \in (x, x+dx)\right)$$

$$= \frac{\partial^n}{\partial z^n} p_1(z, x)dx \Big|_{z=1}.$$

Differentiating (3.72) and (3.74) $n$ times with respect to $z$ at the point $z = 1$ we get the following equations for the partial factorial moments:

$$(\lambda + n\mu)\Phi_{0n} + \mu\Phi_{0,n+1} = \int_0^\infty \Phi_{1n}(x)b(x)dx, \quad (3.82)$$

$$\Phi_{1n}(0) = \lambda\Phi_{0n} + \mu\Phi_{0,n+1}. \quad (3.83)$$

In order to get an equation corresponding to (3.75), we introduce the generating function

$$\Phi_1(z, x) = \sum_{n=0}^\infty \Phi_{1n}(x)\frac{z^n}{n!}.$$

Obviously,

$$\Phi_1(z, x) = p_1(z+1, x).$$

Thus (3.75) can be rewritten as

$$\Phi_1(z, x) = \Phi_1\left(ze^{-(1-H_2)\mu x}, 0\right)[1 - B(x)]$$

$$\times \exp\left\{\frac{\lambda H_1}{\mu(1-H_2)}\left(1 - e^{-(1-H_2)\mu x}\right)z\right\}.$$

Expanding both sides in a power series in $z$ and equating coefficients of equal powers of $z$ we get:

$$\Phi_{1n}(x) = \sum_{k=0}^n \binom{n}{k}\left[\frac{\lambda H_1}{\mu(1-H_2)}\right]^{n-k}\Phi_{1k}(0)$$

$$\times \quad \left[1 - e^{-\mu(1-H_2)x}\right]^{n-k} \left[e^{-\mu(1-H_2)x}\right]^k$$
$$\times \quad [1 - B(x)]. \tag{3.84}$$

Using (3.84) we may rewrite (3.82) as follows:

$$(\lambda + n\mu)\Phi_{0n} + \mu\Phi_{0,n+1} = \sum_{k=0}^{n} \Phi_{1k}(0) \cdot \alpha_{n,k}, \tag{3.85}$$

where

$$\alpha_{n,k} = \binom{n}{k} \left[\frac{\lambda H_1}{\mu(1-H_2)}\right]^{n-k}$$
$$\times \int_0^\infty \left[1 - e^{-\mu(1-H_2)x}\right]^{n-k} \left[e^{-\mu(1-H_2)x}\right]^k dB(x).$$

Eliminating $\Phi_{1k}(0)$ with the help of (3.83) we have from (3.85):

$$\mu\left[1 - \beta(n\mu(1-H_2))\right]\Phi_{0,n+1}$$
$$= -\left[\lambda\left(1 - \beta(n\mu(1-H_2))\right) + \mu\left(n - \alpha_{n,n-1}\right)\right]\Phi_{0n}$$
$$+ \sum_{k=1}^{n-1} \Phi_{0k} \cdot \left(\lambda\alpha_{n,k} + \mu\alpha_{n,k-1}\right) + \lambda\alpha_{n,0}\Phi_{00}. \tag{3.86}$$

Since $\Phi_{00} = p_0$, we have

$$\Phi_{0,0} = A_0 + B_0 p_1,$$

where

$$A_0 = 1, B_0 = -1,$$

and since $\Phi_{01} = N_0$, we have from equation (3.77) that

$$\Phi_{01} = A_1 + B_1 p_1,$$

where

$$A_1 = -\frac{\lambda}{\mu}, B_1 = \frac{1+\rho}{\mu\beta_1}.$$

Thus from (3.86) by induction we have that

$$\Phi_{0n} = A_n + B_n p_1,$$

where coefficients $A_n, B_n$ can be calculated recursively with the help of the following relations:

$$A_{n+1} = -\left[\frac{\lambda}{\mu} + \frac{n - \alpha_{n,n-1}}{1 - \beta(n\mu(1-H_2))}\right] A_n$$

$$+ \sum_{k=1}^{n-1} A_k \frac{\lambda \alpha_{n,k} + \mu \alpha_{n,k-1}}{\mu \left[1 - \beta(n\mu(1 - H_2))\right]}$$

$$+ \frac{\lambda \alpha_{n,0}}{\mu[1 - \beta(n\mu(1 - H_2))]},$$

$$B_{n+1} = - \left[\frac{\lambda}{\mu} + \frac{n - \alpha_{n,n-1}}{1 - \beta(n\mu(1 - H_2))}\right] B_n$$

$$+ \sum_{k=1}^{n-1} B_k \frac{\lambda \alpha_{n,k} + \mu \alpha_{n,k-1}}{\mu \left[1 - \beta(n\mu(1 - H_2))\right]}$$

$$- \frac{\lambda \alpha_{n,0}}{\mu[1 - \beta(n\mu(1 - H_2))]},$$

with initial conditions

$$A_0 = 1, A_1 = -\frac{\lambda}{\mu}, B_0 = -1, B_1 = \frac{1 + \rho}{\mu \beta_1}.$$

Now from (3.83) we have:

$$\Phi_{1n}(0) = (\lambda A_n + \mu A_{n+1}) + (\lambda B_n + \mu B_{n+1})p_1$$

and from (3.84):

$$\Phi_{1n} = C_n + D_n p_1,$$

where

$$C_n = \sum_{k=0}^{n} \binom{n}{k} \left[\frac{\lambda H_1}{\mu(1 - H_2)}\right]^{n-k} (\lambda A_k + \mu A_{k+1})$$

$$\times \int_0^\infty \left[1 - e^{-\mu(1 - H_2)x}\right]^{n-k} \left[e^{-\mu(1 - H_2)x}\right]^k [1 - B(x)]dx,$$

$$D_n = \sum_{k=0}^{n} \binom{n}{k} \left[\frac{\lambda H_1}{\mu(1 - H_2)}\right]^{n-k} (\lambda B_k + \mu B_{k+1})$$

$$\times \int_0^\infty \left[1 - e^{-\mu(1 - H_2)x}\right]^{n-k} \left[e^{-\mu(1 - H_2)x}\right]^k [1 - B(x)]dx.$$

And finally we have:

$$\Phi_n = \Phi_{0n} + \Phi_{1n} = (A_n + C_n) + (B_n + D_n)p_1 \equiv E_n + F_n p_1.$$

### Embedded Markov chain

As for the main model, the sequence $N_i$ of the number of customers in the orbit at the time of the $i$th departure forms a Markov chain. Its stochastic dynamics is described by the following recursive equation:

$$N_i = N_{i-1} - B_i - L_i + \nu_i, \qquad (3.87)$$

where $B_i$ is the number of sources which enter service at time $\xi_i$ (i.e. $B_i = 1$ if the $i$th call is a repeated call and $B_i = 0$ if the $i$th call is a primary call), $L_i$ is the number of customers from orbit which are lost during the service time of the $i$th customer and $\nu_i$ is the number of primary calls which arrive in the system during the service time $S_i$ of the $i$th call and do not leave the system after subsequent blocking.

The Bernoulli random variable $B_i$ depends on the history of the system before time $\eta_{i-1}$ only through $N_{i-1}$; its conditional distribution is given by

$$
\begin{aligned}
\mathrm{P}\{B_i = 0 \mid N_{i-1} = n\} &= \frac{\lambda}{\lambda + n\mu}, \\
\mathrm{P}\{B_i = 1 \mid N_{i-1} = n\} &= \frac{n\mu}{\lambda + n\mu}.
\end{aligned}
\qquad (3.88)
$$

To describe the stochastic structure of the random variables $L_i$ and $\nu_i$ we note that during the service time the orbit can be thought of as an $M/M/\infty$ queueing system. Input flow into this system is formed by those primary calls which are not lost after blocking and thus its rate is $\lambda H_1$. 'Service' in this system means loss of a customer from the orbit. Its duration has an exponential distribution with parameter $\mu(1 - H_2)$. Thus, given that the number of customers in orbit at the beginning of the $i$th service time is equal to $m$ and the service time of the $i$th customer is $t$, each such customer is still present in the orbit at time $t$ independently of other customers with probability $e^{-\mu(1-H_2)t}$. The number of new customers in orbit at time $t$ is equal to the number of busy servers in the $M/M/\infty$ queue (which starts functioning at time $t = 0$ from the empty state) and thus has Poisson distribution with parameter $\frac{\lambda H_1}{\mu(1-H_2)}\left(1 - e^{-\mu(1-H_2)t}\right)$. Thus the random vector $(L_i, \nu_i)$ depends on events which have occurred before epoch $\xi_i$ only through the number of customers in orbit at this epoch, $N_{i-1} - B_i$ and its conditional distribution is given by:

$$
\begin{aligned}
&\mathrm{P}\left(L_i = k, \nu_i = n \mid N_{i-1} - B_i = m\right) \\
&= \int_0^\infty \binom{m}{k} \left[e^{-\mu(1-H_2)t}\right]^{m-k} \left[1 - e^{-\mu(1-H_2)t}\right]^k
\end{aligned}
$$

$$\times \frac{\left[\frac{\lambda H_1}{\mu(1-H_2)}\left(1-e^{-\mu(1-H_2)t}\right)\right]^n}{n!}$$

$$\times \exp\left\{-\frac{\lambda H_1}{\mu(1-H_2)}\left(1-e^{-\mu(1-H_2)t}\right)\right\} dB(t)$$

with generating function

$$E\left(x^{L_i}y^{\nu_i} \,|\, N_{i-1} - B_i = m\right)$$
$$= \int_0^\infty \left[x + (1-x)e^{-\mu(1-H_2)t}\right]^m$$
$$\times \exp\left\{\frac{\lambda H_1}{\mu(1-H_2)}\left(1 - e^{-\mu(1-H_2)t}\right)(y-1)\right\} dB(t).$$

This yields that

$$E\left(L_i | N_{i-1} - B_i = m\right)$$
$$= \int_0^\infty m\left[1 - e^{-\mu(1-H_2)t}\right] dB(t)$$
$$= m\left[1 - \beta(\mu(1-H_2))\right];$$

random variable $\nu_i$ does not depend on $N_{i-1}, B_i$ and

$$E\left(\nu_i\right) = \int_0^\infty \frac{\lambda H_1}{\mu(1-H_2)}\left[1 - e^{-\mu(1-H_2)t}\right] dB(t) \quad (3.89)$$
$$= \frac{\lambda H_1}{\mu(1-H_2)}\left[1 - \beta(\mu(1-H_2))\right].$$

Now investigate ergodicity of the chain.

Because of the recursive structure of the equation (3.87) we will use criteria based on mean drift. For the Markov chain under consideration we have:

$$
\begin{aligned}
x_n &\equiv E(N_i - N_{i-1}|N_{i-1} = n) \\
&= E(-B_i - L_i + \nu_i|N_{i-1} = n) \\
&= -E(B_i|N_{i-1} = n) - E(L_i|N_{i-1} = n) + E(\nu_i|N_{i-1} = n) \\
&= -\frac{n\mu}{\lambda + n\mu} + \frac{\lambda H_1}{\mu(1-H_2)}\left[1 - \beta(\mu(1-H_2))\right] \\
&\quad - E(L_i|N_{i-1} = n).
\end{aligned}
$$

But

$$E(L_i|N_{i-1} = n)$$
$$= E(L_i|N_{i-1} - B_i = n) \cdot P\left(B_i = 0|N_{i-1} = n\right)$$

$$+\mathrm{E}(L_i|N_{i-1} - B_i = n - 1) \cdot \mathrm{P}\,(B_i = 1|N_{i-1} = n)$$

$$= n\,[1 - \beta(\mu(1 - H_2))]\,\frac{\lambda}{\lambda + n\mu}$$

$$+(n - 1)\,[1 - \beta(\mu(1 - H_2))]\,\frac{n\mu}{\lambda + n\mu}$$

$$= \left[n - \frac{n\mu}{\lambda + n\mu}\right][1 - \beta(\mu(1 - H_2))]. \qquad (3.90)$$

Thus

$$
\begin{aligned}
x_n &= -\left[n - \frac{\lambda H_1}{\mu(1 - H_2)}\right][1 - \beta(\mu(1 - H_2))] \\
&\quad - \frac{n\mu}{\lambda + n\mu}\beta(\mu(1 - H_2))
\end{aligned}
$$

As $n \to \infty$ there exists $\lim x_n = -\infty$. Thus the chain $\{N_i\}$ is always ergodic.

In fact, this result is obvious and follows from general theorems about ergodicity of queueing processes.

The above analysis of the structure of the embedded Markov chain allows us to get an equation for the generating function of the stationary distribution $\pi_n$ of the chain. However this equation is so cumbersome that probably a closed form solution does not exist. Nevertheless, we may get a very simple bound for the mean value of this distribution. As we will see this bound implies a bound for the server utilization, which is a key to calculation of all stationary performance characteristics of the system.

Consider the system in the steady state and take expectations of both sides of (3.87):

$$0 = -\mathrm{E}B_i - \mathrm{E}L_i + \mathrm{E}\nu_i.$$

Using (3.88), (3.89) and (3.90) we have:

$$\frac{\beta(\mu(1 - H_2))}{1 - \beta(\mu(1 - H_2))} \cdot \mathrm{E}\left(\frac{\mu N_i}{\lambda + \mu N_i}\right) = \frac{\lambda H_1}{\mu(1 - H_2)} - \mathrm{E}N_i. \qquad (3.91)$$

From Jensen's inequality we have

$$\mathrm{E}\left(\frac{\mu N_i}{\lambda + \mu N_i}\right) \le \frac{\mu \mathrm{E}N_i}{\lambda + \mu \mathrm{E}N_i}$$

and so we have (below $N = \mathrm{E}N_i$):

$$N^2 - \left[\frac{\lambda H_1}{\mu(1 - H_2)} - \frac{\lambda}{\mu} - \frac{\beta(\mu(1 - H_2))}{1 - \beta(\mu(1 - H_2))}\right] N - \frac{\lambda^2 H_1}{\mu^2(1 - H_2)} \geq 0.$$

Since the left-hand side of this inequality is negative for $N = 0$, it implies that

$$N \geq N_+,$$

where $N_+$ is the positive root of the corresponding quadratic equation.

To get a result for the server utilization we note that the queueing process consists of alternating service periods and subsequent idle periods. The mean length of the service period is $\beta_1$ and the mean length of the idle period is $\sum_{n=0}^{\infty} \pi_n \frac{1}{\lambda + n\mu} = \mathrm{E}\left(\frac{1}{\lambda + \mu N_i}\right)$. Thus the fraction of time when the server is busy, i.e. the server utilization $p_1$, is equal to

$$\frac{\beta_1}{\beta_1 + \mathrm{E}\left(\frac{1}{\lambda + \mu N_i}\right)} = \frac{\rho}{\rho + \mathrm{E}\left(\frac{\lambda}{\lambda + \mu N_i}\right)}.$$

Using (3.91) we have:

$$\begin{aligned}
p_1 &= \frac{\rho}{\rho + \mathrm{E}\left(1 - \frac{\mu N_i}{\lambda + \mu N_i}\right)} = \frac{\rho}{1 + \rho - \mathrm{E}\left(\frac{\mu N_i}{\lambda + \mu N_i}\right)} \\
&= \frac{\rho}{1 + \rho + \left(N - \frac{\lambda H_1}{\mu(1 - H_2)}\right) \cdot \frac{1 - \beta(\mu(1 - H_2))}{\beta(\mu(1 - H_2))}} \\
&\leq \frac{\rho}{1 + \rho + \left(N_+ - \frac{\lambda H_1}{\mu(1 - H_2)}\right) \cdot \frac{1 - \beta(\mu(1 - H_2))}{\beta(\mu(1 - H_2))}} \\
&= \frac{\rho + \mu\beta_1 N_+}{1 + \rho + \mu\beta_1 N_+}.
\end{aligned}$$

To get information about the quality of this bound consider the case of exponential service time distribution (for which exact formula (3.71) for $p_1$ is available). It is easy to show that in this case $\mu N_+$ and thus the upper bound for $p_1$ do not depend on $\mu$. Table 3.1 contains values of the server utilization and the upper bound for various values of $\lambda$ and $\mu$ and fixed values $H_1 = 0.9, H_2 = 0.8$, $\nu = 1$. As comparison of the numerical data shows, the bound is quite close to the exact value of the server utilization.

Table 3.1 *Exact value and an upper bound for the server utilization for the M/M/1 retrial queue with nonpersistent subscribers; $H_1 = 0.9$, $H_2 = 0.8$, the mean service time equals 1*

|  | $\lambda = 0.2$ | $\lambda = 0.5$ | $\lambda = 0.8$ | $\lambda = 2.0$ |
|---|---|---|---|---|
| upper bound | 0.194497 | 0.449194 | 0.632674 | 0.880367 |
| $\mu = 0.2$ | 0.193122 | 0.442432 | 0.623047 | 0.877806 |
| $\mu = 0.5$ | 0.191346 | 0.434089 | 0.610629 | 0.873727 |
| $\mu = 1.0$ | 0.188930 | 0.423269 | 0.593977 | 0.866445 |
| $\mu = 5.0$ | 0.179465 | 0.384062 | 0.530987 | 0.813452 |

## 3.4 A single-server multiclass retrial queue

### 3.4.1 Model description

In the main model it is assumed that the input process is homogeneous from the point of view of such characteristics as the service time and the inter-retrial time distributions. In practice, however, these characteristics may differ widely for different subscriber groups. This leads us to multiclass retrial queues.

The simplest multiclass retrial queue can be described as follows. There is a single server which is used to serve $n$ different types of customers. Primary customers of the $i$th type (later on we shall call them $i$-customers) arrive in accordance with a Poisson process with rate $\lambda_i$. We assume that these flows are independent and so one could assume that there is a single input flow with rate $\lambda = \lambda_1 + \ldots + \lambda_n$, but an arriving customer with probability $\frac{\lambda_i}{\lambda}$ has type $i$.

If an arriving $i$-customer finds the server free, it immediately occupies the server and leaves the system after service. Otherwise, if the server is busy, the customer forms a source of repeated calls of $i$th type. Every such a source produces a Poisson flow of repeated calls with rate $\mu_i$, until one of the repeated calls finds the server free. Then the call starts to be served and the source is eliminated. Service times, both for primary and repeated $i$-calls, have the same distribution function $B_i(x)$. As usual we assume that interarrival periods, retrial times and service times are mutually independent.

Let $b_i(x) = B_i'(x)/(1 - B_i(x))$ be the instantaneous service intensity of $i$-calls, $\beta_i(s) = \int_0^\infty e^{-sx} dB_i(x)$ be the Laplace–Stieltjes transform of the service time distribution function $B_i(x)$, $\beta_{i,k} =$

$(-1)^k \beta_i^{(k)}(0)$ be the $k$th moment of the $i$-calls service time about the origin, $\lambda = \lambda_1 + \cdots + \lambda_n$, $\rho_i = \lambda_i \beta_{i,1}$ be the system load due to primary $i$-calls, $B(x) = \sum_{i=1}^{n} \frac{\lambda_i}{\lambda} B_i(x)$ be the distribution function of the service time of a randomly chosen arriving primary customer, $\beta(s) = \int_0^\infty e^{-sx} dB(x)$, $\beta_k = (-1)^k \beta^{(k)}(0)$, $\rho = \sum_{i=1}^{n} \rho_i$.

Let $C(t) = 0$ if at time $t$ the server is free; $C(t) = i$ if at time $t$ the server is occupied by some $i$-call; $N_i(t)$ is the number of $i$-sources at time $t$. If $C(t) \neq 0$ then we define $\xi(t)$ as the elapsed service time of the call being served.

We shall consider the system in the steady state, which exists if and only if $\rho < 1$. This condition is assumed to hold from now on.

### 3.4.2 Stationary performance characteristics

The model under consideration is far more difficult for mathematical analysis than single class models because now the joint queue length process $\mathbf{N}(t) \equiv (N_1(t), \ldots, N_n(t))$ is a random walk on the multidimensional integer lattice $Z_+^n$ rather than on the integer lattice $Z_+$. Thus it is not too surprising that a routine approach (write down Kolmogorov equations for the stationary distribution, transform them with the help of generating functions and solve resulting equations) does not lead to success. So we will use another way of analysis. Namely, we will look for closed sets of equations for moments of the process $\mathbf{N}(t)$ rather than for explicit formulas for the distribution of random variables $N_i(t)$ or for their mean values $N_i = \mathbf{E} N_i(t)$. As we will see, such sets really exist, i.e. the process $\mathbf{N}(t)$ is **momently closed**. The fact that the process $\mathbf{N}(t)$ is momently closed explains why a direct routine approach does not lead to success; as a matter of fact this means that we are trying to solve in an explicit form general sets of linear equations.

**Theorem 3.5** *The mean number of $i$-customers in orbit is*

$$N_i = \frac{\lambda_i \rho}{\mu_i (1 - \rho)} + \frac{\lambda_i \lambda \beta_2}{2} x_i,$$

*where the values $x_i$ can be found as the solution of the system of linear equations*

$$\sum_{j=1}^{n} \frac{\mu_j \rho_j}{\mu_i + \mu_j}(x_i + x_j) = x_i - 1.$$

*Proof.* Let

$$\mathbf{m} = (m_1, \ldots, m_n),$$
$$\mathbf{z} = (z_1, \ldots, z_n),$$

and let

$$\mathbf{e}_i = (0, \ldots, 1, \ldots, 0)$$

be the $n$-dimensional vector with $i$th coordinate equal to 1 and the rest equal to 0,

$$\mathbf{e} = (1, \ldots, 1)$$

be the $n$-dimensional vector which has all coordinates equal to 1. Denote:

$$p_{0,\mathbf{m}} = \mathrm{P}\{C(t) = 0, \mathbf{N}(t) = \mathbf{m}\}$$
$$p_{i,\mathbf{m}}(x) = \frac{d}{dx}\mathrm{P}\{C(t) = i, \xi(t) < x, \mathbf{N}(t) = \mathbf{m}\},$$
$$i = 1, \ldots, n.$$

In a general way we obtain the following equations of statistical equilibrium:

$$\left(\lambda + \sum_{i=1}^{n} \mu_i m_i\right) p_{0,\mathbf{m}} = \sum_{i=1}^{n} \int_0^\infty p_{i,\mathbf{m}}(x) b_i(x) dx,$$

$$\frac{d}{dx} p_{j,\mathbf{m}}(x) = -[\lambda + b_j(x)] p_{j,\mathbf{m}}(x)$$
$$+ \sum_{i=1}^{n} \lambda_i p_{j,\mathbf{m}-\mathbf{e}_i}(x),$$

$$p_{j,\mathbf{m}}(0) = \lambda_j p_{0,\mathbf{m}} + \mu_i(m_j + 1) p_{0,\mathbf{m}+\mathbf{e}_j}.$$

For the generating functions

$$p_0(\mathbf{z}) = \sum_{m_1=0}^{\infty} \cdots \sum_{m_n=0}^{\infty} z_1^{m_1} \cdots z_n^{m_n} p_{0,\mathbf{m}},$$

$$p_i(\mathbf{z}, x) = \sum_{m_1=0}^{\infty} \cdots \sum_{m_n=0}^{\infty} z_1^{m_1} \cdots z_n^{m_n} p_{i,\mathbf{m}}(x)$$

these equations give

$$\sum_{i=1}^{n} \mu_i z_i \frac{\partial p_0(\mathbf{z})}{\partial z_i} + \lambda p_0(\mathbf{z}) = \sum_{i=1}^{n} \int_0^\infty p_i(\mathbf{z}, x) b_i(x) dx, \quad (3.92)$$

$$\frac{\partial}{\partial x}p_j(\mathbf{z}, x) = -\left(\sum_{i=1}^{n}\lambda_i(1 - z_i) + b_j(x)\right)p_j(\mathbf{z}, x), \qquad (3.93)$$

$$p_j(\mathbf{z}, 0) = \lambda_j p_0(\mathbf{z}) + \mu_j\frac{\partial p_0(\mathbf{z})}{\partial z_i}. \qquad (3.94)$$

From (3.93) we find that $p_j(\mathbf{z}, x)$ depends upon $x$ as follows:

$$p_j(\mathbf{z}, x) = p_j(\mathbf{z}, 0)\left[1 - B_j(x)\right]e^{-sx}, \qquad (3.95)$$

where

$$s = \sum_{i=1}^{n}\lambda_i(1 - z_i).$$

From (3.95) it follows that

$$p_j(\mathbf{z}) \equiv \int_0^\infty p_j(\mathbf{z}, x)dx = p_j(\mathbf{z}, 0)\frac{1 - \beta_j(s)}{s}. \qquad (3.96)$$

Now with the help of (3.95) and (3.96), equations (3.92) and (3.94) can be rewritten as follows:

$$\lambda p_0(\mathbf{z}) + \sum_{i=1}^{n}\mu_i z_i\frac{\partial p_0(\mathbf{z})}{\partial z_i} = \sum_{i=1}^{n}\frac{s\beta_i(s)}{1 - \beta_i(s)}p_i(\mathbf{z}), \qquad (3.97)$$

$$\lambda_j p_0(\mathbf{z}) + \mu_j\frac{\partial p_0(\mathbf{z})}{\partial z_j} = \frac{s}{1 - \beta_j(s)}p_j(\mathbf{z}). \qquad (3.98)$$

In order to find the distribution of the server state we multiply (3.98) by $z_j$, then sum over $j = 1, \ldots, n$ and substract from (3.97); after some transformations we get

$$p_0(\mathbf{z}) = \sum_{i=1}^{n}p_i(\mathbf{z})\frac{\beta_i(s) - z_i}{1 - \beta_i(s)}. \qquad (3.99)$$

Fixing some $j$ and putting $z_i = 1$ for all $i \neq j$, we have:

$$p_0(\mathbf{z}) + \sum_{i=1}^{n}p_i(\mathbf{z}) = \frac{1 - z_j}{1 - \beta_j(\lambda_j - \lambda_j z_j)}p_j(\mathbf{z}).$$

Setting $z_j = 1$ and taking into account the normalization condition $\sum_{i=0}^{n}p_i(\mathbf{e}) = 1$ we get

$$p_j(\mathbf{e}) = \rho_j, \qquad (3.100)$$

$$p_0(\mathbf{e}) = 1 - \sum_{j=1}^{n}p_j(\mathbf{e}) = 1 - \rho.$$

Also, with $\mathbf{z} = \mathbf{e}$ equations (3.98) and (3.100) yield

$$\frac{\partial p_0(\mathbf{e})}{\partial z_1} = \frac{\lambda_j \rho}{\mu_j}.$$

Summing up (3.99) over $j = 1, \ldots, n$ and subtracting from (3.97) we have:

$$\sum_{i=1}^n \lambda_i (z_i - 1) p_0(\mathbf{z}) + \sum_{i=1}^n \mu_i(z_i - 1)\frac{\partial p_0(\mathbf{z})}{\partial z_i} = \sum_{i=1}^n \lambda_i(z_i - 1)p(\mathbf{z}),$$

(3.101)

where

$$p(\mathbf{z}) = \sum_{i=0}^n p_i(\mathbf{z}).$$

Differentiating (3.101) with respect to $z_i, z_j$ at the point $\mathbf{z} = \mathbf{e}$ we obtain, after some algebra,

$$(\mu_i + \mu_j) \frac{\partial^2 p_0(\mathbf{e})}{\partial z_i \partial z_j} = \lambda_i N_i + \lambda_j N_j - \lambda_i \lambda_j \rho \frac{\mu_i + \mu_j}{\mu_i \mu_j}. \qquad (3.102)$$

Now differentiate (3.98) with respect to $z_i$ at the point $\mathbf{z} = \mathbf{e}$ :

$$\frac{\partial p_j(\mathbf{e})}{\partial z_i} = \mu_j \beta_{j,1} \frac{\partial^2 p_0(\mathbf{e})}{\partial z_i \partial z_j} + \frac{\lambda_i \rho \rho_j}{\mu_i} + \lambda_i \lambda_j \frac{\beta_{j,2}}{2}$$

and then sum this relation over $j = 1, \ldots n$ :

$$\sum_{j=1}^n \mu_j \beta_{j,1} \frac{\partial^2 p_0(\mathbf{e})}{\partial z_i \partial z_j} = N_i - \frac{\lambda_i \rho(1 + \rho)}{\mu_i} - \lambda_i \lambda \frac{\beta_2}{2}.$$

Using (3.102) we obtain from this equality the following set of linear equations for the mean queue lengths $N_i = EN_i(t)$ :

$$\sum_{j=1}^n \frac{\mu_j \beta_{j,1}}{\mu_i + \mu_j} (\lambda_i N_j + \lambda_j N_i) = N_i - \frac{\lambda_i \lambda \beta_2}{2} - \frac{\lambda_i \rho}{\mu_i}. \qquad (3.103)$$

Introducing the variables $x_i$ by the formula

$$N_i = \frac{\lambda_i \rho}{\mu_i(1 - \rho)} + \frac{\lambda_i \lambda \beta_2}{2} x_i$$

completes the proof. $\qquad\qquad\qquad\qquad\qquad\qquad\qquad\qquad\square$

For any specific $n$ it is easy to obtain the solution in explicit form. For example, if $n = 2$ then we have a system of two linear equations with two unknown variables and so after some algebra we get the following result:

$$N_1 = \frac{\lambda_1 \rho}{\mu_1(1-\rho)}$$

$$+ \frac{\lambda_1 \lambda \beta_2}{2(1-\rho)} \cdot \frac{(1-\rho)\mu_1 + \mu_2}{(1-\rho_1)\mu_1 + (1-\rho_2)\mu_2}, \quad (3.104)$$

$$N_2 = \frac{\lambda_2 \rho}{\mu_2(1-\rho)}$$

$$+ \frac{\lambda_2 \lambda \beta_2}{2(1-\rho)} \cdot \frac{\mu_1 + (1-\rho)\mu_2}{(1-\rho_1)\mu_1 + (1-\rho_2)\mu_2}. \quad (3.105)$$

For large values of $n$ a computer should be used to carry out the calculations.

Let $n = 2$, $\mu_1 \to \infty$, $\mu_2 = \mu$. Then the model under consideration can be thought of as the model with priority subscribers studied in Section 3.2. Using equations (3.104), (3.105) we get for the model with priority subscribers:

$$N_1 = \frac{\lambda_1 \lambda \beta_2}{2(1-\rho_1)},$$

$$N_2 = \frac{\lambda_2 \rho}{\mu(1-\rho)} + \frac{\lambda_2 \lambda \beta_2}{2(1-\rho)(1-\rho_1)},$$

which coincide with formulas (3.34) and (3.35) obtained in Section 3.2.

**Theorem 3.6** *The second moments of queue lengths*

$$N_{ij} = \frac{\partial^2 p(\mathbf{e})}{\partial z_i \partial z_j} = \mathrm{E}\left(N_i(t)N_j(t)\right) - \delta_{i,j}\mathrm{E}N_i(t)$$

*are given by*

$$N_{ij} = \lambda_i \lambda_j \left[ x_{ij} + \frac{\lambda \beta_2}{2} \cdot \frac{x_i + x_j}{\mu_i + \mu_j} + \frac{1}{\mu_i \mu_j} \cdot \frac{\rho^2}{(1-\rho)^2} \right],$$

*where the variables $x_i$ were defined in Theorem 3.5 and the values $x_{ij}$ can be found as a solution of the system of linear equations:*

$$\sum_{k=1}^{n} \mu_k \rho_k \frac{x_{ij} + x_{ik} + x_{kj}}{\mu_i + \mu_j + \mu_k} = x_{ij} - \frac{\lambda \rho \beta_2}{2} \frac{x_i + x_j}{\mu_i + \mu_j}$$

$$- \frac{\lambda \beta_3}{3} - \frac{\lambda \beta_2}{2} \frac{\rho}{1-\rho} \frac{\mu_i + \mu_j}{\mu_i \mu_j}$$

$$-\frac{\lambda\beta_2}{4} \sum_{k=1}^{n} \lambda_k\mu_k\beta_{k,2} \left( \frac{x_i + x_k}{\mu_i + \mu_k} + \frac{x_j + x_k}{\mu_j + \mu_k} \right).$$

The proof is along the lines of Theorem 3.5, but now we have to differentiate equation (3.101) with respect to $z_i z_j z_k$ (instead of differentiating it with respect to $z_i z_j$ as we did earlier in obtaining equation (3.102)) and differentiate equation (3.98) with respect to $z_i z_k$ (instead of differentiating it with respect to $z_i$ as we did earlier in obtaining equation (3.103)).

# CHAPTER 4

# Advanced multiserver models

## 4.1 A multiserver model with priority subscribers

### 4.1.1 Model description

In this section we consider a multiserver version of the retrial queue with priority subscribers studied in section 3.2. This is a $c$-server queueing system in which two different types of primary customers arrive according to independent Poisson flows with rates $\lambda_1$ and $\lambda_2$ respectively. Let $\lambda = \lambda_1 + \lambda_2$ be the rate of the joint flow of primary calls. If there is a free server at the time of any primary call arrival, this call begins to be served immediately and leaves the system after service completion.

However, behaviour of a blocked customer, i.e. one who finds all servers occupied at the time of arrival, depends on its type. Customers from the first flow are queued after blocking and then are served in some discipline such as FIFO. Customers from the second flow who find all servers busy upon arrival cannot be queued and leave the service area, but after some random delay repeat attempts to get service. We assume that intervals between retrials are exponentially distributed with parameter $\mu$.

As in the single server case, a second type customer can be admitted for service only if there is no queue of first type customers. Thus the first type customers have a priority which is usually based on the fact that they have a direct access to the trunk group, and therefore can detect the epoch of a server's release and immediately enter service. This type of priority is similar to the standard head-of-the-line priority discipline. Moreover, as $\mu$ tends to infinity the model under consideration can be thought of as the standard multiserver queueing system with the head-of-the-line priority discipline. Taking this into account we shall refer to the first type customers as priority customers and to the second type customers as nonpriority (or low priority) customers.

In this section we assume that service times are exponentially distributed with mean $1/\nu = 1$ both for priority and low priority customers. The assumption that the statistical properties of service times of both priority and nonpriority calls are identical is quite natural from an applied point of view and technically simplifies further analysis.

As usual, the input flows of primary calls, intervals between repetitions, and service times are mutually independent.

Let $C(t)$ be the total number of customers in service and in the priority queue, and $N(t)$ be the number of sources of (nonpriority) calls. The above assumptions imply that the bivariate process $(C(t), N(t))$ is Markovian. It should be noted that now, as opposed to the main multiserver model, component $C(t)$ takes values in the set $Z_+$ rather than in the finite set $\{0, 1, ..., c\}$, so that the state space of the process $(C(t), N(t))$ is the two-dimensional integer lattice $S = Z_+^2$. The infinitesimal transition rates $q_{(ij)(nm)}$ of the process $(C(t), N(t))$ are given by:

1. for $0 \le i \le c - 1$

$$q_{(ij)(nm)} = \begin{cases} \lambda, & \text{if } (n, m) = (i + 1, j), \\ i, & \text{if } (n, m) = (i - 1, j), \\ j\mu, & \text{if } (n, m) = (i + 1, j - 1), \\ -(\lambda + i + j\mu), & \text{if } (n, m) = (i, j), \\ 0, & \text{otherwise.} \end{cases}$$

2. for $i \ge c$

$$q_{(ij)(nm)} = \begin{cases} \lambda_1, & \text{if } (n, m) = (i + 1, j), \\ \lambda_2, & \text{if } (n, m) = (i, j + 1), \\ c, & \text{if } (n, m) = (i - 1, j), \\ -(\lambda + c), & \text{if } (n, m) = (i, j), \\ 0, & \text{otherwise.} \end{cases}$$

From a practical point of view the most important characteristics of the quality of service to subscribers are:

- the stationary blocking probability $B \equiv \lim_{t \to \infty} \mathrm{P}\{C(t) \ge c\}$;

- the mean queue length of priority subscribers in the steady state $Q \equiv \lim_{t \to \infty} \mathrm{E} \max(C(t) - c, 0)$;

- the mean number of nonpriority subscribers in orbit in the steady state $N \equiv \lim_{t \to \infty} \mathrm{E} N(t)$;

- the stationary carried traffic (which is equal to the mean number of busy servers) $Y \equiv \lim_{t \to \infty} \mathrm{E} \min(c, C(t))$.

### 4.1.2 Ergodicity

Consider a Lyapunov function $\varphi(i,j) = ai + j$, where $a$ is a non-negative parameter which will be determined later on. Then the mean drift $y_{ij} = \sum_{(nm) \neq (ij)} q_{(ij)(nm)} \left( \varphi(n,m) - \varphi(i,j) \right)$ is given by:

$$y_{ij} = \begin{cases} (\lambda_1 + \lambda_2)a + j\mu(a-1) + a \cdot (-i), & \text{if } 0 \leq i \leq c-1, \\ \lambda_2 + \lambda_1 a - ac, & \text{if } i \geq c. \end{cases}$$

Since for $i = 0, 1, \ldots, c-1$ there exist $\lim_{j \to \infty} y_{ij} = (a-1) \cdot \infty$, and for $i \geq c$ variables $y_{ij}$ do not depend on $j$, the process $(C(t), N(t))$ is ergodic if parameter $a$ satisfies the following set of inequalities:

$$\begin{cases} a - 1 < 0 \\ \lambda_2 + \lambda_1 a - ac < 0 \end{cases}.$$

These conditions mean that $a$ belongs to the interval $\left( \frac{\lambda_2}{c - \lambda_1}, 1 \right)$. Such an $a$ can be found iff this interval is not empty, i.e. $\lambda_1 + \lambda_2 < c$. Thus $\lambda_1 + \lambda_2 < c$ is sufficient for ergodicity of $(C(t), N(t))$.

In fact this condition is necessary for ergodicity. Indeed, in the steady state the mean number of busy servers $Y$ equals the intensity of carried traffic, which in turn equals the intensity of offered traffic, i.e. $\lambda_1 + \lambda_2$. On the other hand $Y$ is less than the total number of servers, i.e. $c$.

Below we shall consider the system in the steady state, so that the condition $\lambda_1 + \lambda_2 < c$ will be assumed to hold.

### 4.1.3 Explicit formulas for the main performance characteristics

**Theorem 4.1** For the main performance characteristics the following relations hold:

$$Y = \lambda_1 + \lambda_2, \tag{4.1}$$

$$Q = \frac{\lambda_1}{c - \lambda_1} B, \tag{4.2}$$

$$N = \frac{(1 + \mu)(\lambda - D) - \lambda_1 \left( 1 + \frac{c - \lambda}{c - \lambda_1} \mu \right) B}{\mu(c - \lambda)}, \tag{4.3}$$

where $D$ is the variance of the number of busy servers.

*Proof.* Let $p_{ij} = \mathrm{P}\{C(t) = i, N(t) = j\}$ be the stationary distribution of the process $(C(t), N(t))$. These probabilities satisfy the following set of Kolmogorov equations:

$$
\begin{aligned}
(\lambda + i + j\mu)p_{ij} &= \lambda p_{i-1,j} + (j+1)\mu p_{i-1,j+1} \\
&+ (i+1)p_{i+1,j},\ 0 \le i \le c-1, \\
(\lambda + c)p_{cj} &= \lambda p_{c-1,j} + \lambda_2 p_{c,j-1} \\
&+ (j+1)\mu p_{c-1,j+1} + cp_{c+1,j}, \\
(\lambda + c)p_{ij} &= \lambda_1 p_{i-1,j} + \lambda_2 p_{i,j-1} \\
&+ cp_{i+1,j},\ i \ge c+1.
\end{aligned}
$$

For generating functions

$$
p_i(z) = \sum_{j=0}^{\infty} z^j p_{ij},\ 0 \le i < \infty,
$$

these equations become:

$$
\begin{aligned}
(\lambda + i)p_i(z) + \mu z p_i'(z) &= \lambda p_{i-1}(z) + \mu p_{i-1}'(z) \\
&+ (i+1)p_{i+1}(z), \\
&\quad 0 \le i \le c-1, \qquad (4.4) \\
(\lambda + c)p_c(z) &= \lambda p_{c-1}(z) + \lambda_2 z p_c(z) \\
&+ \mu p_{c-1}'(z) + cp_{c+1}(z), \qquad (4.5) \\
(\lambda + c)p_i(z) &= \lambda_1 p_{i-1}(z) + \lambda_2 z p_i(z) \\
&+ cp_{i+1}(z),\ i \ge c+1. \qquad (4.6)
\end{aligned}
$$

Now introduce the bivariate generating function

$$
g(x,z) = \sum_{i=c+1}^{\infty} x^i p_i(z).
$$

Then equation (4.6) gives:

$$
\begin{aligned}
\left[ \lambda_1 x^2 - (\lambda - \lambda_2 z + c)x + c \right] g(x,z) \\
= x^{c+1} \left[ cp_{c+1}(z) - \lambda_1 x p_c(z) \right]. \qquad (4.7)
\end{aligned}
$$

The coefficient $\lambda_1 x^2 - (\lambda - \lambda_2 z + c)x + c$ in the left-hand side has two zeros:

$$
\begin{aligned}
x_-(z) &= \frac{\lambda - \lambda_2 z + c - \sqrt{(\lambda - \lambda_2 z + c)^2 - 4\lambda_1 c}}{2\lambda_1}; \\
x_+(z) &= \frac{\lambda - \lambda_2 z + c + \sqrt{(\lambda - \lambda_2 z + c)^2 - 4\lambda_1 c}}{2\lambda_1}.
\end{aligned}
$$

If $z \in [0,1]$ then the root $x_-(z) \in [0,1]$ and the root $x_+(z) \in$

$(1, +\infty)$. Note also that

$$x_-(1) = 1;$$
$$x'_-(1) = \frac{\lambda_2}{c - \lambda_1};$$
$$x''_-(1) = \frac{2\lambda_2^2 c}{(c - \lambda_1)^3}.$$

Putting $x = x_-(z)$ in equation (4.7) we get that

$$cp_{c+1}(z) = \lambda_1 x_-(z) p_c(z),$$

which in turn allows us to rewrite equation (4.7) as follows:

$$g(x, z) = \frac{x^{c+1}}{x_+(z) - x} p_c(z).$$

Expanding both sides as a power series in $x$ and equating coefficients of equal powers of $x$ we get:

$$p_i(z) = \left( \frac{1}{x_+(z)} \right)^{i-c} p_c(z)$$
$$= \left( \frac{\lambda_1}{c} x_-(z) \right)^{i-c} p_c(z), \quad c \le i < \infty.$$

Putting here $z = 1$ yields:

$$p_i = \left( \frac{\lambda_1}{c} \right)^{i-c} p_c, \quad c \le i < \infty,$$

where $p_i = \mathrm{P}\left( C(t) = i \right)$. Now for the mean number of customers in the priority queue $Q$ we have:

$$Q = \sum_{i=c}^{\infty} (c - i) p_i = \frac{\lambda_1 c}{(c - \lambda_1)^2} p_c.$$

Since

$$B = \sum_{i=c}^{\infty} p_i = \frac{c}{c - \lambda_1} p_c,$$

the probability $p_c$ can be expressed in terms of the stationary blocking probability:

$$p_c = \left( 1 - \frac{\lambda_1}{c} \right) B,$$

which proves equation (4.2).

To express the mean number of customers in the low priority

queue in terms of the stationary distribution of the number of busy servers we introduce generating functions

$$P_i(z) = \mathrm{E}\left(z^{N(t)}; \text{the number of busy servers equals } i\right),$$
$$0 \le i \le c.$$

Obviously,

$$P_i(z) = \begin{cases} p_i(z), & \text{if } 0 \le i \le c-1, \\ \sum_{k=c}^{\infty} p_k(z) = \dfrac{c}{c - \lambda_1 x_-(z)} p_c(z), & \text{if } i = c. \end{cases}$$

Thus from equations (4.4), (4.5) for generating functions $p_i(z)$ we get the following set of equations for generating functions $P_i(z)$ :

$$\begin{aligned}
(\lambda + i)P_i(z) + \mu z P_i'(z) &= \lambda P_{i-1}(z) + \mu P_{i-1}'(z) \\
&\quad + (i+1)P_{i+1}(z), \\
&\quad 0 \le i \le c-2, \\
(\lambda + c - 1)P_{c-1}(z) + \mu z P_{c-1}'(z) &= \lambda P_{c-2}(z) + \mu P_{c-2}'(z) \\
&\quad + (c - \lambda_1 x_-(z)) P_c(z), \\
(\lambda_2 + c)P_c(z) &= \lambda P_{c-1}(z) + \lambda_2 z P_c(z) \\
&\quad + \mu P_{c-1}'(z) + \lambda_1 x_-(z) P_c(z).
\end{aligned}$$

Introducing generating function $P(x, z) = \sum_{i=0}^{c} x^i P_i(z)$ we transform these equations to

$$\begin{aligned}
\lambda(1 - x)P(x, z) + \mu(z - x)P_z'(x, z) & \\
+(x - 1)P_x'(x, z) + \lambda x^c(x - z)P_c(z) & \\
+\lambda_1 x^{c-1} \left[ x_-(z)(1 - x) + x(z - 1) \right] P_c(z) & \\
-\mu x^c(z - x)P_c'(z) &= 0.
\end{aligned}$$

Differentiating this equation with respect to $z, x, xx, xz, zz$ at the point $x = 1, z = 1$ we get the following equations:

$$\begin{aligned}
\mu N - \lambda_2 B - \mu N_c &= 0, \quad (4.8) \\
\lambda + \mu N - Y - \lambda_2 B - \mu N_c &= 0, \\
\lambda Y + \mu P_{xz}'' - P_{xx}'' - (\lambda_2 c + \lambda_1 \alpha)B - \mu c N_c &= 0, \\
-\lambda N - \mu P_{zz}'' + (1 + \mu)P_{xz}'' + (\lambda_2 - \mu c)N_c & \\
-\lambda_2 \left( c + \frac{\lambda_1}{c - \lambda_1} \right)B + \mu P_{czz}'' &= 0, \\
\mu P_{zz}'' - \lambda_2 N_c - \mu P_{czz}'' &= 0.
\end{aligned}$$

where $N = P'_z(1,1)$, $B = P_c(1)$, $Y = P'_x(1,1)$, $N_c = P'_c(1)$.

Eliminating from these equations variables $N_c, B, P''_{xz}, P''_{zz}, P''_{czz}$ we get relations (4.1) and (4.3).                                                                □

### 4.1.4 Limit theorems

#### High rate of retrials

Intuitive arguments suggest that as $\mu \to \infty$ the retrial queue under consideration can be thought of as an $(M_1, M_2)/M/c/(\infty, \infty)$ queueing system with head-of-the-line priority discipline, which serves two independent Poisson flows of calls; priority customers arrive at rate $\lambda_1$ and nonpriority customers arrive at rate $\lambda_2$. Stochastic behaviour of this limit system can be described with the help of the bivariate Markov process $(C_\infty(t), N_\infty(t))$, where $C_\infty(t)$ is the total number of customers in service (both priority and nonpriority) plus the number of customers in priority queue, and $N_\infty(t)$ is the length of nonpriority queue. The stationary distribution $p_{ij}(\infty)$ of the process $(C_\infty(t), N_\infty(t))$ is known in terms of generating functions $p_i^{(\infty)}(z) = \sum_{j=0}^{\infty} z^j p_{ij}(\infty)$ (Davis, R. (1966) Waiting-time distribution of a multiserver priority queue system. *Operations Research*, **14**):

$$
p_i^{(\infty)}(z) = \begin{cases} \dfrac{\lambda^i}{i!} p_{00}(\infty), & \text{if } 0 \leq i \leq c-1, \\[3mm] \dfrac{(1-z)x_-(z)}{x_-(z) - z} \left[ \dfrac{\lambda_1 x_-(z)}{c} \right]^{i-c} \dfrac{\lambda^c}{c!} p_{00}(\infty), & \text{if } i \geq c, \end{cases}
$$

where

$$
p_{00}(\infty) = \left[ \sum_{i=0}^{c-1} \frac{\lambda^i}{i!} + \frac{c}{c-\lambda} \frac{\lambda^c}{c!} \right]^{-1}
$$

and the function $x_-(z)$ was introduced earlier. It should be noted that probabilities $p_{ij}(\infty) = 0$ if $0 \leq i \leq c-1$, $j \geq 1$.

In particular, the stationary distribution $P_i(\infty)$ of the number of busy servers in the limit system is given by

$$
P_i(\infty) = \begin{cases} \dfrac{\lambda^i}{i!} P_0(\infty), & \text{if } 0 \leq i \leq c-1, \\[3mm] \dfrac{c}{c-\lambda} \dfrac{\lambda^c}{c!} P_0(\infty) & \text{if } i = c, \end{cases}
$$

where $P_0(\infty) = p_{00}(\infty)$, so that the stationary blocking probability in the limit system is

$$B(\infty) \equiv P_c(\infty) = \frac{\lambda^c}{(c-1)!} \Bigg/ \left( (c-\lambda) \sum_{i=0}^{c-1} \frac{\lambda^i}{i!} + \frac{\lambda^c}{(c-1)!} \right).$$

The later formulas are obvious since when only the total number of customers in the system is observed then it can be viewed as the standard $M/M/c/\infty$ queue with offered traffic $\lambda = \lambda_1 + \lambda_2$.

Using the method developed in section 2.7 for the main model we can prove rigorously that $\lim_{\mu \to \infty} p_{ij} = p_{ij}(\infty)$ and get the second terms of expansions of probabilities $p_{ij}$ in a power series in $\frac{1}{\mu}$. Taking into account that the main performance characteristics are expressed in terms of the stationary distribution of the number of busy servers, we give asymptotic expansions only for these probabilities.

**Theorem 4.2** *Under $\mu \to \infty$ the stationary distribution of the number of busy servers can be represented as*

$$P_i = P_i(\infty) \cdot \left[ 1 + \frac{1}{\mu} \cdot I \cdot A_i + o\left(\frac{1}{\mu}\right) \right],$$

*where*

$$
\begin{aligned}
I &= \frac{\lambda_1 - \lambda_2 - c + R}{2\lambda} - \ln \frac{2(c-\lambda)}{c - \lambda + R} \\
&+ \frac{c\lambda_1}{\lambda^2} \ln \frac{(c-\lambda)(\lambda + c + R)}{c(c - \lambda + R)},
\end{aligned}
$$

$$
A_i = \begin{cases}
-B(\infty), & \text{if } 0 \le i \le c-2, \\
\lambda - B(\infty), & \text{if } i = c-1, \\
1 - (c-\lambda) - B(\infty), & \text{if } i = c,
\end{cases}
$$

*and*

$$R = \sqrt{(\lambda + c)^2 - 4\alpha c}.$$

*Low rate of retrials*

Intuitive arguments suggest that as $\mu \to 0$ the retrial queue under consideration can be viewed as an $(M_1, M_2)/M/c/(\infty, 0)$ queueing system, i.e. as a $c$-server system which serves two independent Poisson flows of calls; customers from the first flow arrive in Poisson flow with rate $\lambda_1$ and are queued in the case of blocking, whereas customers from the second flow arrive in Poisson flow with rate $\lambda_2 + r$ (additional load $r$ is formed by repeated calls) and are lost in the case of blocking.

Stochastic behaviour of this limit system can be described with the help of the process of the total number of customers in the service and in the priority queue. This process is a birth and death process with rates of birth

$$
\Lambda_i = \begin{cases} \lambda + r, & \text{if } 0 \leq i \leq c - 1, \\ \\ \alpha, & \text{if } i \geq c, \end{cases}
$$

and rates of death $\mu_i = \min(i, c)$. Thus its stationary distribution $p_i$ is given by

$$
p_i = \begin{cases} \dfrac{(\lambda + r)^i}{i!} p_0, & \text{if } 0 \leq i \leq c - 1, \\ \\ \dfrac{(\lambda + r)^c}{c!} \left(\dfrac{\lambda_1}{c}\right)^{i-c} p_0, & \text{if } i \geq c, \end{cases} \tag{4.9}
$$

where

$$
p_0 = \left[ \sum_{i=0}^{c-1} \frac{(\lambda + r)^i}{i!} + \frac{(\lambda + r)^c}{c!} \frac{c}{c - \lambda_1} \right]^{-1}.
$$

In particular, the stationary distribution $P_i$ of the number of busy servers in the limit system is given by

$$
P_i = \begin{cases} \dfrac{(\lambda + r)^i}{i!} P_0, & \text{if } 0 \leq i \leq c - 1, \\ \\ \dfrac{(\lambda + r)^c}{c!} \dfrac{c}{c - \lambda_1} P_0, & \text{if } i = c, \end{cases} \tag{4.10}
$$

where $P_0 = p_0$.

The parameter $r$ can be found as a unique positive solution of the equation

$$
r \sum_{i=0}^{c-1} \frac{(\lambda + r)^i}{i!} = \lambda_2 \frac{(\lambda + r)^c}{c!} \frac{c}{c - \lambda_1}, \tag{4.11}
$$

which is a limit version of the relation (4.1).

Using the methods developed in section 2.7 for the main multiserver model, we can make the above consideration more precise. For example, the following result holds.

**Theorem 4.3** *Let $r = r(c, \lambda_1, \lambda_2)$ be the positive root of the equation (4.11) and*

$$D = r + r \frac{\frac{\lambda_2^2 \lambda_1}{(\lambda_2+r)(c-\lambda_1)^2} + \frac{r}{\lambda+r} \sum_{j=0}^{c-1} \frac{1}{P_j} \left( \sum_{i=0}^{j} P_i \right)^2}{\frac{\lambda_2}{\lambda_2+r} - \frac{r(c-\lambda)}{\lambda+r}}.$$

*Then as $\mu \to 0$*

$$\mathrm{E} \left\{ \exp \left( it \frac{\mu N(u) - r}{\sqrt{\mu}} \right) ; C(u) = n \right\} \to p_n \exp \left( -\frac{Dt^2}{2} \right),$$

*i.e. asymptotically*

- *the scaled number of customers in the nonpriority queue and the total number of customers in the priority queue and in service are independent;*
- *the number of customers in the nonpriority queue is Gaussian with mean $r/\mu$ and variance $D/\mu$;*
- *the variable $C(u)$ has distribution (4.9); in particular, the distribution of the number of busy servers is given by (4.10).*

### 4.1.5 Approximations

Very accurate approximations for the main performance characteristics of the retrial queue under consideration can be obtained with the help of interpolation between two extreme cases: $\mu \to 0$ and $\mu \to \infty$.

First consider the mean number of customers in the nonpriority queue. Since there exist

$$\lim_{\mu \to 0} \mu N = r$$

and

$$\lim_{\mu \to \infty} N = N(\infty) = \frac{\lambda_2 c}{(c - \lambda)(c - \lambda_1)} B(\infty),$$

following arguments similar to those used in section 2.8 we consider as an approximation for $N$ the following relation:

$$\mu N \approx r + \mu N(\infty),$$

or equivalently

$$N \approx \frac{r}{\mu} + \frac{\lambda_2 c}{(c - \lambda)(c - \lambda_1)} B(\infty). \qquad (4.12)$$

Now consider the conditional mean queue length given that all servers are busy, $E(N(t)|C(t) \geq c) = N_c/B$. Since there exist

$$\lim_{\mu \to 0} \mu E(N(t)|C(t) \geq c) = r$$

and

$$\lim_{\mu \to \infty} E(N(t)|C(t) \geq c) = \frac{N(\infty)}{B(\infty)} = \frac{\lambda_2 c}{(c-\lambda)(c-\lambda_1)},$$

we will approximate $E(N(t)|C(t) \geq c)$ as follows:

$$\mu E(N(t)|C(t) \geq c) \approx r + \mu \frac{\lambda_2 c}{(c-\lambda)(c-\lambda_1)}. \tag{4.13}$$

Using approximations (4.12) and (4.13) we can approximate the stationary blocking probability $B$. With this goal rewrite equation (4.8) as

$$B = \frac{\mu N}{\lambda_2 + \mu E(N(t)|C(t) \geq c)},$$

so that

$$B \approx \frac{r(c-\lambda)(c-\lambda_1) + \lambda_2 c \mu B(\infty)}{(\lambda_2 + r)(c-\lambda)(c-\lambda_1) + \lambda_2 c \mu}. \tag{4.14}$$

Using relation (4.2) we now can approximate the mean number of customers in the priority queue:

$$Q \approx \lambda_1 \frac{r(c-\lambda_1-\lambda_2)(c-\lambda_1) + \lambda_2 c \mu B(\infty)}{(\lambda_2 + r)(c-\lambda_1-\lambda_2)(c-\lambda_1) + \lambda_2 c \mu}.$$

The key part of the above approximate analysis is numerical solution of the algebraic equation (4.11). This can be done easily with the help of the following inequality for the paramenter $r$ :

$$0 < r < \frac{\lambda}{c-\lambda},$$

which follows from (4.3).

In Table 4.1 we give values of the blocking probability $B(\mu)$ and the approximate blocking probability $B_{\mathrm{appr}}(\mu)$ (calculated with the help of the right-hand side of equation (4.14)) in the retrial queue under consideration with $c = 10$ servers in the case $\lambda_1 = \lambda_2 = \frac{\lambda}{2}$.

Table 4.1 *Exact and approximate values of the blocking probability in* $(M_1, M_2)/M/c$ *type retrial queue with priority subscribers in the case* $c = 10$, $\lambda_1 = \lambda_2 = \lambda/2$

|  |  | $\mu = 0+$ | $\mu = 1$ | $\mu = 5$ | $\mu = 10$ |
|---|---|---|---|---|---|
| $\lambda = 5$ | $B(\mu)$ | 0.02605 | 0.02843 | 0.03165 | 0.03303 |
|  | $B_{\mathrm{appr}}(\mu)$ | 0.02605 | 0.02812 | 0.03173 | 0.03331 |
| $\lambda = 8$ | $B(\mu)$ | 0.29197 | 0.32470 | 0.36355 | 0.37858 |
|  | $B_{\mathrm{appr}}(\mu)$ | 0.29197 | 0.33547 | 0.37951 | 0.39219 |

## 4.2 A multiserver model with impatient subscribers

### 4.2.1 Model description

In this section we consider a multiserver version of the retrial queue with nonpersistent subscribers studied in section 3.3. This is a queueing system with $c$ servers in which a Poisson flow of primary customers with rate $\lambda$ arrives.

If there is a free server at time of a primary call arrival, this call begins to be served immediately and leaves the system after service completion. However if all servers are busy at the time of arrival of a primary call, then with probability $1 - H_1$ the call leaves the system without service and with probability $H_1 > 0$ forms a source of repeated calls. Every such source produces a Poisson process of repeated calls with intensity $\mu$. If an incoming repeated call finds a free server, it is served and leaves the system after service, while the source which produced this repeated call disappears. Otherwise, i.e. if all servers are occupied at the time of a repeated call arrival, with probability $1 - H_2$ the source leaves the system without service and with probability $H_2$ retries for service again.

We assume that service times are exponentially distributed with parameter $\nu = 1$. As usual, we suppose that the input flow of primary calls, intervals between repetitions, service times and decisions whether or not to retry for service are mutually independent.

The functioning of the system can be described by means of a bivariate process $(C(t), N(t))$, where $C(t)$ is the number of busy servers and $N(t)$ is the number of sources of repeated calls at time $t$. Under the above assumptions process $(C(t), N(t))$ is Markovian with the lattice semi-strip $S = \{0, 1, ..., c\} \times Z_+$ as the state space. Its infinitesimal transition rates $q_{(ij)(nm)}$ are given by:

1. for $0 \leq i \leq c - 1$

$$
q_{(ij)(nm)} = \begin{cases}
\lambda, & \text{if } (n, m) = (i + 1, j), \\
i, & \text{if } (n, m) = (i - 1, j), \\
j\mu, & \text{if } (n, m) = (i + 1, j - 1), \\
-(\lambda + i + j\mu), & \text{if } (n, m) = (i, j), \\
0, & \text{otherwise.}
\end{cases}
$$

2. for $i = c$

$$
q_{(cj)(nm)} = \begin{cases}
\lambda H_1, & \text{if } (n, m) = (c, j + 1), \\
j\mu(1 - H_2), & \text{if } (n, m) = (c, j - 1), \\
c, & \text{if } (n, m) = (c - 1, j), \\
-(\lambda H_1 + j\mu(1 - H_2) + c), & \text{if } (n, m) = (c, j), \\
0, & \text{otherwise.}
\end{cases}
$$

From a practical point of view the most important characteristics of the quality of service to subscribers are:

- the stationary blocking probability $B \equiv \lim_{t \to \infty} P\{C(t) = c\}$;

- the mean queue length in the steady state $N \equiv \lim_{t \to \infty} EN(t)$;

- the stationary carried traffic (which is equal to the mean number of busy servers) $Y \equiv \lim_{t \to \infty} EC(t)$;

- the fraction of lost primary calls $L \equiv 1 - \frac{Y}{\lambda}$.

### 4.2.2 Ergodicity

Consider the Lyapunov function $\varphi(i, j) = ai + j$, where $a$ is a positive parameter which will be determined later on. Then the mean drift $y_{ij} = \sum_{(nm) \neq (ij)} q_{(ij)(nm)} (\varphi(n, m) - \varphi(i, j))$ is given by:

$$
y_{ij} = \begin{cases}
\lambda a + j\mu(a - 1) + a \cdot (-i), & \text{if } 0 \leq i \leq c - 1, \\
\lambda H_1 - ac - j\mu(1 - H_2), & \text{if } i = c.
\end{cases}
$$

Thus for all $i = 0, 1, \ldots, c$ there exist

$$
\lim_{j \to \infty} y_{ij} \equiv L_i = \begin{cases}
(a - 1) \cdot \infty, & \text{if } 0 \leq i \leq c - 1, \\
-\infty, & \text{if } i = c \text{ and } H_2 < 1, \\
\lambda H_1 - ac, & \text{if } i = c \text{ and } H_2 = 1.
\end{cases}
$$

By Statement 8, section 2.2, the process $(C(t), N(t))$ is ergodic if all variables $L_i$ are negative. For $H_2 < 1$ this means that nonnegative parameter $a$ must be less than 1. Since such a value of $a$ can always be found, the process $(C_t, N_t)$ is ergodic for any arrival rate if $H_2 < 1$. For $H_2 = 1$ the condition $L_c < 0$ means that $a$ must belong to the interval $\left(\frac{\lambda H_1}{c}, 1\right)$. Such $a$ can be found iff this interval is not empty, i.e. $\lambda H_1 < 1$. This is a sufficient condition for ergodicity in the case $H_2 = 1$. In fact this condition is necessary for ergodicity. This follows from formula (4.25) for the mean number of busy servers, which will be proved later on.

### 4.2.3 Explicit formulas for the main performance characteristics

Let $p_{ij} = P\{C(t) = i, N(t) = j\}$ be the joint distribution of the number of busy servers and the queue length in the steady state. These probabilities satisfy the following set of Kolmogorov equations:

$$
\begin{aligned}
(\lambda + i + j\mu)p_{ij} &= \lambda p_{i-1,j} + (j+1)\mu p_{i-1,j+1} \\
&+ (i+1)p_{i+1,j}, \ 0 \le i < c, \quad (4.15) \\
(\lambda H_1 + j\mu(1 - H_2) + c)p_{cj} &= \lambda p_{c-1,j} \\
&+ (j+1)\mu p_{c-1,j+1} + \lambda H_1 p_{c,j-1} \\
&+ (j+1)\mu(1 - H_2)p_{c,j+1}. \quad (4.16)
\end{aligned}
$$

For generating functions

$$
p_i(z) = \sum_{j=0}^{\infty} z^j p_{ij}, \ 0 \le i \le c,
$$

these equations become

$$
\begin{aligned}
(\lambda + i)p_i(z) + \mu z p_i'(z) &= \lambda p_{i-1}(z) + \mu p_{i-1}'(z) \\
&+ (i+1)p_{i+1}(z), \\
&\quad 0 \le i \le c - 1, \quad (4.17) \\
(\lambda H_1(1 - z) + c)p_c(z) &= \lambda p_{c-1}(z) + \mu p_{c-1}'(z) \\
&+ \mu(1 - H_2)(1 - z)p_c'(z). \quad (4.18)
\end{aligned}
$$

Now introduce the bivariate generating function

$$
p(x, z) = \sum_{i=0}^{c} x^i p_i(z).
$$

Then equations (4.17), (4.18) give:

$$\lambda(1-x)p(x,z) + \mu(z-x)p'_z(x,z) + (x-1)p'_x(x,z)$$
$$+\lambda x^c(x-1)p_c(z) + \lambda H_1 x^c(1-z)p_c(z)$$
$$+\mu(1-H_2)x^c(z-1)p'_c(z) - \mu x^c(z-x)p'_c(z) = 0.$$

Differentiating this equation with respect to $z, x, xx, xz, zz$ at the point $x = 1, z = 1$ we get the following equations:

$$\mu N - \lambda H_1 B - \mu H_2 N_c = 0, \qquad (4.19)$$

$$\lambda + \mu N - Y - \lambda B - \mu N_c = 0, \qquad (4.20)$$

$$\lambda Y + \mu p''_{xz} - p''_{xx} - \lambda c B - \mu c N_c = 0, \qquad (4.21)$$

$$-\lambda N - \mu p''_{zz} + (1+\mu)p''_{xz} + \lambda N_c$$
$$-\mu H_2 c N_c - \lambda H_1 c B + \mu p''_{czz} = 0, \qquad (4.22)$$

$$\mu p''_{zz} - \lambda H_1 N_c - \mu H_2 p''_{czz} = 0, \qquad (4.23)$$

where $N = EN(t) = p'_z(1,1)$, $B = P\{C(t) = c\} = p_c(1)$, $Y = EC(t) = p'_x(1,1)$, $N_c = E\{N(t); C(t) = c\} = p'_c(1)$. Eliminating from equations (4.19) and (4.20) the variable $N_c$ we get:

$$\mu(1-H_2)N = \lambda H_2 + \lambda(H_1 - H_2)B - H_2 Y.$$

In the case $H_2 < 1$ this allows us to express the mean queue length $N$ in terms of the distribution of the number of busy servers:

$$N = \frac{\lambda H_2 + \lambda(H_1 - H_2)B - H_2 Y}{\mu(1-H_2)}. \qquad (4.24)$$

To get a parallel formula in the case $H_2 = 1$ one should eliminate from equations (4.19), (4.20), (4.21), (4.22) and (4.23) the variables $N_c, p''_{xz}, p''_{zz}, p''_{czz}$. Taking into account that $p''_{xx}(1,1) = EC^2(t) - EC(t)$ we get:

$$Y = \lambda - \lambda(1-H_1)B, \qquad (4.25)$$

$$N = \frac{1+\mu}{\mu(c-\lambda H_1)}\Big\{\lambda + \lambda^2 - EC^2(t)$$
$$-\lambda(1-H_1)\Big[\lambda + c + 1 - \frac{\lambda H_1}{1+\mu}\Big]B\Big\}. \qquad (4.26)$$

As for the main model, the steady state distribution of the number of busy servers $p_i = P(C(t) = i)$ can be written in the form of the stationary distribution of a birth and death process. With this goal put $z = 1$ in equations (4.17), (4.18):

$$\lambda p_i + \mu N_i - (i+1)p_{i+1} = \lambda p_{i-1} + \mu N_{i-1} - ip_i,$$

$$0 \leq i \leq c-1,$$

$$\lambda p_{c-1} + \mu N_{c-1} - c p_c = 0,$$

where $N_i = \mathrm{E}\left(N(t); C(t) = i\right)$. These equations yield that

$$\lambda p_i + \mu N_i - (i+1)p_{i+1} = 0, \; 0 \leq i \leq c-1. \qquad (4.27)$$

Denote the ratio $\mu N_i / p_i = \mathrm{E}\left(\mu N(t) | C(t) = i\right)$, which equals the rate of flow of repeated calls given that the number of busy servers is $i$, as $r_i$. Then equation (4.27) can be rewritten as

$$p_{i+1} = \frac{\lambda + r_i}{i+1} p_i, \; 0 \leq i \leq c-1.$$

From this we recursively have:

$$p_i = \frac{(\lambda + r_{i-1})\dots(\lambda + r_0)}{i!} p_0, \; 0 \leq i \leq c, \qquad (4.28)$$

and from the normalizing condition $\sum_{k=0}^{c} p_k = 1$ :

$$p_0 = \left( \sum_{k=0}^{c} \frac{(\lambda + r_{k-1})\dots(\lambda + r_0)}{k!} \right)^{-1}. \qquad (4.29)$$

Thus the steady state distribution of the number of busy servers in the retrial queue is identical to the steady state distribution of the number of busy servers in the Erlang loss model with state dependent arrival rate $\Lambda_i = \lambda + r_i$. The extra load $r_i$ is formed by repeated calls.

It should be noted that for the parameters $r_i$ an additional equation holds: eliminating from (4.19), (4.20) variable $N$ we get:

$$Y = \lambda - \lambda(1 - H_1)B - \mu(1 - H_2)N_c,$$

which can be rewritten as

$$\sum_{k=0}^{c} k \frac{(\lambda + r_{k-1})\dots(\lambda + r_0)}{k!} = \lambda \sum_{k=0}^{c} \frac{(\lambda + r_{k-1})\dots(\lambda + r_0)}{k!}$$

$$- (\lambda(1 - H_1) + (1 - H_2)r_c) \frac{(\lambda + r_{c-1})\dots(\lambda + r_0)}{c!}. \qquad (4.30)$$

Although parameters $r_i$, $0 \leq i \leq c$, are unknown and thus equations (4.28), (4.29), (4.30) do not give a closed form solution, these equations provide some insight into the problem and will be used later on.

### 4.2.4  Truncated model

*Model description*

In this model as opposed to the main model, the number of sources of repeated calls is bounded by a given constant $M$. If the number of sources equals $M$ then the blocked calls are lost and have no influence on the further functioning of the system. The stochastic dynamics of the system can be described by means of a process of two variables $(C^{(M)}(t), N^{(M)}(t))$, where $C^{(M)}(t)$ is the number of busy servers and $N^{(M)}(t)$ is the number of sources of repeated calls (queue length) at time $t$. Under the above assumptions, process $(C^{(M)}(t), N^{(M)}(t))$ is Markovian with the finite lattice semi-strip $S^{(M)} = \{0, 1, ..., c\} \times \{0, 1, ..., M\}$ as the state space. Its infinitesimal transition rates $q^{(M)}_{(ij)(nm)}$ are given by:

1. for $0 \le i \le c - 1$, $0 \le j \le M$

$$q^{(M)}_{(ij)(nm)} = \begin{cases} \lambda, & \text{if } (n,m) = (i+1,j), \\ i, & \text{if } (n,m) = (i-1,j), \\ j\mu, & \text{if } (n,m) = (i+1,j-1), \\ -(\lambda + i + j\mu), & \text{if } (n,m) = (i,j), \\ 0, & \text{otherwise.} \end{cases}$$

2. for $i = c$, $0 \le j \le M - 1$

$$q^{(M)}_{(cj)(nm)} = \begin{cases} \lambda H_1, & \text{if } (n,m) = (c, j+1), \\ j\mu(1 - H_2), & \text{if } (n,m) = (c, j-1), \\ c, & \text{if } (n,m) = (c-1, j), \\ -(\lambda H_1 + j\mu(1 - H_2) + c), & \text{if } (n,m) = (c,j), \\ 0, & \text{otherwise.} \end{cases}$$

3. for $i = c$, $j = M$

$$q^{(M)}_{(cM)(nm)} = \begin{cases} c, & \text{if } (n,m) = (c-1, M), \\ M\mu(1 - H_2), & \text{if } (n,m) = (c, M-1), \\ -(c + M\mu(1 - H_2)), & \text{if } (n,m) = (c, M), \\ 0, & \text{otherwise.} \end{cases}$$

Thus the rates $q^{(M)}_{(ij)(nm)}$ are the same as those of the initial model except for the boundary state $i = c, j = M$.

Since the state space of the process $(C^{(M)}(t), N^{(M)}(t))$ is finite, the process is always ergodic. Its stationary distribution $p^{(M)}_{ij} = P\{C^{(M)}(t) = i, N^{(M)}(t) = j\}$ may be found as a solution of the following set of linear equations:

$$(\lambda + i + j\mu)p_{ij}^{(M)} = \lambda p_{i-1,j}^{(M)} + (j+1)\mu p_{i-1,j+1}^{(M)}$$
$$+ \quad (i+1)p_{i+1,j}^{(M)},$$
$$0 \le i < c, 0 \le j < M,$$

$$(\lambda + i + M\mu)p_{iM}^{(M)} = \lambda p_{i-1,M}^{(M)} + (i+1)p_{i+1,M}^{(M)},$$
$$0 \le i \le c-1,$$

$$(\lambda H_1 + j\mu(1 - H_2) + c)p_{cj}^{(M)} = \lambda p_{c-1,j}^{(M)} + (j+1)\mu p_{c-1,j+1}^{(M)}$$
$$+ \quad \lambda H_1 p_{c,j-1}^{(M)}$$
$$+ \quad (j+1)\mu(1 - H_2)p_{c,j+1}^{(M)},$$
$$0 \le j \le M-1,$$

$$(c + M\mu(1 - H_2))p_{cM}^{(M)} = \lambda p_{c-1,M}^{(M)} + \lambda H_1 p_{c,M-1}^{(M)},$$

which satisfies the normalizing condition

$$\sum_{i=0}^{c} \sum_{j=0}^{M} p_{ij}^{(M)} = 1.$$

*Explicit formulas for the main performance characteristics*

Using the same approach as for the main $M/M/c$ type model we get the following formulas for the mean queue length in terms of the stationary distribution of the number of busy servers:

1. if $H_2 < 1$, then

$$N^{(M)} = \frac{\lambda H_2 + \lambda(H_1 - H_2)B^{(M)} - H_2 Y^{(M)} - \lambda H_1 p_{cM}^{(M)}}{\mu(1 - H_2)},$$

2. if $H_2 = 1$, then

$$N^{(M)} = \frac{1 + \mu}{\mu(c - \lambda H_1)} \left\{ \lambda + \lambda^2 - \mathrm{E}\left(C^{(M)}(t)\right)^2 \right.$$
$$- \quad \lambda(1 - H_1)\left[\lambda + c + 1 - \frac{\lambda H_1}{1 + \mu}\right]B^{(M)}$$
$$+ \quad \left.\left[\lambda + c + 1 + \frac{\mu M + \lambda(1 - H_1)}{1 + \mu}\right]\lambda H_1 p_{cM}^{(M)}\right\}.$$

*An algorithm for numerical calculation of the stationary distribution in the truncated system*

The stationary distribution $p_{ij}^{(M)}$ for the model under consideration can be calculated numerically with the help of a recursive algorithm similar to that used for the main model.

First introduce new variables $r_{ij}^{(M)}$, $0 \leq i \leq c, 0 \leq j \leq M$, by

$$r_{ij}^{(M)} = \frac{p_{ij}^{(M)}}{p_{0M}^{(M)}},$$

so that

$$p_{ij}^{(M)} = \frac{r_{ij}^{(M)}}{\sum\limits_{i=0}^{c} \sum\limits_{j=0}^{M} r_{ij}^{(M)}}.$$

Variables $r_{ij}^{(M)}$ satisfy the following set of equations, which follow from Kolmogorov equations for probabilities $p_{ij}^{(M)}$:

$$
\begin{aligned}
r_{0M}^{(M)} &= 1, && (4.31) \\
(\lambda + i + j\mu)r_{ij}^{(M)} &= \lambda r_{i-1,j}^{(M)} + (j+1)\mu r_{i-1,j+1}^{(M)} \\
&\quad + (i+1)r_{i+1,j}^{(M)}, \\
&\qquad 0 \leq i < c, 0 \leq j < M, && (4.32) \\
(\lambda + i + M\mu)r_{iM}^{(M)} &= \lambda r_{i-1,M}^{(M)} + (i+1)r_{i+1,M}^{(M)}, \\
&\qquad 0 \leq i \leq c-1, && (4.33) \\
\lambda H_1 r_{cj}^{(M)} &= (j+1)\mu \sum_{i=0}^{c-1} r_{i,j+1}^{(M)} \\
&\quad + (j+1)\mu(1 - H_2)r_{c,j+1}^{(M)}, \\
&\qquad 0 \leq j \leq M-1 && (4.34)
\end{aligned}
$$

As for the main model, we will calculate variables $r_{ij}^{(M)}$ by groups, each of size $c+1$; first calculate $r_{0M}^{(M)}, ..., r_{cM}^{(M)}$, then $r_{0M-1}^{(M)}, ..., r_{cM-1}^{(M)}$ and so on, until we find $r_{00}^{(M)}, ..., r_{c0}^{(M)}$. To be more exact,

1. Put $j = M$. Calculate variables $r_{0M}^{(M)}, ..., r_{cM}^{(M)}$ recursively from equation (4.33) using equation (4.31) as initial condition.

2. Put $j = j - 1$.

2.1. From equation (4.34) calculate $r_{cj}^{(M)}$.

2.2 From equation (4.32) with the help of the 'forward elimination, back substitution' algorithm calculate $r_{c-1,j}^{(M)}, ..., r_{0,j}^{(M)}$ .

Note that since equation (4.32) does not contain parameters $H_1$ and $H_2$, this step of the algorithm is identical to that for the main model.

3. Repeat step 2 while $j \geq 0$.

The following Pascal program calculates the joint stationary distribution $p_{ij}^{(M)}$ of the number of busy servers and the queue length, blocking probability $B^{(M)}$, mean number of busy servers $Y^{(M)}$, second moment of the number of busy servers $\mathrm{E}\left(C^{(M)}(t)\right)^2$ and mean number of customers in the queue $N^{(M)}$.

```pascal
Program retrial(Input,Output);
Uses Crt;
Var
i,j,c,M :integer;
lambda,mu,H1,H2,sum,bl,Y,V,N:extended;
r,p :array[0..20,0..100] of extended;
b,D : array[0..20] of extended;
Begin
writeln('input the number of servers '); read(c);
writeln('input the trancation limit '); read(M);
writeln('input the arrival rate '); read(lambda);
writeln('input the retrial rate '); read(mu);
writeln('input the probability H_1 '); read(H1);
writeln('input the probability H_2 '); read(H2);
r[0,M]:=1; r[1,M]:=lambda+M*mu;
for i:=2 to c do
r[i,M]:=((lambda+i-1+M*mu)*r[i-1,M]
-lambda*r[i-2,M])/i;
for j:=M-1 downto 0 do
begin
r[c,j]:=0;
for i:=0 to c-1 do r[c,j]:=r[c,j]+r[i,j+1];
r[c,j]:=r[c,j]+(1-H2)*r[c,j+1];
r[c,j]:=(j+1)*mu*r[c,j]/(lambda*H1);
b[0]:=0; D[0]:=0;
for i:=1 to c-1 do
begin
b[i]:=i*(j*mu+b[i-1])/(lambda+j*mu+b[i-1]);
```

```
D[i]:=(j+1)*mu*r[i-1,j+1]
+lambda*D[i-1]/(lambda+j*mu+b[i-1]);
end;
for i:=c-1 downto 0 do
r[i,j]:=(D[i]+(i+1)*r[i+1,j])/(lambda+j*mu+b[i]);
end;
sum:=0; bl:=0; Y:=0; V:=0; N:=0;
for i:=0 to c do
for j:=0 to M do
begin
sum:=sum+r[i,j];
if i=c then bl:=bl+r[i,j];
Y:=Y+i*r[i,j];
V:=V+i*i*r[i,j];
N:=N+j*r[i,j];
end;
bl:=bl/sum; Y:=Y/sum; V:=V/sum; V:=V-Y*Y; N:=N/sum;
for i:=0 to c do
for j:=0 to M do
p[i,j]:=r[i,j]/sum;
writeln('blocking probability=',bl:6:4);
writeln('the mean number of busy servers=',Y:8:4);
writeln('the variance of
the number of busy servers=',V:8:4);
writeln('the mean number of sources=',N:8:4);
End.
```

*Relation with the initial system*

First note that for the model with nonpersistent subscribers the process $(C(t), N(t))$ is a migration process with the following infinitesimal characteristics:

$$\lambda^1_{(n,m)} = \begin{cases} \lambda, & \text{if } 0 \leq n \leq c-1; \\ 0, & \text{if } n = c; \end{cases}$$

$$\lambda^2_{(n,m)} = \begin{cases} 0, & \text{if } 0 \leq n \leq c-1; \\ \lambda H_1, & \text{if } n = c; \end{cases}$$

$$\mu^1_{(n,m)} = n;$$

$$\mu^2_{(n,m)} = \begin{cases} 0, & \text{if } 0 \leq n \leq c-1; \\ m\mu(1 - H_2), & \text{if } n = c; \end{cases}$$

$$a_{(n,m)}^{1,2} = 0;$$

$$a_{(n,m)}^{2,1} = \begin{cases} m\mu, & \text{if } 0 \le n \le c-1; \\ 0, & \text{if } n = c. \end{cases}$$

Similarly, the process $(C^{(M)}(t), N^{(M)}(t))$ for the truncated variant of the model with nonpersistent subscribers is a migration process with the following infinitesimal characteristics:

$$\lambda_{(n,m)}^{1(M)} = \begin{cases} \lambda, & \text{if } 0 \le n \le c-1; \\ 0, & \text{if } n = c; \end{cases}$$

$$\lambda_{(n,m)}^{2(M)} = \begin{cases} 0, & \text{if } 0 \le n \le c-1; \\ \lambda H_1, & \text{if } n = c, 0 \le m \le M-1; \\ 0, & \text{if } n = c, m = M; \end{cases}$$

$$\mu_{(n,m)}^{1(M)} = n;$$

$$\mu_{(n,m)}^{2(M)} = \begin{cases} 0, & \text{if } 0 \le n \le c-1; \\ m\mu(1-H_2), & \text{if } n = c; \end{cases}$$

$$a_{(n,m)}^{1,2(M)} = 0;$$

$$a_{(n,m)}^{2,1(M)} = \begin{cases} m\mu, & \text{if } 0 \le n \le c-1; \\ 0, & \text{if } n = c. \end{cases}$$

Taking into account the description of both main and truncated retrial models as migration processes and applying Statement 13 from section 2.4 we get the following theorems.

**Theorem 4.4** *If*

$$(C^{(M)}(0), N^{(M)}(0)) \le_{st} (C(0), N(0)),$$

*then for all $t \ge 0$ we have:*

$$(C^{(M)}(t), N^{(M)}(t)) \le_{st} (C(t), N(t)).$$

*In particular, for the corresponding stationary distributions we get:*

$$\{p_{ij}^{(M)}\} \le_{st} \{p_{ij}\}.$$

**Theorem 4.5** *If*

$$(C^{(M)}(0), N^{(M)}(0)) \le_{st} (C^{(M+1)}(0), N^{(M+1)}(0)),$$

*then for all $t \ge 0$ we have:*

$$(C^{(M)}(t), N^{(M)}(t)) \le_{st} (C^{(M+1)}(t), N^{(M+1)}(t)).$$

*In particular, for the corresponding stationary distributions we get:*

$$\{p_{ij}^{(M)}\} \leq_{st} \{p_{ij}^{(M+1)}\}.$$

Verbatim repetition of arguments used in section 2.6 allows us to prove that stationary performance characteristics of the truncated model converge to the corresponding characteristics of the initial model as $M \to \infty$.

Errors in approximating stationary performance characteristics of the model with nonpersistent subscribers with the help of the corresponding characteristics of the truncated model can be estimated as follows:

$$0 \leq B - B^{(M)} \leq \frac{\lambda H_1}{1 + \lambda(1 - H_1)} p_{cM}^{(M)},$$

$$0 \leq Y - Y^{(M)} \leq \lambda H_1 p_{cM}^{(M)},$$

$$0 \leq N - N^{(M)} \leq F \cdot \lambda H_1 p_{cM}^{(M)},$$

where

$$F = \begin{cases} \dfrac{1}{\mu(1 - H_2)}, & \text{if } H_2 < 1 \text{ and } H_2 \geq H_1, \\[2ex] \dfrac{1 + \lambda(1 - H_2)}{\mu(1 - H_2)(1 + \lambda(1 - H_1))}, & \text{if } H_2 < 1 \text{ and } H_2 < H_1, \\[2ex] \dfrac{1 + \mu}{\mu} \cdot \dfrac{\lambda + c + \frac{\mu M + \lambda(1 - H_1)}{1 + \mu}}{c - \lambda H_1}, & \text{if } H_2 = 1. \end{cases}$$

The boundary probability $p_{cM}^{(M)}$ can be estimated as follows:

$$p_{cM}^{(M)} \leq \frac{\dfrac{(\lambda H_1)^M}{M! \mu^M} \displaystyle\prod_{i=0}^{M} \dfrac{\lambda + i\mu}{c + (\lambda + i\mu)(1 - H_2)}}{\displaystyle\sum_{j=0}^{M} \dfrac{c + \lambda + j\mu}{c + (\lambda + i\mu)(1 - H_2)} \dfrac{(\lambda H_1)^j}{j! \mu^j} \prod_{i=0}^{j-1} \dfrac{\lambda + i\mu}{c + (\lambda + i\mu)(1 - H_2)}}$$

The analysis is similar to that used for the main model and thus is omitted.

*Some numerical results*

Table 4.2 contains values of blocking probability $B$ and carried traffic $Y$ for the model with $c = 10$ servers, offered traffic $\lambda = 9$ and $\lambda = 12$, rate of retrials $\mu = 5$, $H_1 = H_2 = H$ for various values of the repetition probability $H$. These numerical results show

Table 4.2 *Dependence of the blocking probability B and the carried traffic Y in the M/M/c type retrial queue with impatient subscribers on the repetition probability H*

| H | $\lambda = 9$ | | $\lambda = 12$ | |
| | B | Y | B | Y |
|---|---|---|---|---|
| 0.1 | 0.1738 | 7.5180 | 0.3133 | 8.4183 |
| 0.2 | 0.1806 | 7.5521 | 0.3265 | 8.4660 |
| 0.3 | 0.1887 | 7.5919 | 0.3424 | 8.5215 |
| 0.4 | 0.1985 | 7.6391 | 0.3616 | 8.5873 |
| 0.5 | 0.2107 | 7.6965 | 0.3858 | 8.6671 |
| 0.6 | 0.2265 | 7.7686 | 0.4171 | 8.7671 |
| 0.7 | 0.2482 | 7.8636 | 0.4604 | 8.8981 |
| 0.8 | 0.2806 | 7.9988 | 0.5257 | 9.0827 |
| 0.9 | 0.3387 | 8.2242 | 0.6443 | 9.3814 |
| 1.0 | 0.5824 | 9.0000 | 1.0000 | 10.000 |

high sensitivity of the performance characteristics to the repetition probability in the case of very persistent subscribers when $H$ is close to 1, espesially for overloaded systems when $\lambda > c$.

### 4.2.5 Limit theorems

#### High rate of retrials

Limit behaviour of performance characteristics of the model under consideration is different in the cases $H_2 = 1$ and $H_2 < 1$.

Intuitive arguments suggest that in the case $H_2 < 1$ and $\mu \to \infty$ the retrial queue may be identified with the corresponding Erlang loss model. Indeed, if a primary call is blocked then it makes all attempts instantly, so that all repeated calls will be blocked as well. Thus the probability that the primary call after all joins the orbit is equal to $H_1 \prod_{i=1}^{\infty} H_2 = 0$. For this limit system, the joint stationary distribution of the number of busy servers and the number of customers in orbit is given by

$$p_{ij}(\infty) = \begin{cases} \dfrac{\lambda^i}{i!} \Big/ \sum\limits_{k=0}^{c} \dfrac{\lambda^k}{k!}, & \text{if } 0 \le i \le c, \ j = 0; \\ \\ 0, & \text{if } 0 \le i \le c, \ j \ge 1. \end{cases}$$

Rigorous proof of the fact that $\lim_{\mu \to \infty} p_{ij} = p_{ij}(\infty)$ can be given with the help of the method developed in section 2.7 for the main model.

First note that because $0 < Y < c$, $0 < B < 1$, the numerator of the right-hand side of equation (4.24) is bounded when $\mu$ varies over the half-line $0 < \mu < \infty$. Thus

$$\lim_{\mu \to \infty} N = 0 \qquad (4.35)$$

and since for $j \geq 1$ probability $p_{ij} = \frac{1}{j}jp_{ij} \leq \frac{1}{j}N$, we get that $\lim_{\mu \to \infty} p_{ij} = 0 \equiv p_{ij}(\infty)$ if $0 \leq i \leq c$, $j \geq 1$.

Moreover, using this fact and (4.15) we have by induction on $i$ that

$$\lim_{\mu \to \infty} \mu p_{ij} = 0, \text{ if } 0 \leq i \leq c-1, \ j \geq 1, \qquad (4.36)$$

which in turn implies that

$$\lim_{\mu \to \infty} \mu^2 p_{ij} = 0, \text{ if } 0 \leq i \leq c-2, \ j \geq 1.$$

Now express from equation (4.15) with $j = 0$ all probabilities $p_{i0}$ in terms of probability $p_{00}$ :

$$p_{i0} = \frac{\lambda^i}{i!}p_{00} - \frac{\lambda^i}{i!}\sum_{l=1}^{i}\frac{(l-1)!}{\lambda^l}\sum_{k=0}^{l-2}\mu p_{k1}, \ 0 \leq i \leq c. \qquad (4.37)$$

Besides rewrite the normalizing condition as follows:

$$\sum_{i=0}^{c}p_{i0} + \sum_{i=0}^{c}\sum_{j=1}^{\infty}p_{ij} = 1.$$

Using (4.37) and solving for $p_{00}$ we get:

$$p_{00} = \left(1 + \sum_{i=0}^{c}\frac{\lambda^i}{i!}\sum_{l=1}^{i}\frac{(l-1)!}{\lambda^l}\sum_{k=0}^{l-2}\mu p_{k1} - \sum_{i=0}^{c}\sum_{j=1}^{\infty}p_{ij}\right)$$

$$\times \left(\sum_{i=0}^{c}\frac{\lambda^i}{i!}\right)^{-1}. \qquad (4.38)$$

There are two terms containing unknown probabilities $p_{ij}$ in the numerator of the right-hand side of this relation. However as $\mu \to \infty$ the limits of these terms equal 0. For the first term this follows from (4.36) and for the second term from (4.35). Thus equation

(4.38) yields that there exists

$$\lim_{\mu \to \infty} p_{00} = 1 \Big/ \sum_{i=0}^{c} \frac{\lambda^i}{i!} \equiv p_{00}(\infty).$$

Now from (4.37) and (4.36) it follows that there exists

$$\lim_{\mu \to \infty} p_{i0} = \frac{\lambda^i}{i!} \lim_{\mu \to \infty} p_{00} = p_{i0}(\infty), \ 1 \le i \le c.$$

Of course, the relation $\lim_{\mu \to \infty} p_{ij} = p_{ij}(\infty)$ is obvious intuitively, but the above analysis is of interest because it can be generalized in order to get next terms of the expansions of the probabilities $p_{ij}$ in powers of $1/\mu$ and get estimates of remainders.

**Theorem 4.6**

$$0 < \frac{B - B(\infty)}{B} < \frac{\lambda H_1}{\mu(1 - H_2)}(1 - B(\infty)),$$

*where*

$$B(\infty) \equiv E_c(\lambda) \equiv \frac{\lambda^c}{c!} \Big/ \sum_{i=0}^{c} \frac{\lambda^i}{i!}$$

*is the stationary blocking probability in the classical $M/M/c/0$ Erlang loss queueing system.*

*Proof.* From equation (4.27) we get:

$$p_i < \frac{i+1}{\lambda} p_{i+1}, \ 0 \le i \le c - 1,$$

so that

$$p_i < \frac{\lambda^i}{i!} \frac{c!}{\lambda^c} B, \ 0 \le i \le c - 1.$$

Summing this inequality with respect to $i$ we get:

$$1 - B < \sum_{i=0}^{c-1} \frac{\lambda^i}{i!} \frac{c!}{\lambda^c} B,$$

or equivalently $B > B(\infty)$.

Next, from (4.37) and (4.38) we have:

$$B - B(\infty) = p_{c0} - p_{c0}(\infty) + \sum_{j=1}^{\infty} p_{cj}$$

$$= \frac{\lambda^c}{c!} (p_{00} - p_{00}(\infty)) - \frac{\lambda^c}{c!} \sum_{l=1}^{c} \frac{(l-1)!}{\lambda^l} \sum_{k=0}^{l-2} \mu p_{k1} + \sum_{j=1}^{\infty} p_{cj}$$

$$= B(\infty) \left[ \sum_{i=0}^{c} \frac{\lambda^i}{i!} \sum_{l=1}^{i} \frac{(l-1)!}{\lambda^l} \sum_{k=0}^{l-2} \mu p_{k1} - \sum_{i=0}^{c} \sum_{j=1}^{\infty} p_{ij} \right]$$

$$- \frac{\lambda^c}{c!} \sum_{l=1}^{c} \frac{(l-1)!}{\lambda^l} \sum_{k=0}^{l-2} \mu p_{k1} + \sum_{j=1}^{\infty} p_{cj}$$

$$= B(\infty) \left[ -\sum_{i=0}^{c} \frac{\lambda^i}{i!} \sum_{l=i+1}^{c} \frac{(l-1)!}{\lambda^l} \sum_{k=0}^{l-2} \mu p_{k1} - \sum_{i=0}^{c-1} \sum_{j=1}^{\infty} p_{ij} \right]$$

$$+ (1 - B(\infty)) \sum_{j=1}^{\infty} p_{cj}$$

$$< (1 - B(\infty)) \sum_{j=1}^{\infty} p_{cj}.$$

Summing equations (4.15), (4.16) with respect to $i$ we get:

$$\lambda H_1 p_{cj} - (j+1)\mu \sum_{i=0}^{c} p_{ij+1} + (j+1)\mu H_2 p_{cj+1} = 0.$$

This implies that

$$\lambda H_1 p_{cj} + (j+1)\mu H_2 p_{cj+1} = (j+1)\mu \sum_{i=0}^{c} p_{ij+1} > (j+1)\mu p_{cj+1},$$

so that

$$p_{cj+1} < \frac{\lambda H_1}{(j+1)\mu(1-H_2)} p_{cj}.$$

Summing with respect to $j$ we have:

$$\sum_{j=1}^{\infty} p_{cj} < \frac{\lambda H_1}{\mu(1-H_2)} \sum_{j=0}^{\infty} \frac{p_{cj}}{j+1}$$

$$< \frac{\lambda H_1}{\mu(1-H_2)} \sum_{j=0}^{\infty} p_{cj}$$

$$= \frac{\lambda H_1}{\mu(1-H_2)} B,$$

which yields the desired statement. $\quad\Box$

Analysis similar to that used in section 2.7 for the main model allows us to get the following result.

**Theorem 4.7** *If $H_2 < 1$, then as $\mu \to \infty$ the stationary blocking probability $B$ and the stationary mean queue length $N$ are*

$$
\begin{aligned}
B \;=\;& B(\infty) \\
& + \frac{\lambda H_1}{\mu(1 - H_2)} B(\infty)(1 - B(\infty)) \\
& + o\left(\frac{1}{\mu}\right), \\
N \;=\;& \frac{\lambda H_1}{\mu(1 - H_2)} B(\infty) \\
& + \frac{\lambda H_1}{\mu^2(1 - H_2)^2} B(\infty) \left[\lambda H_1(1 - B(\infty)) - H_2 c\right] \\
& + o\left(\frac{1}{\mu^2}\right).
\end{aligned}
$$

Now consider the case $H_2 = 1$. Intuitive arguments suggest that if $\mu \to \infty$ then the retrial queue may be considered as the corresponding $M/M/c/\infty$ queue where arriving customers with probability $H_1$ decline to join the queue and leave the system without service. For the limit system, the stationary distribution of the number of busy servers is given by

$$
p_i(\infty) = 
\begin{cases}
\dfrac{\lambda^i}{i!} p_0(\infty), & \text{if } 0 \le i \le c - 1, \\[2ex]
\dfrac{\lambda^c}{(c-1)!} \dfrac{1}{c - \lambda H_1} p_0(\infty), & \text{if } i = c,
\end{cases}
$$

where

$$
p_0(\infty) = 1 \left/ \left( \sum_{i=0}^{c-1} \frac{\lambda^i}{i!} + \frac{\lambda^c}{(c-1)!} \frac{1}{c - \lambda H_1} \right) \right. .
$$

Applying a method similar to that used for asymptotic analysis of the main model in the case $\mu \to \infty$ we get the following result.

**Theorem 4.8** *If $H_2 = 1$, then as $\mu \to \infty$ the stationary distribution of the number of busy servers in the retrial queue with nonpersistent subscribers can be represented as*

$$
p_i = p_i(\infty) \left( 1 + \frac{1}{\mu} A_i \ln\left(1 - \frac{\lambda H_1}{c}\right) + o\left(\frac{1}{\mu}\right) \right),
$$

*where*

$$A_i = \begin{cases} p_c(\infty)\,(1 + \lambda(1 - H_1)), & \text{if } 0 \le i \le c - 2, \\ -\lambda + (1 + \lambda(1 - H_1))\,p_c(\infty), & \text{if } i = c - 1, \\ c - 1 - \lambda + (1 + \lambda(1 - H_1))\,p_c(\infty), & \text{if } i = c. \end{cases}$$

Using this theorem one can estimate the mean number of customers in orbit with the help of (4.26).

*Low rate of retrials*

Let $r$ be a positive root of the equation

$$r \sum_{k=0}^{c} \frac{(\lambda + r)^k}{k!} = (\lambda H_1 + r H_2) \frac{(\lambda + r)^c}{c!}, \qquad (4.39)$$

and $\Lambda = \lambda + r$. It is easy to show that this root exists and is unique if $H_2 < 1$ or $H_2 = 1$, $\lambda H_1 < c$.

Using the same approach as for the main $M/M/c$ type model, we get that if the model under consideration is in the steady state and $\mu \to 0$ then asymptotically

- the scaled queue length $(\mu N(t) - r)/\sqrt{\mu}$ and the number of busy servers are independent;

- the number of busy servers has Erlang loss distribution

$$p_n = \frac{\Lambda^n}{n!} \Bigg/ \sum_{k=0}^{c} \frac{\Lambda^k}{k!};$$

- the number of sources of repeated calls is Gaussian with mean $r/\mu$ and variance $D/\mu$, where

$$D = r + \frac{r^2}{\Lambda(r + 1) - cr - \Lambda(r + H_2)p_c} \sum_{k=0}^{c-1} \frac{1}{p_k} \left( \sum_{n=0}^{k} p_n \right)^2.$$

Taking into account this result we may approximate the stationary distribution of the number of busy servers with the help of the corresponding distribution in the Erlang loss model with offered traffic $\Lambda = \lambda + r$, where the parameter $r$ is a positive solution to the equation (4.39).

The same approximation can be obtained as a consequence of the assumption that conditional rate of flow of repeated calls $r_i = \mathrm{E}\,(\mu N(t)|C(t) = i)$ does not depend on $i$ :

$$r_i \equiv r, \ 0 \le i \le c.$$

Under this assumption equations (4.28), (4.29) yield that the number of busy servers has Erlang loss distribution with parameter $\lambda + r$, and equation (4.30) becomes equation (4.39).

## 4.3 A multiserver retrial queue with a finite number of sources of primary calls

### 4.3.1 Model description

So far we have assumed that the flow of primary calls is Poisson. For this model of input flow, the probability of arrival of a new call during interval $(t, t + dt)$ is equal to $\lambda dt$ no matter how many calls are already present in the system at time $t$. Usually this means that primary calls are generated by a very large number of sources and each of them generates primary calls very seldom. From this point of view, a model with Poisson input flow is a model with an infinite number of sources of primary calls. In the present section we consider a model with a finite number of sources of primary calls. This allows us to take into account decrease of the rate of flow of primary calls when the number of busy circuits and/or number of sources of repeated calls increases.

Thus, consider a $c$-server queueing system where (contrary to the main model) primary calls are generated by $K$, $c < K < \infty$, sources. Each source can be in one of three states

- service
- sending repeated calls (i.e. waiting for service)
- free

If a source is free at time $t$ (i.e if it is not being served and is not waiting for service) then it may generate a primary call during interval $(t, t + dt)$ with probability $\alpha \cdot dt$.

If there is a free server at the time of arrival of a primary call then the call (or equivalently the source which produced the call) starts to be served. During service the source cannot generate new primary calls. After service the source moves into the free state and can generate a new primary call.

If all servers are busy at time of arrival of a primary call, then the source starts generation of repeated calls at exponential intervals with mean $1/\mu$ until it finds a free server, at which time the source starts to be served. As before, after service the source becomes free and can generate a new primary call.

The service time has an exponential distribution with a finite mean $1/\nu = 1$ both for primary calls and repeated calls.

The functioning of the system can be described by means of process $(C(t), N(t))$, where $C(t)$ is the number of busy servers and $N(t)$ is the number of sources of repeated calls (queue length) at time $t$. Under the above assumptions process $(C(t), N(t))$ is Markovian with finite state space $S = \{0, 1, ..., c\} \times \{0, 1, ..., K - c\}$. Its infinitesimal transition rates $q_{(ij)(nm)}$ are given by:

1. for $0 \leq i \leq c - 1$

$$q_{(ij)(nm)} = \begin{cases} (K - i - j)\alpha, & \text{if } (n, m) = (i + 1, j), \\ i, & \text{if } (n, m) = (i - 1, j), \\ j\mu, & \text{if } (n, m) = (i + 1, j - 1), \\ -((K - i - j)\alpha & \\ \quad + i + j\mu), & \text{if } (n, m) = (i, j), \\ 0 & \text{otherwise.} \end{cases}$$

2. for $i = c$

$$q_{(cj)(nm)} = \begin{cases} (K - c - j)\alpha, & \text{if } (n, m) = (c, j + 1), \\ c, & \text{if } (n, m) = (c - 1, j), \\ -((K - c - j)\alpha + c), & \text{if } (n, m) = (c, j), \\ 0 & \text{otherwise.} \end{cases}$$

Since the state space of the process $(C(t), N(t))$ is finite, the process is ergodic for all values of the rate of generation of new primary calls, and from now on we will assume that the system is in the steady state.

From a practical point of view the most important characteristics of the quality of service to subscribers are:

- the mean rate of generation of primary calls

$$\overline{\lambda} = \alpha E(K - C(t) - N(t))$$

(which equals the mean number of busy servers $Y = EC(t)$);
- the mean number of sources of repeated calls $N = EN(t)$;
- the fraction of primary calls which were blocked (i.e. met all servers busy)

$$B = \frac{\alpha E(K - C(t) - N(t); C(t) = c)}{\alpha E(K - C(t) - N(t))};$$

- the mean waiting time $W = N/\overline{\lambda}$ and the mean conditional waiting time $W_B = W/B$.

*4.3.2 The outside observer's distribution of the server state and the queue length*

Let $p_{ij} = \mathrm{P}\{C(t) = i, N(t) = j\}$ be the joint distribution of the number of busy servers and the queue length in the steady state. These probabilities satisfy the following set of Kolmogorov equations (below $p_{ij} = 0$ if $(i,j)$ does not belong to the state space of the process $(C(t), N(t))$):

$$
\begin{aligned}
((K - i - j)\alpha + i + j\mu)p_{ij} &= (K - i + 1 - j)\alpha p_{i-1,j} \\
&+ (j + 1)\mu p_{i-1,j+1} \\
&+ (i + 1)p_{i+1,j}, \\
&\text{if} \quad 0 \le i \le c - 1, \qquad (4.40) \\
((K - c - j)\alpha + c)p_{cj} &= (K - c + 1 - j)\alpha p_{c-1,j} \\
&+ (K - c - j + 1)\alpha p_{c,j-1} \\
&+ (j + 1)\mu p_{c-1,j+1}. \qquad (4.41)
\end{aligned}
$$

Summing these equations with respect to $i = 0, 1, \ldots, c$ we get the following useful relation:

$$
(K - c - j)\alpha p_{cj} = (j + 1)\mu \sum_{i=0}^{c-1} p_{i,j+1}, 0 \le j \le K - c. \qquad (4.42)
$$

The set of equations (4.40), (4.41) is finite and thus we do not need to truncate it to get a numerical solution. It is easy to see that the structure of the set is very similar to the structure of the Kolmogorov equations for the truncated variant of the main model (variable $K - c$ plays the role of the truncation limit $M$, so that below we will use notation $M = K - c$) and thus it can be solved with the help of a recursive algorithm similar to that used for the main model.

First introduce variables $r_{ij}$ by

$$
r_{ij} = \frac{p_{ij}}{p_{0,M}},
$$

so that

$$
p_{ij} = \frac{r_{ij}}{\sum_{i=0}^{c} \sum_{j=0}^{M} r_{ij}}.
$$

Variables $r_{ij}$ satisfy the set of equations (4.40)-(4.42) and equation:

$$
r_{0M} = 1. \qquad (4.43)
$$

As for the main model we will calculate variables $r_{ij}$ by groups, each of size $c + 1$; first calculate $r_{0,M}, \ldots, r_{c,M}$, then calculate $r_{0,M-1}, \ldots, r_{c,M-1}$ and so on, until we find $r_{0,0}, \ldots, r_{c,0}$. The detailed algorithm looks as follows.

1. Put $j = M$.

1.2 Calculate variables $r_{0,M}, \ldots, r_{c,M}$ recursively from equation (4.40) (with $j = M$) using equation (4.43) as initial condition:

$$r_{1M} = (c\alpha + M\mu)r_{0M} = c\alpha + M\mu,$$

$$
\begin{aligned}
r_{iM} &= \frac{(c - i + 1)\alpha + i - 1 + M\mu}{i} r_{i-1,M} \\
&\quad - \frac{(c - i + 2)\alpha}{i} r_{i-2,M}, \text{ if } i = 2, \ldots, c.
\end{aligned}
$$

2. Put $j = j - 1$.

2.1. From equation (4.42) calculate $r_{cj}$ :

$$r_{cj} = \frac{(j + 1)\mu}{(M - j)\alpha} \sum_{i=0}^{c-1} r_{i,j+1}.$$

2.2. Consider the difference equation (4.40) as a set of linear equations with respect to unknowns $x_i = r_{ij}$, $0 \le i \le c - 1$. This set of equations has the form

$$\alpha_i x_{i-1} + \beta_i x_i + \gamma_i x_{i+1} = \delta_i, \ 0 \le i \le c - 1,$$

where

$$
\begin{aligned}
x_i &= r_{ij}, \\
\alpha_i &= -(K - i + 1 - j)\alpha, \\
\beta_i &= (K - i - j)\alpha + i + j\mu, \\
\gamma_i &= -(i + 1), \\
\delta_i &= (j + 1)\mu r_{i-1,j+1},
\end{aligned}
$$

and values

$$x_{-1} = 0, \ x_c = r_{cj}$$

are known.

Thus we can apply the 'forward elimination, back substitution' algorithm in order to calculate $r_{c-1,j}, \ldots, r_{0,j}$ :

• calculate variables $b_i, D_i$, $0 \le i \le c - 1$, with the help of equations:

$$b_0 = 0,$$

$$D_0 = 0;$$

$$b_i = i\frac{b_{i-1} + j\mu}{b_{i-1} + (K - i + 1 - j)\alpha + j\mu}, \quad 1 \le i \le c - 1,$$

$$D_i = (j + 1)\mu r_{i-1,j+1} + \frac{(K - i + 1 - j)\alpha D_{i-1}}{b_{i-1} + (K - i + 1 - j)\alpha + j\mu},$$
$$1 \le i \le c - 1.$$

- then recursively calculate $r_{ij}, 0 \le i \le c - 1$, (in reverse order, starting with $r_{cj}$ known from step 2.1) with the help of the equation

$$r_{ij} = \frac{D_i + (i + 1)r_{i+1,j}}{b_i + (K - i - j)\alpha + j\mu}, \quad i = c - 1, c - 2, ..., 1, 0.$$

3. Repeat step 2 while $j \ge 0$ (i.e. successively for $j = M - 1, j = M - 2, ..., j = 0$).

Since $p_{ij} = r_{ij} \cdot p_{0M}$ , we have:

$$p_{0M} = \frac{1}{\sum_{i=0}^{c} \sum_{j=0}^{M} r_{ij}}.$$

Now we can calculate probabilities $p_{ij}$ as products $r_{ij} \cdot p_{0M}$ and correspondingly performance characteristics, which can be expressed in terms of the distribution $p_{ij}$.

The following Pascal program calculates the mean number of busy servers $Y$, the mean number of customers in the queue $N$, blocking probability $B$ and the mean conditional waiting time $W_B$.

```
Program retrial(Input,Output);
{finite source retrial queue}
Uses Crt;
Var
i,j,j1,j2,c,M,K :integer;
a,mu,sum,pc,Y,N,Nc,bl,W:extended;
r :array[0..100,0..1] of extended;
b,D : array[0..100] of extended;
Begin
writeln('input the number of servers '); read(c);
writeln('input the number of sources '); read(K);
writeln('input the new calls generation rate');
```

```
read(a);
writeln('input the retrial rate '); read(mu);
M:=K-c;
j1:=0;j2:=1;j:=M;
sum:=0; pc:=0; Y:=0; N:=0; Nc:=0;
r[0,j1]:=1;
sum:=sum+r[0,j1];N:=N+j*r[0,j1];
r[1,j1]:=c*a+M*mu;
sum:=sum+r[1,j1];N:=N+j*r[1,j1];Y:=Y+r[1,j1];
if c=1 then
begin pc:=pc+r[1,j1];Nc:=Nc+j*r[1,j1] end;
for i:=2 to c do
begin
r[i,j1]:=(((c-i+1)*a+i-1+M*mu)*r[i-1,j1]
-(c-i+2)*a*r[i-2,j1])/i;
sum:=sum+r[i,j1]; N:=N+j*r[i,j1]; Y:=Y+i*r[i,j1];
if i=c then begin pc:=pc+r[i,j1];Nc:=Nc+j*r[i,j1] end;
end;
for j:=M-1 downto 0 do
begin
j1:=1-j1;j2:=1-j2;
r[c,j1]:=0;
for i:=0 to c-1 do r[c,j1]:=r[c,j1]+r[i,j2];
r[c,j1]:=r[c,j1]*(j+1)*mu/((M-j)*a);
sum:=sum+r[c,j1]; N:=N+j*r[c,j1];
Y:=Y+c*r[c,j1]; pc:=pc+r[c,j1];Nc:=Nc+j*r[c,j1];
b[0]:=0;D[0]:=0;
for i:=1 to c-1 do
begin
b[i]:=i*(j*mu+b[i-1])/((K-i+1-j)*a+j*mu+b[i-1]);
D[i]:=(j+1)*mu*r[i-1,j2]
+(K-i+1-j)*a*D[i-1]/((K-i+1-j)*a+j*mu+b[i-1]);
end;
for i:=c-1 downto 0 do
begin
r[i,j1]:=(D[i]+(i+1)*r[i+1,j1])/((K-i-j)*a+j*mu+b[i]);
sum:=sum+r[i,j1]; N:=N+j*r[i,j1]; Y:=Y+i*r[i,j1];
end;
end;
pc:=pc/sum; Y:=Y/sum; N:=N/sum; Nc:=Nc/sum;
W:=N/Y;bl:=a*(M*pc-Nc)/Y;
```

```
writeln('the mean number of busy servers=',Y:8:4);
writeln('the mean number
of sources of repeated calls=',N:8:4);
writeln('blocking probability=',bl:6:4);
writeln('the mean conditional waiting time=',
W/bl:9:4);
End.
```

### 4.3.3 The arriving customer's distribution of the server state and the queue length

Probabilities $p_{ij}$, which we obtained before, give the proportion of time the system spends in the state $(C(t) = i, N(t) = j)$. Thus they can be thought of as **an outside observer's distribution** of the system state. However, for evaluating the quality of service of calls, it is more important to know the state of the system at the time when a particular source generates a new primary call. Let us denote this **arriving customer's distribution** by $\pi_{ij}$, i.e. $\pi_{ij}$ is the probability that the given source finds the system in the state $(C(t) = i, N(t) = j)$ when placing a primary request.

An important feature of the finite-source model is that the arriving customer's distribution differs from the outside observer's distribution. Indeed, it is clear that for $i = 0, \ldots, c - 1$ probabilities $\pi_{ij}$ are positive for $j = 0, 1, \ldots, K - c$, and probabilities $\pi_{cj}$ are positive for $j = 0, 1, \ldots, K - c - 1$ (but $\pi_{c,K-c} = 0$), whereas probabilities $p_{ij}$ are positive for $j = 0, 1, \ldots, K - c$. Thus, at least $\pi_{c,K-c} \neq p_{c,K-c}$.

It is well known that for birth and death type queueing models with $K$ sources of calls, the arriving customer's distribution of the total number of customers in the queue and in service, $\pi_x = \pi_x(K)$, is the same as the outside observer's distribution of the total number of customers in the queue and in service, $p_x(K - 1)$, in the corresponding model with $K - 1$ sources of calls: $\pi_x(K) = p_x(K - 1)$. Such a result does not hold for the retrial queue with finite source input. Indeed, as we noted, $\pi_{0,K-c}(K) > 0$, whereas $p_{0,K-c}(K - 1) = 0$.

Nevertheless, for the retrial model the arriving customer's distribution $\pi_{ij}$ can be easily related to the outside observer's distribution $p_{ij}$. With this goal let us characterize the state of the system with the help of vector $x = (x_1, ..., x_K)$, where

$$x_i = \begin{cases} 0, & \text{if } i\text{th source is free,} \\ s, & \text{if } i\text{th source is served,} \\ r, & \text{if } i\text{th source is in orbit.} \end{cases}$$

For any such microscopic state $x$ denote by $C(x)$ the number of served calls (i.e. $C(x) = \sum_{i=1}^{K} I(x_i = s)$), and by $N(x)$ the number of sources of repeated calls (i.e. $N(x) = \sum_{i=1}^{K} I(x_i = r)$). Let $X_{ij}$ be the set of all microscopic states $x$ such that $C(x) = i$, $N(x) = j$; $i = 0, \ldots, c$; $j = 0, 1, ..., K - c$. It is easy to see that the cardinality of the set $X_{ij}$ is $\binom{K}{i} \cdot \binom{K-i}{j}$.

By symmetry, all states $x \in X_{ij}$ have the same probability $p_{ij}^*$. On the other hand, probability of the set $X_{ij}$ (i.e. sum of probabilities of all states $x \in X_{ij}$) is the outside observer's probability $p_{ij}$. Thus,

$$p_{ij}^* = \frac{p_{ij}}{\binom{K}{i} \cdot \binom{K-i}{j}}.$$

Now fix some source $i_0$ and denote by $\widehat{p}_{ij}$ the probability that in the steady state this source is free, the number of busy servers equals $i$ and there are $j$ sources of repeated calls. It is clear that this event is formed by states $x \in X_{ij}$ such that $x_{i_0} = 0$. The total number of such states is $\binom{K-1}{i} \cdot \binom{K-1-i}{j}$. Thus,

$$\widehat{p}_{ij} = \binom{K - 1}{i} \cdot \binom{K - 1 - i}{j} \cdot p_{ij}^* = \frac{K - i - j}{K} p_{ij}.$$

Therefore the stationary probability that the fixed source is free equals

$$\widehat{p} = \sum_i \sum_j \widehat{p}_{ij} = \frac{1}{K} \sum_i \sum_j (K - i - j) p_{ij} = \frac{\overline{\lambda}}{K\alpha},$$

and the conditional distribution of the system state provided the fixed source $i_0$ is free is given by:

$$\widetilde{p}_{ij} = \frac{\widehat{p}_{ij}}{\widehat{p}} = \frac{(K - i - j)\alpha}{\overline{\lambda}} p_{ij}.$$

This is the outside observer's distribution of the system state during periods when the $i_0$th source is free. But since the period from the time when the source becomes free until the time when it generates a new primary call has an exponential distribution, the PASTA property holds (Wolff, R. W. (1989) *Stochastic modeling and the theory of queues*, Prentice Hall, Inc., section 5-16), i.e. $\tilde{p}_{ij}$ is identical to the arriving customer's distribution $\pi_{ij}$:

$$\pi_{ij} = \frac{(K - i - j)\alpha}{\bar{\lambda}} p_{ij}. \qquad (4.44)$$

### 4.3.4  Waiting time

#### The method of analysis

Consider some fixed source $i_0$ and assume that at time $t$ it places a primary request (which means, in particular, that the source was free just prior to the time $t$). Then the period until the source starts to be served is its virtual waiting time $W(t)$.

As is usually the case for retrial queues, analysis of the waiting time process is much more difficult than analysis of the number in the system. Although the mean waiting time $W$ can be derived easily from the outside observer's distribution of the system state with the help of Little's formula $W = \frac{N}{\lambda}$, and the probability $B$ that a customer must wait for service can be calculated from this distribution as the fraction of blocked primary calls $\frac{1}{\lambda} \sum_{j=0}^{K-c-1} (K - c - j)\alpha p_{cj}$, detailed and accurate analysis requires more than the outside observer's distribution. The blocking probability $B$ is the probability that all servers are occupied from the point of view of a source which places a primary request; thus $B$ should be defined as $\sum_{j=0}^{K-c-1} \pi_{cj}$, where $\pi_{ij}$ is the arriving customer's distribution. Using formula (4.44) for $\pi_{ij}$ we can show rigorously that

$$B \equiv \sum_{j=0}^{K-c-1} \pi_{cj} = \frac{1}{\bar{\lambda}} \sum_{j=0}^{K-c-1} (K - c - j)\alpha p_{cj},$$

i.e. as one could expect, the probability $B$ that a customer must wait for service is equal to the fraction of blocked primary calls.

Calculation of the distribution and higher moments of the waiting time can be done with the help of the approach developed in Chapter 1 for the main single-server model.

Assume that at time $t = 0$ there are $j$ customers in orbit and $i$ customers in service, $j = 1, 2, ..., K - c$; $i = 0, ..., c$. Mark one of the customers in orbit and denote by $f_{ij}(t)$ the probability that by the time $t$ this customer has not been served, i.e. the residual waiting time of the tagged customer, $\tau_{ij}$, is greater than $t$

$$f_{ij}(t) = \mathrm{P}\left(\tau_{ij} > t\right).$$

In terms of these probabilities the complementary waiting time distribution function $\overline{F}(t)$ can be expressed as follows:

$$\overline{F}(t) = \sum_{j=0}^{K-c-1} \pi_{cj} \cdot f_{c,j+1}(t).$$

Using (4.44) we can rewrite this formula in terms of the outside observer's distribution:

$$\overline{F}(t) = \frac{1}{\lambda} \sum_{j=0}^{K-c-1} (K - c - j)\alpha p_{cj} \cdot f_{c,j+1}(t). \tag{4.45}$$

To get equations for probabilities $f_{ij}(t)$ we introduce an auxiliary Markov process $\zeta(t)$ with state space $\{0,\ldots,c\} \times \{1,\ldots,K-c\} \cup \{s\}$. State $(i,j)$ can be thought of as the presence in the system of $i$ customers in service and $j$ customers in orbit, including the tagged one. The special state $s$ is an absorbing state, and transition into this state means that the tagged customer starts to be served. Thus the residual waiting time of the tagged customer, $\tau_{ij}$, is simply the time until absorption in the state $s$:

$$f_{ij}(t) = \mathrm{P}(\zeta(t) \neq s \mid \zeta(0) = (i,j)) = 1 - \mathrm{P}(\zeta(t) = s \mid \zeta(0) = (i,j)).$$

It is easy to see that if $0 \leq i \leq c - 1$ then the process $\zeta(t)$ can move from state $(i,j)$ into one of the following states:

- $(i+1,j)$ with rate $(K - i - j)\alpha$;

- $(i+1,j-1)$ with rate $(j - 1)\mu$;

- $s$ with rate $\mu$;

- $(i-1,j)$ with rate $i$.

If $i = c$ then the process $\zeta(t)$ can move from state $(c,j)$ into one of the following states:

- $(c,j+1)$ with rate $(K - c - j)\alpha$;

- $(c-1,j)$ with rate $c$.

Therefore from the Kolmogorov backward equations for the Markov chain $\zeta(t)$ we get:

$$f'_{ij}(t) = -[(K - i - j)\alpha + j\mu + i]f_{ij}(t)$$

$$+ (K - i - j)\alpha f_{i+1,j}(t)$$

$$+ (j - 1)\mu f_{i+1,j-1}(t) + i f_{i-1,j}(t),$$

$$\text{if } 0 \le i \le c - 1, \ 1 \le j \le K - c;$$

$$f'_{cj}(t) = -[(K - c - j)\alpha + c]f_{cj}(t) + (K - c - j)\alpha f_{c,j+1}(t)$$

$$+ c f_{c-1,j}(t), \ \text{if } i = c, \ 1 \le j \le K - c.$$

For Laplace transforms $\varphi_{ij}(s) = \int_0^\infty e^{-st} f_{ij}(t) dt$ these equations become:

$$
\begin{aligned}
-1 + s\varphi_{ij}(s) &= -[(K - i - j)\alpha + j\mu + i]\varphi_{ij}(s) \\
&+ (K - i - j)\alpha\varphi_{i+1,j}(s) \\
&+ (j - 1)\mu\varphi_{i+1,j-1}(s) + i\varphi_{i-1,j}(s), \\
&\text{if } 0 \le i \le c - 1, \ 1 \le j \le K - c; \quad (4.46) \\
-1 + s\varphi_{cj}(s) &= -[(K - c - j)\alpha + c]\varphi_{cj}(s) \\
&+ (K - c - j)\alpha\varphi_{c,j+1}(s) \\
&+ c\varphi_{c-1,j}(s), \\
&\text{if } i = c, \ 1 \le j \le K - c. \quad (4.47)
\end{aligned}
$$

Now we obtain a useful formula for the Laplace transform of the virtual waiting time, which can be thought of as a generalization of Little's formula for the mean waiting time.

With this goal, multiply equation (4.46) by $p_{ij}$ and equation (4.47) by $p_{cj}$, replace terms $[(K - i - j)\alpha + j\mu + i]p_{ij}$ and $[(K - c - j)\alpha + c]p_{cj}$ in the right-hand sides of the resulting equations with the help of the Kolmogorov equations (4.40), (4.41) for the probabilities $p_{ij}$, and sum these equations with respect to $i = 0, \dots, c$. After some algebra we get:

$$\sum_{j=0}^{K-c} (K - c - j)\alpha p_{cj}\varphi_{c,j+1}(s) = N - s\sum_{i=0}^{c}\sum_{j=0}^{K-c} j p_{ij}\varphi_{ij}(s), \quad (4.48)$$

where $N$ is the mean number of customers in orbit.

Introducing the Laplace transform of the virtual waiting time,

$$\mathrm{E}e^{-sW(t)} = 1 - s \int\limits_0^\infty e^{-st}\overline{F}(t)dt,$$

and Laplace transforms of the conditional waiting times $\tau_{ij}$,

$$\mathrm{E}e^{-s\tau_{ij}} = 1 - s\varphi_{ij}(s),$$

and using (4.45) we can rewrite equation (4.48) as follows:

$$\mathrm{E}e^{-sW(t)} = 1 - s\frac{1}{\lambda} \sum_{i=0}^{c} \sum_{j=1}^{K-c} jp_{ij}\mathrm{E}e^{-s\tau_{ij}}. \qquad (4.49)$$

Differentiating (4.49) with respect to $s$ at the point $s = 0$ we get Little's formula for the mean waiting time:

$$\mathrm{E}W(t) = \frac{\mathrm{E}N(t)}{\overline{\lambda}}.$$

Thus the mean waiting time can be calculated with the help of the algorithm for calculation of the stationary distribution of the number in the system which was described above.

Differentiating (4.49) twice with respect to $s$ at the point $s = 0$ we get another interesting relation:

$$\mathrm{E}\left(W(t)\right)^2 = \frac{2}{\overline{\lambda}} \sum_{i=0}^{c} \sum_{j=1}^{K-c} jp_{ij}\mathrm{E}\tau_{ij}. \qquad (4.50)$$

Thus to calculate variance of the waiting time, we need to know only the mean conditional waiting times $m_{ij} \equiv \mathrm{E}\tau_{ij}$ (and the stationary distribution of the number in the system $p_{ij}$).

*Calculation of the mean conditional waiting times*

Putting $s = 0$ in equations (4.46), (4.47) we get the following equations for the mean conditional waiting times $m_{ij} \equiv \mathrm{E}\tau_{ij}$:

$$
\begin{aligned}
-1 &= -((K - i - j)\alpha + j\mu + i)m_{ij} + (K - i - j)\alpha m_{i+1,j} \\
&+ (j - 1)\mu m_{i+1,j-1} + im_{i-1,j}, \\
&\quad \text{if } 0 \le i \le c - 1,\ 1 \le j \le K - c; \\
-1 &= -((K - c - j)\alpha + c)m_{cj} + (K - c - j)\alpha m_{c,j+1} \\
&+ cm_{c-1,j},\ \text{if } i = c,\ 1 \le j \le K - c.
\end{aligned}
$$

Rewrite these equations as follows:

$$m_{i+1,j} = \frac{(K-i-j)\alpha + j\mu + i}{(K-i-j)\alpha} m_{ij} - \frac{i}{(K-i-j)\alpha} m_{i-1,j}$$

$$- \frac{(j-1)\mu}{(K-i-j)\alpha} m_{i+1,j-1} - \frac{1}{(K-i-j)\alpha},$$

$$\text{if } 0 \leq i \leq c-1, \ 1 \leq j \leq K-c; \qquad (4.51)$$

$$m_{c,j+1} = \frac{(K-c-j)\alpha + c}{(K-c-j)\alpha} m_{cj} - \frac{c}{(K-c-j)\alpha} m_{c-1,j}$$

$$- \frac{1}{(K-c-j)\alpha}, \ \text{if } i = c, \ 1 \leq j \leq K-c-1; \quad (4.52)$$

$$-1 = -cm_{c,K-c} + cm_{c-1,K-c}$$

$$(\text{case } i = c, j = K-c). \qquad (4.53)$$

This set of equations can be solved with the help of the following algorithm.

1. Put $j = 1$ in equation (4.51):

$$m_{i+1,1} = \frac{((K-i-1)\alpha + \mu + i)\, m_{i1} - im_{i-1,1} - 1}{(K-i-1)\alpha},$$

$$\text{if } 0 \leq i \leq c-1. \qquad (4.54)$$

From this equation we can recursively express variables $m_{i,1}$, $0 \leq i \leq c$, in terms of variable $m_{0,1}$ by the linear equation:

$$m_{i,1} = u_{i,1} \cdot m_{0,1} + v_{i,1}.$$

Coefficients $u_{i,1}, v_{i,1}$ can be calculated numerically with the help of the following recursive procedure:

$$u_{0,1} = 1; \qquad\qquad v_{0,1} = 0;$$

$$u_{1,1} = \frac{(K-1)\alpha + \mu}{(K-1)\alpha}; \quad v_{1,1} = -\frac{1}{(K-1)\alpha};$$

$$u_{i+1,1} = \frac{((K-i-1)\alpha + \mu + i)u_{i1} - iu_{i-1,1}}{(K-i-1)\alpha},$$

$$\text{if } 1 \leq i \leq c-1;$$

$$v_{i+1,1} = \frac{((K-i-1)\alpha + \mu + i)\, v_{i1} - iv_{i-1,1} - 1}{(K-i-1)\alpha},$$

$$\text{if } 1 \leq i \leq c-1.$$

2. Increase $j$ by 1 and assume that variables $m_{i,k}$, $0 \leq i \leq c$, $1 \leq k \leq j-1$, are already expressed linearly in terms of $m_{0,1}$ as $u_{i,k} \cdot m_{0,1} + v_{i,k}$. Then from equation (4.51) we can recursively express variables $m_{ij}$, $0 \leq i \leq c$, linearly in terms of variables $m_{0,j}$ and $m_{0,1}$:

$$m_{ij} = x_{ij}m_{0j} + y_{ij}m_{01} + z_{ij}.$$

Coefficients $x_{ij}, y_{ij}, z_{ij}$ can be calculated numerically with the help of the following recursive procedure:

$$x_{0,j} = 1, \quad x_{1,j} = \frac{(K-j)\alpha + j\mu}{(K-j)\alpha},$$

$$y_{0,j} = 0, \quad y_{1,j} = -\frac{(j-1)\mu}{(K-j)\alpha}u_{1,j-1},$$

$$z_{0,j} = 0, \quad z_{1,j} = -\frac{(j-1)\mu v_{1,j-1} + 1}{(K-j)\alpha},$$

and for $i = 1, \ldots, c-1$

$$x_{i+1,j} = \frac{((K-i-j)\alpha + i + j\mu)\, x_{ij} - ix_{i-1,j}}{(K-i-j)\alpha},$$

$$y_{i+1,j} = \frac{((K-i-j)\alpha + i + j\mu)\, y_{ij} - iy_{i-1,j}}{(K-i-j)\alpha}$$
$$- \frac{(j-1)\mu u_{i+1,j-1}}{(K-i-j)\alpha},$$

$$z_{i+1,j} = \frac{((K-i-j)\alpha + i + j\mu)\, z_{ij} - iz_{i-1,j}}{(K-i-j)\alpha}$$
$$- \frac{(j-1)\mu v_{i+1,j-1} - 1}{(K-i-j)\alpha}.$$

On the other hand, from equation (4.52) we have:

$$m_{c,j} = u_{c,j} \cdot m_{0,1} + v_{c,j},$$

where

$$u_{c,j} = \frac{((K-c-j+1)\alpha + c)\, u_{c,j-1} - cu_{c-1,j-1}}{(K-c-j+1)\alpha};$$

$$v_{c,j} = \frac{((K-c-j+1)\alpha + c)\, v_{c,j-1} - cv_{c-1,j-1} - 1}{(K-c-j+1)\alpha}.$$

From the two relations for $m_{c,j}$, $m_{c,j} = u_{c,j} \cdot m_{0,1} + v_{c,j}$ and $m_{c,j} = x_{c,j}m_{0j} + y_{c,j}m_{01} + z_{c,j}$, we can express $m_{0,j}$ as a linear function of $m_{0,1}$ :

$$m_{0,j} = \frac{u_{c,j} - y_{c,j}}{x_{c,j}} m_{0,1} + \frac{v_{c,j} - z_{c,j}}{x_{c,j}} \equiv u_{0,j} \cdot m_{0,1} + v_{0,j},$$

which in turn allows us to express all variables $m_{ij}$, $0 \leq i \leq c$, as linear functions of $m_{0,1}$ :

$$m_{i,j} = u_{i,j} \cdot m_{0,1} + v_{i,j}, \ 0 \leq i \leq c,$$

where

$$
\begin{aligned}
u_{0,j} &= \frac{u_{c,j} - y_{c,j}}{x_{c,j}}, \\
v_{0,j} &= \frac{v_{c,j} - z_{c,j}}{x_{c,j}}, \\
u_{ij} &= x_{ij} u_{0j} + y_{ij}, \text{ if } 1 \leq i \leq c, \\
v_{ij} &= x_{ij} v_{0j} + z_{ij}, \text{ if } 1 \leq i \leq c.
\end{aligned}
$$

3. Repeat step 2 until $j$ becomes equal to $K - c$. As a result we express all variables $m_{ij}$, $0 \leq i \leq c, 1 \leq j \leq K - c$, in terms of $m_{01}$ as follows:

$$m_{i,j} = u_{i,j} \cdot m_{0,1} + v_{i,j}, \ 0 \leq i \leq c, 1 \leq j \leq K - c.$$

4. From equation (4.53) we have:

$$c \cdot [u_{c-1,K-c} \cdot m_{0,1} + v_{c-1,K-c}] - c \cdot [u_{c,K-c} \cdot m_{0,1} + v_{c,K-c}] = -1.$$

This allows us to calculate the main variable $m_{0,1}$ :

$$m_{0,1} = \frac{v_{c-1,K-c} - v_{c,K-c} + 1/c}{u_{c,K-c} - u_{c-1,K-c}}.$$

5. Using relation

$$m_{ij} = u_{ij} m_{01} + v_{ij}, \ 0 \leq i \leq c, \ 1 \leq j \leq K - c,$$

we can now calculate all mean conditional waiting times $m_{ij}$.

The following Pascal program realizes this algorithm.

```
Program retrial(Input,Output);
{the mean conditional waiting times
for finite source retrial queue}
Uses Crt;
Var
```

```
i,j,c,K :integer;
a,mu :extended;
m,u,v,x,y,z :array[0..10,0..20] of extended;
Begin
writeln('input the number of servers '); read(c);
writeln('input the number of sources '); read(K);
writeln('input the new calls generation rate');
read(a);
writeln('input the retrial rate '); read(mu);
u[0,1]:=1; v[0,1]:=0;
u[1,1]:=((K-1)*a+mu)/((K-1)*a); v[1,1]:=-1/((K-1)*a);
for i:=1 to c-1 do
begin
u[i+1,1]:=(((K-i-1)*a+i+mu)*u[i,1]
-i*u[i-1,1])/((K-i-1)*a);
v[i+1,1]:=(((K-i-1)*a+i+mu)*v[i,1]
-i*v[i-1,1]-1)/((K-i-1)*a);
end;
for j:=2 to K-c do
begin
x[0,j]:=1; y[0,j]:=0; z[0,j]:=0;
x[1,j]:=((K-j)*a+j*mu)/((K-j)*a);
y[1,j]:=-(j-1)*mu*u[1,j-1]/((K-j)*a);
z[1,j]:=-((j-1)*mu*v[1,j-1]+1)/((K-j)*a);
for i:=1 to c-1 do
begin
x[i+1,j]:=(((K-i-j)*a+i+j*mu)*x[i,j]
-i*x[i-1,j])/((K-i-j)*a);
y[i+1,j]:=(((K-i-j)*a+i+j*mu)*y[i,j]
-i*y[i-1,j]-(j-1)*mu*u[i+1,j-1])/((K-i-j)*a);
z[i+1,j]:=(((K-i-j)*a+i+j*mu)*z[i,j]-
i*z[i-1,j]-(j-1)*mu*v[i+1,j-1]-1)/((K-i-j)*a);
end;
u[c,j]:=(((K-c-j+1)*a+c)*u[c,j-1]
-c*u[c-1,j-1])/((K-c-j+1)*a);
v[c,j]:=(((K-c-j+1)*a+c)*v[c,j-1]
-c*v[c-1,j-1]-1)/((K-c-j+1)*a);
u[0,j]:=(u[c,j]-y[c,j])/x[c,j];
v[0,j]:=(v[c,j]-z[c,j])/x[c,j];
for i:=1 to c do
begin
```

```
u[i,j]:=x[i,j]*u[0,j]+y[i,j];
v[i,j]:=x[i,j]*v[0,j]+z[i,j];
end;
end;
m[0,1]:=(v[c-1,K-c]-v[c,K-c]+1/c)
/(u[c,K-c]-u[c-1,K-c]);
for i:=0 to c do
for j:=1 to K-c do
m[i,j]:=u[i,j]*m[0,1]+v[i,j];
End.
```

# CHAPTER 5

# Bibliographical remarks

## 5.1 The main single-server model

The $M/M/1$ retrial queue in the steady state was studied long ago by Cohen (1957) (in fact he considered the more general $M/M/c$ retrial queue with impatient customers). The first results on $M/G/1$ retrial queues are due to Keilson, Cozzolino and Young (1968) who used the method of supplementary variable to investigate the joint distribution of the channel state and the number of customers in orbit in the steady state. Later in the case of exponential service time, Jonin and Sedol (1970a, 1970b, 1970c) independently obtained explicit formulas for $p_{0n}$, $p_{1n}$ (but $p_{00}$ was given in a form of some series only) as well as expressions for $p_1$ and $EN(t)$. In the case of Erlangian service times, Jonin (1971), Jonin and Brezgunova (1972) gave expressions for $p_1$ and $EN(t)$. Aleksandrov (1974) and Falin (1975) considered the case of arbitrarily distributed service times and obtained the results given in section 1.2, but by methods different from that used in Keilson *et al.* (1968).

The property of stochastic decomposition for the $M/G/1$ retrial queue (and many of its variants) was observed by many authors. The idea of using this property for deriving an explicit formula for moments of the number of customers in orbit was suggested by Artalejo and Falin (1994b). In recent papers Yang *et al.* (1994b) and Yang and Li (1995a) established the property for single-server retrial queues with general distribution of inter-repetition intervals and for discrete time retrial queues.

The Markov chain embedded at service completion times was considered by Falin (1975). The influence of retrial rate on ergodicity properties in the case $\rho = 1$ was investigated by Falin (1987a). Liang and Kulkarni (1993b) investigated ergodicity of the single server retrial queue with nonexponential interarrival times and retrial times and gave an example when the condition $\rho < 1$ is not sufficient for ergodicity.

The limit theorems stated in section 1.4 were proved by Falin (1979b, 1983b).

Stochastic inequalities for the distribution of the number of customers in the $M/G/1$ retrial queue and related results were obtained by Falin (1986b), Khalil and Falin (1994). Liang and Kulkarni (1993a) investigated monotonicity properties of a single-server retrial queue with finite capacity, arbitrary arrival process, arbitrary service time distribution and phase-type retrial times.

The busy period and the functioning of the system in a nonstationary regime were studied by Falin (1979b). But there was an error in the finiteness condition of the busy period in the case $\rho = 1$ (it appeared in the corresponding statements of the papers Falin (1979a), (1981b) too). The structure of the busy period was discussed qualitatively in the paper of Aleksandrov (1974), but some statements were wrong. Choo and Conolly (1979) studied the busy period for exponentially distributed service times. The explicit formulas for the mean and the variance of the busy period length were obtained. In a paper of Falin (1990), a direct method of calculation of the Laplace transform of the busy period was suggested. Inequalities for the mean characteristics of the busy period were obtained by Falin (1986b). Variables $A^{(k)}$ and $B^{(k)}$ were studied by Artalejo and Falin (1996a).

The method for obtaining the distribution of the virtual waiting time was proposed in a paper of Falin (1976), and an expression for $EW(t)$ was given. Falin (1977) applied this method to finding $E(W(t))^2$, but the expression obtained was very complicated. It was simplified by Falin (1986b). Asymptotic properties of the virtual waiting time distribution under heavy traffic were studied in Falin (1980b). Later Choo and Conolly (1979) suggested another method of calculating the distribution of $W(t)$, however this method was wrong (Conolly (1982), Kulkarni (1982)). Furthermore the method given by Kulkarni (1982) coincides with the method of Falin (1976) completely. An important advance on this problem was made by Falin (1988a), Falin and Fricker (1991b); in section 1.8 we followed these papers. The description of the waiting time process in retrial queues with the help of the number of retrials $R(t)$ was suggested and studied in full detail for exponential service times in Falin (1986a). The case of arbitrary service times was studied by Falin (1988a), Falin and Fricker (1991b). The phase transitions associated with the virtual waiting time were discovered by Falin (1988a, 1989a). Grishechkin (1990, 1991, 1992) obtained interesting results concerning the waiting time via the theory of branching processes.

The departure process was studied by Falin (1978, 1979c). Later the same formulas for the moments of the random variables $T_i$ were independently obtained by Choo and Conolly (1979). The quasi-input process for the $M/G/1$ retrial queue was considered by Falin (1990).

The problem of estimation of the retrial rate was analysed by Hoffman and Harris (1986) and by Falin (1995a). The nonstandard Markovian description of retrial queues which was introduced by Falin (1995a) is closely connected with the mean number of unsuccessful retrials made during a service period; this performance characteristic of single channel retrial queues was considered by Kulkarni (1983a).

Methods of numerical calculation of the steady state distribution were developed by de Kok (1982), (1984).

There are several papers on methods of approximate analysis of single-server retrial queues. Falin (1984f) suggested fluid flow and diffusion approximations for the single-server retrial queue. The mathematical theory of diffusion approximations for retrial queues was developed by Lukashuk (1990), Falin (1991c) and Anisimov and Atadzhanov (1991a, 1991b, 1991c). Information theoretic approximations for retrial queues were suggested by Artalejo (1992). The approach from this paper was further developed by Falin, Martin and Artalejo (1994).

## 5.2 The main multichannel model

Necessary conditions for ergodicity of retrial queues can be obtained quite easily from the fact that the mean number of busy channels (which in the steady state equals the intensity of carried traffic and can usually be expressed in terms of the system parameters and performance characteristics) must be less than the total number of channels which are available to calls. Often the conditions obtained in this manner are sufficient for ergodicity, but a proof of this is much more difficult. A direct approach which is based on explicit (in some sense) solution of the Kolmogorov equations for the stationary distribution leads to very cumbersome arguments and does not seem to be useful. However, it was applied by Cohen (1957), section 2.3, case (ii), who obtained the solution with the help of a truncated system. Shortly after this paper the first criteria based on mean drifts were published, so that at present the theory of Lyapunov functions gives the most natural way for analysis of ergodicity. Deul (1980) investigated the ergodicity of the main multichannel model with the help of an embedded Markov chain which differs from the standard embedded chain for a continuous time Markov process. Correspondingly, the Lyapunov function he used is very complicated and it is difficult to apply this approach to more complex retrial queues. Falin (1984a) used the standard embedded chain (considering the initial continuous time process only at times when it changes its states) and suggested as Lyapunov function a linear combination of coordinates of a vector Markov process, which describes the functioning of a system. This gives a simple unified method of investigating various retrial queues. Another class of Lyapunov functions was introduced by Hanschke (1987). Later on, Falin (1987a) applied the theory of Lyapunov functions to investigate nonergodicity and transience.

Considerable efforts have been made to find a closed form solution of Kolmogorov equations for the steady state distribution of the number of busy channels and the queue length. For the model with $c = 2$ servers, Jonin and Sedol (1970a), (1970b), (1970c) found a solution. Falin (1984b) and Hanschke (1987) simplified this solution and obtained further results. In the general case, Cohen (1957) obtained the steady state distribution of the number of busy channels and the queue length for the truncated model in terms of Laguerre polynomials and some functions which can be found as a solution of a linear set of equations (with order equal to the level of truncation). Hanschke (1978) gave a solution in terms of generalized continued fractions. A similar approach was developed by Pearce (1989). In Pearce (1987), for a truncated $M/M/c$ type model which slightly generalizes the model with nonpersistent subscribers, the steady state joint distribution of the channel state and the queue length is obtained in terms of sigma polynomials (when $H_2 < 1$) and Stirling numbers of the first kind (when $H_2 = 1$) and some variables $c_0, \ldots, c_{M-1}$ which are given by a recursive formula. However, it should be noted that no satisfactory analytical solution of the Kolmogorov equations for the steady state distribution of the number of busy channels and the queue length was obtained; perhaps such a solution does not exist at all. Useful 'semi-explicit' formulas (which reduce calculation of performance characteristics to calculation of the moments of the number of busy channels) were obtained by Falin (1984d).

The use of a truncated model for numerical solution of the Kolmogorov equations for the original model with an unlimited number of sources was suggested by Wilkinson (1956). He also noted that stationary probabilities for the truncated model can be calculated recursively from the Kolmogorov equations, but did not give a detailed algorithm. The first algorithm based on this idea was published by Jonin and Sedol (1970a, 1970b, 1970c). In the book Jonin and Sedol (1970c), they published tables of performance characteristics of the multiserver retrial queue with nonpersistent subscribers (Cohen's model). An approximation with the help of the model where the retrial rate equals infinity when the number of customers in orbit exceeds some level was suggested by Falin (1983c). An approximation with the help of the model where the retrial rate stays constant when the number of customers in orbit exceeds some level was suggested by Greenberg (1986). Detailed analysis of this approximating model is carried out by Neuts and Rao (1990).

Attempts to calculate the error of the approximation of retrial queue by the corresponding truncated model, or equivalently, evaluate the truncation limit $M$ which guarantees the preassigned error, have been made for a long time. Wilkinson (1956) suggested that the truncation limit $M$ must be chosen in such a way that the probability of ultimate loss $p_{cM}^{(M)}$ in the truncated model equals the probability $1 - H_1$ that a calling subscriber makes no further attempt after failure of the primary attempt. This recommendations is unnatural since, in the original model, calling subscribers are assumed to be absolutely persistent. Cohen (1957) investigated the error numerically and derived some properties of the dependence of the error upon the parameters of the model. However, his investigation is limited and does not give any recommendation concerning the choice of the truncation limit. For a long time in engineering practice, in order to calculate performance characteristics with given error $\varepsilon$, it has been recommended to take the value $M$ such that $p_{cM}^{(M)} < \varepsilon$. Falin (1983d) showed that in some domains of parameters (heavy traffic or low retrial rate) this method is not correct. Some estimates of the error were obtained by Stepanov (1983a). Estimates given in section 2.6 were obtained by Falin (1985c, 1987c, 1989b).

The method of asymptotic analysis of multiserver retrial queues under high retrial intensity was suggested by Falin (1984d). The case of the low retrial rate was considered by Cohen (1957). It is shown that the stationary distribution of the number of busy channels can be approximated by the Erlang loss distribution with increased arrival rate. Falin (1985c), Falin and Sukharev (1988e) obtained the asymptotics of the shifted and scaled queue length with the help of the theory of singular perturbed differential equations. The functional limit theorem about convergence of the shifted and scaled queue legth process to the Ornstein–Uhlenbeck process is proved in Falin (1991c).

An approximation of retrial queues with the help of the corresponding loss model with properly increased offered traffic was suggested by Cohen (1957). Riordan (1962) suggested the interpolation between extreme cases $\mu \to 0$ and $\mu \to \infty$. These approximations are first-order approximations (since they take into account only rates of flows). Later on Berry (1987) and Reeser (1989) suggested the use of the classical Equivalent Random Type (ERT) method (as a second order approximation) to take into account peakedness of the flow of repeated calls and its dependence on the number of busy channels. A similar approach is developed in several papers by Pourbabai (1986,1987, 1988a, 1988b, 1989, 1990).

## 5.3 Advanced single-server models

Batch arrival retrial queues were considered for the first time by Falin (1976), who used the embedded Markov chain technique to derive the joint distribution of the channel state and the queue length. Another approach to the problem was proposed by Yang and Templeton (1987). A more detailed analysis of the model was given by Falin (1981b), who studied the nonstationary regime and the busy period. Multiclass batch arrival retrial queues were considered by Kulkarni (1986) and Falin (1988c).

The model with priority customers that can be queued was considered by Falin (1981a), who obtained (in terms of generating functions) the joint distribution of the queue lengths of priority and nonpriority calls in the steady state. A variant of the model with a finite number of sources of primary priority calls was also considered in this paper. Later Choi and Park (1990) independently obtained the same results. More extensive analysis of this model was carried out by Falin, Artalejo and Martin (1993) (existence of the stationary regime, embedded Markov chain, stochastic decomposition, limit theorems under high and low rates of retrials and heavy traffic analysis) and by Choi, Han and Falin (1993b) (waiting time). A model with a limited priority queue was considered by Choi *et al.* (1995). The joint distribution of queue sizes in the steady state and some performance measures were obtained. This model is of interest for performance analysis of a mobile cellular radio communication system. A single server priority retrial queue with preemptive resume priority and server vacations is studied by Langaris and Moutzoukis (1995) and Langaris and Moutzoukis (1996). It is assumed that the customers arrive in batches according to a Poisson process; the numbers of customers of each type in an arriving batch may be dependent random variables. The number in the system is studied in transient and in stationary regimes. Besides, the waiting time process is considered in the steady state.

For the first time a model taking into account nonpersistence of calling subscribers was considered by Cohen (1957) for the $M/M/c$ type retrial queue. It is a generalization of the ordinary model with impatient demands and it supposes that sources of repeated calls leave the system with some intensity $\delta$. As a matter of fact, this supposition is equivalent to dealing with the persistence function $H_1 = H_2 = \mu/(\mu + \delta)$. The $M/M/1$ model with persistence function $(H_1, H_2, H_2, ...)$ was considered by Falin (1980a) who obtained a solution in terms of Kummer confluent functions. Lubacz and Roberts (1984) obtained expressions for the blocking probability and mean queue length in the case of nonpersistent subscribers with the persistence function $(H_1, 1, 1, ...)$. In the cases $H_1 = H_2$ and $H_1 = 1$ this system was considered by Jonin and Sedol (1970b) too. A more sophisticated $M/M/1$ type retrial queue with nonpersistent subscribers is studied by Hanschke (1985a). Yang *et al.* (1990a) considered an $M/G/1$ model with persistence function $H_1 = H_2$ and obtained expressions for the moments of the queue length in terms of the server utilization. They also suggested a numerical algorithm for calculation of the server utilization, but did not prove its convergence rigorously. For the same model Gilbert (1988) and Greenberg (1989) derived bounds for performance measures. Results presented in section 3.3 generalize results of Yang *et al.*(1990a), Gilbert (1988) and Greenberg (1989) to the case of a more general persistence function $H_1, H_2$. Keilson and Servi (1993) considered a wide class of Markovian single-server retrial queues (including the model with nonpersistent subscribers) in the more general context of a matrix $M/M/\infty$ system.

Kornyshev (1974a, 1977a) suggested models of two-way communication (in fact, in the models considered in these papers some additional parameters are also incorporated in order to describe the subscribers' behavior in fuller detail). For exponential service times, Kornyshev obtained explicit formulas for the main performance characteristics in the steady state. In the case of general service time distribution, the model was investigated by Falin (1979a, 1986b).

A retrial model in which outgoing calls may be dispatched by two subscribers (coupled switching) was considered by Kornyshev (1968). In the case of exponentially distributed service times, he obtained explicit formulas for mean characteristics in the steady state. Falin (1979a) studied the problem in the case of a general distribution of service times and derived the joint distribution of channel state and queue length in the steady state, busy period and nonstationary distribution. In Falin (1981a) the model of coupled switching was generalized to the case when a source of outgoing calls has an arbitrary finite or infinite size.

The multiclass repeated orders queueing system was suggested by Kornyshev (1980b), who obtained in the case of exponential service times a set of linear equations for the mean queue lengths in the steady state. The case of equal retrial rates (which is relatively simple) was investigated by Hanschke (1985b). For the case of general distributions of service times with two types of calls, explicit formulas for the mean queue length in the steady state were obtained by Kulkarni (1983a). For an arbitrary number of call classes, Falin (1983a) obtained the set of linear equations for the mean queue lengths in the steady state. Later Falin (1988c) showed that the joint queue length process is moment closed. It should be noted that the models considered by Kulkarni (1983a) and Falin (1988c) in fact deal with batch arrivals. Interesting game theoretic problems for multiclass retrial queues were stated and solved by Kulkarni (1983a, 1983b). An extensive analysis of multiclass batch arrival retrial queues was carried out by Grishechkin (1992) with the help of the theory of branching processes with immigration.

A model which takes into account the possibility of network blocking was considered by Jonin, Sedol and Kibild (1975) under Erlangian service time distribution. Formulas for the blocking probability and the mean queue length were obtained. Kornyshev (1977b) considered two types of blocking, in the network itself and at the destination of a call, and correspondingly two orbits. In this paper it is assumed that subscribers do not move between orbits.

Sukharev (1984) considered the case in which subscribers, after each unsuccessful retrial, move to the orbit corresponding to the type of blocking experienced on that retrial.

Jonin (1980, 1982b) considered models with collisions in the context of double connections. For the model of Jonin (1982b) with general service time distribution, Falin and Sukharev (1985b) found the necessary and sufficient condition of existence of the stationary regime and the steady state distribution of the channel state and the number of sources of repeated calls. For the purpose of describing the functioning of local area computer networks with CSMA/CD protocol, Ruskov *et al.* (1984) considered a more detailed model where the service consists of two phases and collisions can occur only at the first stage. They obtained a set of linear equations for the stationary probabilities and gave its numerical solution. The analytical solution to this and other similar models was later given by Khomichkov (1986, 1987b, 1988c) using the methods due to Falin and Sukharev (1985b). Using time discretization Dimitrov and Ruskov (1985) studied the nonstationary regime of systems with collisions. A retrial queue with collision arising from unslotted CSMA/CD protocol under the retrial control policy in which the retrial rate is inversely proportional to the number of customers in orbit was investigated by Choi, Shin and Ahn (1992a). Asymptotic analysis of retrial queues with collisions was done by Anisimov and Atadzhanov (1991b).

A discrete time single-channel retrial queue with a geometric input flow, geometric retrial times and general service times is considered by Yang and Li (1995). The joint distribution of the channel state and the queue length in the steady state is obtained in terms of infinite products. Based on this solution it is shown that the stochastic decomposition holds for the discrete time retrial queue.

A single-server retrial queue with finite number of sources of primary calls was considered by de Kok (1984). With the help of the theory of regenerative processes, this author obtained a recursive scheme for computing the joint distribution of the channel state and the queue length in the steady state. In fact, the finite source model was treated as a special case of a more general model, where the rate of arrival of a new primary call is a general function $\lambda_{ij}$ of the number of customers in service, $i$, and the number of customers in orbit, $j$. Ohmura (1985) obtained recursive formulas for computing the joint distribution of the channel state and the queue length in the steady state with the help of discrete transformations. He also suggested application of this model to analysis of the waiting time for access to magnetic disc memory. The number in the system and the waiting time process also were considered by Dragieva (1988, 1994). However, the waiting time process was investigated only in the cases of exponential and deterministic service times and only the mean characteristics were obtained. It should be noted that some results seem to be incorrect. Artalejo and Gómez-Corral (1995a) used information theoretic approximation to estimate the stationary distribution of the number of customers in the system.

A model where only one customer from orbit can retry for service was suggested by Fayolle (1986). Under Markovian assumptions he obtained the necessary and sufficient condition for ergodicity. The analysis of the waiting time process is a much more difficult problem, as it usually is. Fayolle (1986) gave a solution of this problem in terms of a meromorphic function, the poles and residues of which are easily computed recursively. He also investigated the tail behavior of the waiting time distribution in the steady state. Sukharev (1987) considered the case of a general distribution of interretrial times and obtained the necessary and sufficient condition for ergodicity, the joint distribution of the channel state and the number of customers in the orbit in the steady state, and also investigated the waiting time in terms of the number of retrials. The same model as in Sukharev (1987) was later independently investigated by Choi, Park and Pearce (1993c) in more detail. Falin (1988b) generalized the model of Fayolle (1986) in another direction; he assumed that the channel holding times are generally distributed whereas retrial times have an exponential distribution. The same model was later independently investigated by Farahmand (1990). A variant of this model with a fixed number of so-called recurrent customers, who after service immediately return to orbit, was studied in papers of Farahmand (1996a, 1996b) (the case of ordinary retrials is also considered there).

A specific finite buffer single-server retrial queue with a single control device was considered by Bocharov and Albores (1983) in the context of a queueing network with reservicing after blocking. These authors obtained the necessary and sufficient condition for ergodicity and an algorithm for calculation of the steady state distribution.

In all models considered so far, retrials are due to limited capacity of the service zone. In some applications queueing models with infinite capacity of the service zone and retrials due to balking or impatience of arriving customers arise. Boyer, Dupuis and Khelladi (1988), Fayolle and Brune (1988) considered a single-server system with unlimited buffer. All arriving customers join the buffer and then are served in accordance with some discipline (such as FIFO). However, the customers are impatient, i.e. their waiting times are bounded. If the waiting time expires then the awaiting customer leaves the buffer. After this with some probability the customer leaves the system forever and with complementary probability tries his luck again after a random delay. Under the usual assumption that all random variables involved in the model description are independent and exponentially distributed, the functioning of the system can be described by a bivariate Markov process $(C(t), N(t))$, where $C(t)$ is the number of customers in the buffer (both awaiting and served) and $N(t)$ is the number of sources of repeated calls. Since the state space of the process $(C(t), N(t))$ is the two dimensional integer lattice $Z_+^2$, it is more difficult for mathematical analysis than the corresponding process in ordinary retrial queues. Fayolle and Brune (1988) derived ergodicity conditions for the process using Lyapunov functions. Calculation of the steady state distribution of the number of customers in the system is reduced to the solution of a homogeneous integral equation of the first kind. Boyer, Dupuis and Khelladi (1988) considered a truncated model and investigated the model numerically.

A variant of the $M/G/1$ retrial queue was considered by Neuts and Ramalhoto (1984). In contrast to the main model, it is assumed that an arriving primary customer first joins the retrial queue and then retries for service at random intervals. The authors obtained the steady state distribution of the channel state and the number of customers in orbit, and discussed some properties of the distribution including the problem of its numerical calculation. In the case of exponential service times, a corresponding model with nonpersistent customers was considered.

A single-server retrial queue with unreliable server was considered by Kulkarni (1990), Yang and Li (1994a), Aissani (1993), Artalejo (1994a) and Aissani (1995) in a more general context of randomly varying service rate.

An $GI/M/1$ retrial queue with intervals between arrivals of primary calls distributed according to either hyper-exponential or Erlang distribution of the second order was studied by Khomichkov (1987a). This author obtained the joint distribution of the channel state and the queue length in the steady state in terms of hypergeometric functions. The case of a Cox input process of second order was considered in Khomichkov (1988a). A multiserver model with a general Coxian input flow was numerically studied by Yang, Posner and Templeton (1992).

Retrial queues with 'negative arrivals' (which delete customers from the orbit) were introduced by Gómez-Corral (1996) and Artalejo and Gómez-Corral (1996b). A necessary and sufficient condition for ergodicity and the steady state distribution of the channel state and the number in orbit were obtained.

## 5.4 Advanced multiserver models

The repeated orders queueing system with priority calls was introduced by Falin (1984a) and Lederman (1985) (as a model of subscriber lines modules). It should be noted that in the model of Lederman (1985) priority calls may leave the priority queue (because of impatience) and then reattempt or leave the system forever. Falin (1984a) investigated ergodicity of the model. Lederman (1985) using iterative methods calculated numerically corresponding stationary probabilities and the main performance characteristics. More detailed analysis (explicit formulas, limit theorems under high and low rate of retrials, heavy traffic, an approximation, a numerical method, generalization to nonpersistent subscribers) was done by Khalil, Falin and Yang (1992).

A multiserver batch arrival retrial queue was introduced by Inamori et al. (1985) as a model of the NTT public facsimile communication network. With the help of an approximate algorithm, performance characteristics are calculated numerically.

Nonpersistence of subscribers was taken into account by Cohen (1957). On the analogy of the ordinary model with impatient customers, it was assumed there that customers from the orbit leave the system forever with some rate. Wilkinson and Radnik (1968) and Bretschneider (1970) take into account this aspect by means of a persistence function. This approach to the impatience of subcribers is commonly accepted now. Later, Zelinskiy and Kornyshev (1978) showed that Cohen's model is equivalent to the model with persistence function of the form $H_1 = H_2$ with a certain correlation of parameters. The influence of the persistence function on performance measures of retrial queues was studied numerically by Deul (1982).

A multiserver retrial queue with a buffer was considered by

Hashida and Kawashima (1979). A numerical algorithm and an approximation with the help of the corresponding loss model similar to those used for the main model were described. A finite source retrial queue with a buffer was considered by Kornyshev (1974b).

Artalejo (1995b), Falin and Artalejo (1995b) considered a multiserver system with $c$ servers and unlimited waiting room. This system serves a Poisson flow of (primary) customers with rate $\lambda$. An arriving customer who finds $i$ other customers in the system (including customers in service) with probability $\alpha_i$ joins the queue and then is served in accordance with some discipline (such as FIFO), and with probability $1 - \alpha_i$ leaves this main service system and joins the orbit. It should be noted that if $i \leq c - 1$ (and $\alpha_i = 1$ for $i \leq c - 1$) then an arriving customer starts to be served immediately. It is assumed that service times are exponentially distributed with the mean $1/\nu$. Each orbiting customer stays there for an exponentially distributed period of time and then arrives to the main service systems. Its further behaviour is identical to that of primary calls, i.e. with probability $\alpha_i$ (which depends on the number of customers in the main queue), the arriving orbiting customer joins the queue and with probability $1 - \alpha_i$ leaves the service system and joins the orbit again. As usual input flow of primary calls, service times, retrial times and decisions whether or not join the queue are assumed to be mutually independent. Similar to the model with retrials due to impatience, the functioning of the system can be described by means of a two-dimensional Markov process $(C(t), N(t))$, where $C(t)$ is the number of customers in the main service system and $N(t)$ is the number of customers in orbit at time $t$. Artalejo (1995b) using the Returning Customers See Time Averages (RTA) approximation studied numerically performance characteristics. This analysis was extended by Falin and Artalejo (1995b) and a limit theorem about convergence of the scaled and shifted queue length process to the Ornstein–Uhlenbeck process was proved.

# References

Afanas'eva, L.G. (1991) On ergodicity condition for queueing systems with repeated calls. In: *Problems of Stability of Stochastic Models. Proceedings of the Seminar of All-Union Institute of System Research*, Moscow, 3–8 (in Russian).

Aissani, A. (1993) Unreliable queueing with repeated orders. *Microelectron. Reliab.*, **33**, No.14, 2093–2106.

Aissani, A. (1995) A retrial queue with redundancy and unreliable server. *Queueing Systems*, **17**, 431–449.

Akyildiz, I.F. (1989) Exact analysis of queueing networks with rejection blocking. *Queueing Networks with Blocking*, Elsevier Science Publishers B.V. (North-Holland), Amsterdam, 19–29.

Aleksandrov, A.M. (1974) A queueing system with repeated orders. *Engineering Cybernetics*, **12**, No.3, 1–4.

Anisimov, V.V. and Atadzhanov Kh.L. (1991a) Diffusion approximation of systems with repeated calls. *Theory of Probability and Mathematical Statistics*, No.44.

Anisimov, V.V. and Atadzhanov, Kh.L. (1991b) Asymptotic analysis of queues in systems with repeated calls and duplex connections. *Preprint, 91-28. Akad. Nauk Ukrain. SSR, Inst. Kibernet.*, Kiev (in Russian).

Anisimov, V.V. and Atadzhanov, Kh.L. (1991c) Asymptotic analysis of highly reliable systems with repeated calls. *Issled. Operatsii i ASU.*, No.37, 24–31 (in Russian).

Arhipov, I.M. (1978) Algorithm of simulation of a circuit switching network with repeated calls. In: *Collection of Algorithms and Programs*, Moscow, No. 6 (in Russian).

Arhipov, I.M. and Shkolny E.I. (1979) On calculation of circuit switching networks with repeated calls. In: *Methods and Structures of Teletraffic Systems*, Nauka, Moscow, 8–14 (in Russian).

Arhipov, I.M. (1981) On influence of repeated calls on effectiveness of transits between three nodes. In: *Information Networks and Automatic Switching. Proceedings of 4th All-Union Conference on Information Networks*, Nauka, Moscow, 78–79 (in Russian).

Artalejo, J.R. (1991) Convolutive methods for queueing systems with repeated attempts. *XV Jornadas Luso-Espanholas de Matematica.*

Vol.IV. Universidade de Evora, Departamento de Matemática, Evora, 67–72 (in Spanish).

Artalejo, J.R. (1992) Information theoretic approximations for retrial queueing systems. *Trans. 11th Prague Conf. Inf. Theory, Statist. Decis. Funct., Random Processes, Prague. Aug. 27-31, 1990*, (eds S.Kubik and J.A.Visek), Kluwer Academic Publishers, Dordrecht, 263–270.

Artalejo, J.R. (1993) Explicit formulae for the characteristics of the $M/H_2/1$ retrial queue.*J. Oper. Res. Soc.*, **44**, No.3, 309–313.

Artalejo, J.R. (1994a) New results in retrial queueing systems with breakdown of the servers. *Statistica Neerlandica*, **48**, No.1, 23–36.

Artalejo, J.R. and Falin G.I. (1994b) Stochastic decomposition for retrial queues. *TOP*, vol.2, No.2, 329–342.

Artalejo, J.R. and Martin, M. (1994c) A maximum entropy analysis of the $M/G/1$ queue with constant repeated attempts. *Selected Topics on Stochastic Modelling*, (eds R. Gutierrez and M.J. Valderrama), World Scientific, Singapore, 181–190.

Artalejo, J.R. and Gómez-Corral, A. (1995a) Information theoretic analysis for queueing systems with quasi-random input. *Mathl. Comput. Modelling*, **22**, No.3, 65–76.

Artalejo, J.R. (1995b) A queueing system with returning customers and waiting line. *Operations Research Letters*, **17**, 191–199.

Artalejo, J.R., Falin, G.I. (1996a) On the orbit characteristics of the $M/G/1$ retrial queue. *Naval Research Logistics*, **43**, No.8, 1147–1161.

Artalejo, J.R., Gómez-Corral, A. (1996b) Stochastic analysis of the departure and quasi-input processes in a versatile single-server queue. *J. Appl. Math. Stochastic Anal.*, **9**, No.2, 171–183.

Artalejo, J.R. (1996c) Stationary analysis of the characteristics of the $M/M/2$ queue with constant repeated attempts. *Opsearch*, **33**, No.2, 83–95.

Arutunyan, A.V. (1977) Investigation of the ideal symmetric systems with repeated calls. In: *Teletraffic Theory and Information Networks*, Nauka, Moscow, 60–67 (in Russian).

Aubakirov, T.U. (1985a) On a queueing system with limited queue and repeated calls. Paper #4701-85, All-Union Institute for Scientific and Technical Information, Moscow (in Russian).

Aubakirov, T.U. (1985b) Error estimation in approximation of a countable Markov chain associated with a general model of subscribers' behaviour in fully available systems of automatic switching. *Paper #4702-85, All-Union Institute for Scientific and Technical Information*, Moscow (in Russian).

Aubakirov, T.U. (1985c) Error estimation in approximation of a countable Markov chain associated with a model of a priority service in automatic switching systems. Paper #7821-B, All-Union Institute for

Scientific and Technical Information, Moscow (in Russian).

Aubakirov, T.U. (1986) On a computational procedure for calculating of a multichannel queue with repeated calls. In: *Numerical Methods in Mathematical Physics*, Moscow, 96–97 (in Russian).

Aubakirov, T.U. (1987) Monotonicity properties and stability of population processes and their use for estimation of error of Wilkinson method. *Bulletin of Kazakh Academy of Sciences*, **10**, 77–80 (in Russian).

Aubakirov, T.U. (1990) Approximations of a priority queueing system with repeated calls by system of a simpler structure. In: *Structural Properties of Algebraic Systems. Collection of Scientific Works*, Karaganda, 101–113 (in Russian).

Berry, L.T.M. (1987) A repeated calls model encompassing non-Poisson traffic streams in alternative routing networks. *Australian Telecommunication Research*, **21**, No.1, 21–28.

Bocharov, P.P. and Albores, F.J. (1983) On two-node exponential queueing network with internal losses or blocking. *Problems of Control and Information Theory*, **12**, No.4, 243–254.

Boyer, P., Dupuis, A. and Khelladi, A. (1988) A simple model for repeated calls due to time-outs. *Teletraffic Science for New Cost-Effective Systems, Networks and Services. ITC-12. Proceedings of the 12 International Teletraffic Congress. Torino, Italy, June 1–8, 1988*, Pt. 1. (ed M.Bonatti), Elsevier Science Publishers B.V., Amsterdam, 356–363.

Bretschneider, G. (1970) Repeated calls with limited repetition probability. A paper presented at the 6th International Teletraffic Congress, Munich.

Choi, B.D. and Park, K.K. (1990) The $M/G/1$ retrial queue with Bernoulli schedule. *Queueing Systems*, **7**, No.2, 219–227.

Choi, D.D., Shin, Y.W. and Ahn, W.C. (1992a) Retrial queues with collision arising from unslotted CSMA/CD protocol. *Queueing Systems*, **11**, No.4, 335–356.

Choi, B.D. and Kulkarni, V.G. (1992b) Feedback retrial queueing systems. *Queueing and related models. Oxford Statist. Sci. Ser., 9*, Oxford Univ. Press, New York, 93–105.

Choi, B.D., Rhee, K.H. and Park, K.K. (1993a) The $M/G/1$ retrial queue with retrial rate control policy. *Probab. Eng. and Inf. Sci.*, **7**, No.1, 29–46.

Choi, B.D., Han, D.H. and Falin, G. (1993b) On the virtual waiting time for an $M/G/1$ retrial queue with two types of calls. *J.Appl. Math. Stochastic Anal.*, **6**, No.1, 11–23.

Choi, B.D., Park, K.K. and Pearce, C.E.M. (1993c) An $M/M/1$ retrial queue with control policy and general retrial times. *Queueing Systems*, **14**, 275–292.

Choi, B.D., Choi, K.B. and Lee, Y.W. (1995) $M/G/1$ retrial queueing systems with two types of calls and finite capacity. *Queueing Systems*, **19**, 215–229.

Choo, Q.H. (1978) The Interaction of Theory and Simulation in Queueing Analysis. Ph.D. Thesis, Chelsea College, University of London.

Choo, Q.H. and Conolly, B. (1979) New results in the theory of repeated orders queueing systems. *Journal of Applied Probability*, **16**, No.3, 631–640.

Cohen, J.W. (1957) Basic problems of telephone traffic theory and the influence of repeated calls. *Philips Telecommunication Review*, **18**, No.2, 49–100.

Conolly, B.W. (1982) Letter to the editor. *Journal of Applied Probability*, **19**, No.4, 904–905.

de Kok, A.G. (1982) Computational methods for single server systems with repeated attempts. Report No.89, Interfaculteit der Actuariele Wetenschappen en Econometrie, Amsterdam.

de Kok, A.G. (1984) Algorithmic methods for single server systems with repeated attempts. *Statistica Neerlandica*, **38**, No.1, 23–32.

Deul, N. (1980) Stationary conditions for multi-server queueing systems with repeated calls. *Elektronische Informationsverarbeitung und Kybernetik (Journal of Information Processing and Cybernetics)*, **16**, No.10–12, 607–613.

Deul, N. (1982) The influence of the perseverance function in queueing systems with repeated calls. *Elektronische Informationsverarbeitung und Kybernetik (Journal of Information Processing and Cybernetics)*, **18**, No.10–11 587–594.

Diamond, J.E. and Alfa, A.S. (1995) Matrix analytical method for $M/PH/1$ retrial queues. *Stochastic Models*, **11**, No.3, 447–470.

Dimitrov, B.N. and Ruskov, P.R. (1985) Discrete model of a single-line queueing system with repeated calls. *Mathematics and Mathematical Education. Proceedings of the Fourteenth Spring Conference of the Union of Bulgarian Mathematicians, Sunny Beach, April 6-9, 1985*, Bulgarian Academy of Science, Sofia, 376–384 (in Russian).

Douz, V.I., Zelinskiy, A.M. and Kornyshev, Yu.N. (1984) A study of traffic handling processes with repeated calls and variable parameters. *Telecommunications and Radio Engineering*, **38**, No.8, 32–37.

Dragieva, V.I. (1988) Queueing system with finite source and repeated demands.*International Seminar on Teletraffic Theory and Computer Modelling. Sofia, Bulgaria, March 21-26, 1988. Proceedings*, Sofia, 135–145 (in Russian).

Dragieva, V.I. (1994) Single-line queue with finite source and repeated calls. *Problems of Information Transmission*, vol.30, No.3, 283–289.

Dudin, A.N. (1985) On a system with repeated calls and changing operating conditions. Paper #293-85, All-Union Institute for Scientific and Technical Information, Moscow (in Russian).

Duffy, F.P. and Mercer, R.A. (1978) A study of network performance and customer behaviour during direct-distance-dialling call attempts in the USA. *The Bell System Technical Journal*, **57**, No.1, 1–33.

Elldin, A. (1966) Description of the mechanism of repeated call attempts in telephone traffic. A paper presented at the III Congr. Venezolano de Ing. Electr. y Mec., Caracas.

Elldin, A. (1967) Approach to the theoretical description of repeated call attempts. *Ericsson Technics*, **23**, No. 3, 346–407.

Elldin, A. (1977) Traffic engineering in developing countries. Some observations from the ESCAP region. *Telecomm.*, **44**, 9, 427–436.

Evers, R. and Meyenberg, E. (1971) The recognition of audible-tones and speech on a telephone-line. *Nachrichtentehn. Ztschr.*, **24**, 536–540.

Evers, R. (1972) A survey of subscriber behaviour including repeated call attempts – results of measurements in two PABX's. A paper presented at the 6th International Symp. on Human Factors in Telecomm., Stockholm.

Evers, R. (1973) Measurement of subscriber reaction to unsuccessful call attempts and the influence of reasons of failure. A paper presented at the 7th International Teletraffic Congress, Stockholm.

Evers, R. (1976) Analysis of traffic flows on subscriber-lines dependent of time and subscriber class. A paper presented at the 8th International Teletraffic Congress, Melbourne.

Evers, R. (1974a) The structure of traffic offered by different groups of users. A paper presented at the 7th International Symposium on Human Factors in Telecomm., Montreal.

Evers, R. and Manz, K. (1974b) Influencing the efficiency rate of the international network by different models of handling ineffective call attempts. A paper presented at the International Switching Symposium, Munich.

Evers, R. (1974c) Über das Verhalten des Fernsprechteilnehmers bei erfolglosen Anrufversuchen und den Einfluss dieses Verhaltens auf den Fernsprechverkehr (Dissertation). Fachbereich Elektrotechnik der Technischen Universität Berlin.

Falin, G.I. (1975) Multi-phase servicing in a single-channel system for automation of experiments with repeated calls. In: *Problems of Automation of Scientific Investigations in Radio Engineering and Electronics*, Institute for Radiotechnics and Electronics of the USSR Academy of Science, Moscow, 30-37 (in Russian).

Falin, G.I. (1976) Aggregate arrival of customers in one-line system with repeated calls. *Ukrainian Math. J.*, **28**, No.4, 437–440.

Falin, G.I. (1977) Waiting time in a single-channel queuing system with repeated calls. *Moscow University Computational Mathematics and Cybernetics*, No.4, 66–69.

Falin, G.I. (1978) The exit flow of a single-line queueing system when there are secondary orders. *Engineering Cybernetics*, **16**, No.5, 64–67.

Falin, G.I. (1979a) Model of coupled switching in presence of recurrent calls. *Engineering Cybernetics*, **17**, No.1, 53–59.

Falin, G.I. (1979b) A single-line system with secondary orders. *Engineering Cybernetics*, **17**, No.2, 76–83.

Falin, G.I. (1979c) Effect of the recurrent calls on output flow of a single channel system of mass service. *Engineering Cybernetics*, **17**, No.4, 99–102.

Falin, G.I. (1979d) Queueing Systems with Repeated Calls. Ph.D.Thesis, Moscow State University, Mechanics and Mathematics Faculty (in Russian).

Falin, G.I. (1980a) An $M/M/1$ queue with repeated calls in presence of persistence function. Paper #1606-80, All-Union Institute for Scientific and Technical Information, Moscow (in Russian).

Falin, G.I. (1980b) $M/G/1$ system with repeated calls in heavy traffic. *Moscow University Mathematics Bulletin*, **35**, No.6, 48–51.

Falin, G.I. (1980c) Not completely accessible schemes with allowance for repeated calls, *Engineering Cybernetics*, **18**, No.5, 56–63.

Falin, G.I. (1980d) Switching systems with allowance for repeated calls. *Problems of Information Transmission*, **16**, 145–151.

Falin, G.I. (1980e) Repeated calls in structurally complex systems. *Engineering Cybernetics*,**18**, No.6, 46–51.

Falin, G.I. (1981a) Calculation of the load on a share-use telephone instrument, *Moscow University Computational Mathematics and Cybernetics*, No.2, 76–80.

Falin, G.I. (1981b) Functioning under nonsteady conditions of a single-channel system with group arrival of requests and repeated calls. *Ukrainian Math.J.*, **33**, No.4, 429–432.

Falin, G.I. (1981c) Investigation of weakly loaded switching systems with repeated calls. *Engineering Cybernetics*, **19**, No.3, 69–73.

Falin, G.I. (1982a) Servicing of a finite number of demands by fully available trunk with repetition of blocked calls. Paper #2112-82, All-Union Institute for Scientific and Technical Information, Moscow (in Russian).

Falin, G.I. (1983a) The influence of inhomogeneity of the composition of subscribers on the functioning of telephone systems with repeated calls. *Engineering Cybernetics*, **21**, No.6, 78–82.

Falin, G.I. (1983b) Asymptotic properties of the number of demands distribution in an $M/G/1/\infty$ queueing system with repeated calls. Paper #5418-83, All-Union Institute for Scientific and Technical Information, Moscow (in Russian).

Falin, G.I. (1983c) Calculation of probability characteristics of a multiline system with repeat calls. *Moscow University Computational*

*Mathematics and Cybernetics*, No.1, 43–49.

Falin, G.I. (1983d) On the accuracy of a numerical method of calculation of characteristics of systems with repeated calls. *Elektrosvyaz*, No.8, 35–36 (in Russian).

Falin, G.I. (1984a) On sufficient conditions for ergodicity of multi-channel queueing systems with repeated calls. *Advances in Applied Probabability*, **16**, 447–448.

Falin, G.I. (1984b) Two-channel queueing system with repeated calls. Paper #4221-84, All-Union Institute for Scientific and Technical Information, Moscow (in Russian).

Falin, G.I. (1984c) Multiserver fully-available systems with repeat calls under conditions of heavy traffic. *Moscow University Computational Mathematics and Cybernetics*, No.3, 82–85.

Falin, G.I. (1984d) Asymptotic investigation of fully available switching systems with high repetition intensity of blocked calls. *Moscow University Mathematics Bulletin*, **39**, No.6, 72–77.

Falin, G.I. (1984e) Calculation of probabilistic characteristics of a fully available trunk in presence of repeated calls. Paper #8310-84, All-Union Institute for Scientific and Technical Information, Moscow (in Russian).

Falin, G.I. (1984f) Continuous approximation for a single server system with an arbitrary service time under repeated calls. *Engineering Cybernetics*, **22**, No.2, 66–71.

Falin, G.I. (1985a) A probabilistic model for investigation of load of subscribers' lines with waiting places. In: *Probability Theory, Stochastic Processes and Functional Analysis*. Moscow State University, Moscow (in Russian).

Falin, G.I. and Sukharev, Yu.I. (1985b) On single-line queue with double connections. Paper #6582-85, All-Union Institute for Scientific and Technical Information, Moscow (in Russian).

Falin, G.I. (1985c) Limit theorems for queueing systems with repeated calls. A paper presented at the 4th Int. Vilnius Conf. on Probability Theory and Mathematical Statistics, Vilnius, USSR.

Falin, G.I. (1985d) Probabilistic model for investigation of load of subscribers' lines with waiting places. In: *Probability Theory, Theory of Stochastic Processes and Functional Analysis*, Moscow State University, 64–66 (in Russian).

Falin, G.I. (1985e) Repeated calls in information transmission systems. In:*Investigation of Ways of Increasing Efficiency of Computer and Communication Networks*, Minsk, 57–58 (in Russian).

Falin, G.I. (1986a) On waiting time process in single-line queue with repeated calls. *Journal of Applied Probability*, **23**, No.1, 185–192.

Falin, G.I. (1986b) Single-line repeated orders queueing systems. *Mathematische Operationsforschung und Statistik. Optimization*, No.5, 649–

667.

Falin, G.I. (1986c) On heavily loaded systems with repeated calls. *Soviet Journal of Computer and Systems Sciences*, **24**, No.4, 124–128.

Falin, G.I. (1987a) On ergodicity of multilinear queueing systems with repeated calls. *Soviet Journal of Computer and Systems Sciences*, **25**, No.1, 60–65.

Falin, G.I. (1987b) Multichannel queueing systems with repeated calls under high intensity of repetition. *Journal of Information Processing and Cybernetics*, No.1, 37–47.

Falin, G.I. (1987c) Error estimates for approximations of countable Markov chains associated with repeated calls models. *Moscow University Mathematics Bulletin*, **42**, No.2, 12–15.

Falin, G.I. (1988a) Virtual waiting time in systems with repeated calls. *Moscow University Mathematics Bulletin*, **43**, No.6, 6–10.

Falin, G.I. (1988b) A system with feedback. *Paper #719-B88, All-Union Institute for Scientific and Technical Information*, Moscow (in Russian).

Falin, G.I. (1988c) On a multiclass batch arrival retrial queue. *Advances in Applied Probability*, **20**, No.2, 483–487.

Falin, G.I. (1988d) Comparability of migration processes. *Theory of Probability and Its Applications*, **33**, No.2, 370–372.

Falin, G.I. and Sukharev, Yu.I. (1988e) Singularly disturbed equations and asymptotic analysis of stationary characteristics of repeat-call-queueing systems. *Moscow University Mathematics Bulletin*, **43**, No.5, 7–10.

Falin, G.I. (1989a) Phase transitions in queueing systems, which are connected with virtual waiting time. *Ukrainian Math. J.*, **41**, No.7, 813–817.

Falin, G.I. (1989b) Ergodicity and stability of systems with repeated calls. *Ukrainian Math. Journal*, **41**, No.5, 559–562.

Falin, G.I. (1990) A survey of retrial queues. *Queueing Systems*, **7**, No.2, 127–167.

Falin, G.I. (1991a) Phase transitions in overloaded computer and communication systems. *Proceedings of KAIST Mathematical Workshop*, Korea Advanced Institute of Science and Technology, Taejon, Korea, vol.6, 21–34.

Falin, G. and Fricker, C. (1991b) On the virtual waiting time in an $M/G/1$ retrial queue. *Journal of Applied Probability*, **28**, No.2, 446–460.

Falin, G.I. (1991c) A diffusion approximation for retrial queueing systems. *Theory of Probability and Its Applications*, **36**, No.1, 149–152.

Falin, G.I., Artalejo, J.R. and Martin, M. (1993) On the single server retrial queue with priority customers. *Queueing Systems*, **14**, 439–455.

Falin, G.I., Artalejo, J.R. and Martin, M. (1994) Information theoretic approximations for the $M/G/1$ retrial queue. *Acta Informatica*, **31**, 559–571.

Falin, G. (1995a) Estimation of retrial rate in a retrial queue. *Queueing Systems*, **19**, 231–246.

Falin, G.I. and Artalejo, J.R. (1995b) Approximations for multiserver queues with balking/retrial discipline. *OR Spectrum*, **17**, 239–244.

Farahmand, K. (1990) Single line queue with repeated demands. *Queueing Systems*, **6**, No.2, 223–228.

Farahmand, K., Smith, N.H. (1996a) Retrial queues with recurrent demand option. *Journal of Applied Mathematics and Stochastic Analysis*, **9**, No.2.

Farahmand, K. (1996b) Single line queue with recurrent repeated demands. *Queueing Systems*, **22**, 425–435.

Fayolle, G. (1986) A simple telephone exchange with delayed feedbacks. *Teletraffic Analysis and Computer Performance Evaluation*, (eds O.J.Boxma, J.W.Cohen and H.C.Tijms), Elsevier Science Publishers B.V., Amsterdam, 245–253.

Fayolle, G. and Brun, M.A. (1988) On a system with impatience and repeated calls. *Queueing Theory and its Applications. Liber Amicorum for J.W.Cohen*, (eds O.J.Boxma and R.Syski), Elsevier Science Publishers B.V., Amsterdam, 283–305.

Feinberg, M. A. (1991) Analytical model of automated call distribution system. Queueing, Performance and Control in ATM. ITC-13 Workshops. Proceedings of 13 International Teletraffic Congress. Copenhagen, Denmark, June 19–26, 1991, (eds J.W. Cohen and C.D. Pack), Elsevier Science Publishers B.V., Amsterdam, 193–197.

Fredericks, A.A. and Reisner, G.A. (1979) Approximations to stochastic service systems, with an application to a retrial model. *The Bell System Technical Journal*, **58**, No.3, 557–576.

Gilbert, E.N. (1988) Retrials and balks. *IEEE Transactions on Information Theory*, **34**, No.6, 1502–1508.

Gómez-Corral, A. (1996) Queueing systems with repeated attempts and negative arrivals. Ph.D. Thesis, Department of Statistics and Operations Research, Complutense University, Madrid, Spain (in Spanish).

Gosztony, G. (1976a) Comparison of calculated and simulated results for trunk groups with repeated attempts. A paper presented at the 8th International Teletraffic Congress, Melbourne.

Gosztony, G. (1976b) Stochastic service systems with two input processes and repeated calls (Traffic engineering of telex stations). In: *Progress in Operations Research*, North-Holland, Amsterdam, 467–492.

Gosztony, G. (1976c) Repeated call attempts and their effect on traffic engineering. *Budavox Telecommunication Review*, **2**, 16–26.

Gosztony, G. (1977) Comparison of calculated and simulated results for trunk groups with repeated attempts. *Budavox Telecommunication Review*, No.1, 1–18.

Gosztony, G., Rahko, K., Chapuis, R. (1979) The grade of service in the world-wide telephone network. Pt.I. *Telecomm. J.*, **46**, No.9, 556–565.

Gosztony, G. (1985) A general (rH$\beta$) formula of call repetition: validity and constraints. *Teletraffic Issues in an Advanced Information Society. ITC-11. Proceedings of the 11 International Teletraffic Congress, Kyoto, September 4-11, 1985*, Pt. 2. (ed M.Akiyama), Elsevier Science Publishers, Amsterdam, 1010–1016.

Grillo, D. (1979) Telephone network behaviour in repeated attempts environment: a simulation analysis. A paper presented at the 9th International Teletraffic Congress, Torremolinos.

Greenberg, B.S. (1986) Queueing systems with returning customers and the optimal order of tandem queues. Ph.D. Thesis, University of California, Berkeley.

Greenberg, B.S. and Wolff, R.W. (1987) An upper bound on the peformance of queues with returning customers. *Journal of Applied Probability*, **24**, 466–475.

Greenberg, B.S. (1989) $M/G/1$ queueing systems with returning customers. *Journal of Applied Probability*, **26**, No.1, 152–163.

Grishechkin, S.A. (1990) Branching processes and systems with repeated calls or random discipline. *Theory Probab. Appl.*, **35**, No.1, 28–53.

Grishechkin, S.A. (1991) Application of branching processes to finding stationary distributions in queueing theory. *Theory Probab. Appl.*, **36**, No.3, 477–493.

Grishechkin, S.A. (1992) Multiclass batch arrival retrial queues analyzed as branching processes with immigration. *Queueing Systems*, **11**, No.4, 395–418.

Guerineau, J.P. and Pellieux, G. (1974) Nouveaux résultats concernant le comportement de l'abonné du réseau téléphonique de Paris. *Commutation et Electronique*, **45**, 53–71.

Hanschke, T. (1978) Die von Bretschneider, Cohen und Schwarzbart /Puri entwickelte Warteschlangenmodelle mit wiederholten Versuchen: eine Methode zur Berechnung der ergodischen Projection ihrer Markovschen Warteprozesse und die Simulation der Wartezeit, Fakultät für Mathematik der Universität Karlsruhe.

Hanschke, T. (1985a) A model for planning switching networks. In: *Operations Research Proceedings 1984*, Springer, Berlin, 555–562.

Hanschke, T. (1985b) The $M/G/1/1$ queue with repeated attempts and different types of feedback effects. *OR Spectrum*, **7**, No.4, 209–215.

Hanschke, T. (1986) A computational procedure for the variance of the waiting time in the $M/M/1/1$ queue with repeated calls. In: *Operations Research Proceedings*, Springer, Berlin, 525–532.

Hanschke, T. (1987) Explicit formulas for the characteristics of the

$M/M/2/2$ queue with repeated attempts. *Journal of Applied Probability*, **24**, 486–494.

Hashida, O. and Kawashima, K. (1979) Buffer behaviour with repeated calls. *Electronics and Comunications in Japan*, **62-B**, No.3, 27–35.

Hauschilat, B. and Iverson, V.B. (1977) On subscriber reactions to the telephone systems. A paper presented at the 8th International Symposium on Human Factors in Telecomm., Cambridge.

Haywarder, W.S. and Wilkinson, R.I. (1970) Human factors in telephone systems and their influence on traffic theory, specially with regard to future facilities. A paper presented at the 6th International Teletraffic Congress, Munich.

Hoffman, K.L. and Harris, C.M. (1986) Estimation of a caller retrial rate for a telephone information system. *European Journal of Operational Research*, **27**, 207–214.

Honi, G. (1975) Some macro-models for discussing repeated call attempts. *Budavox Telecomm. Rev.*, **2**, 21–39.

Honi, G. and Gosztony, G. (1976) Some practical problems of the traffic engineering of overloaded telephone networks. A paper presented at the 8th International Teletraffic Congress, Melbourne.

Inamori, H., Sawai, M., Endo, T., Tanabe, K. (1985) An automatically repeated call model in NTT public facsimile communication systems. *Teletraffic Issues in an Advanced Information Society. ITC-11. Proceedings of the 11 International Teletraffic Congress, Kyoto, September 4-11, 1985*, Pt. 2. (ed M.Akiyama), Elsevier Science Publishers, Amsterdam, 1017–1023.

Iokhvin, B.M. (1982) Determination of the economic indicators of telephone traffic systems with reringing. *Telecommunications and Radio Engineering*, No.3, 28–33.

Iokhvin, B.M., Pevtsov, N.V., Sterlina, E.I., Babasinov, A.I. (1983) Reserves of operating efficiency in long-distance automatic telephone exchanges. *Telecommunications and Radio Engineering*, No.4, 1–5.

Iversen, W.B. (1973) Analysis of real teletraffic processes based on computerized measurements. *Ericsson Technics*, **29**, No.1, 3–64.

Jonin, G.L. and Sedol, J.J. (1970a) Telephone systems with repeated calls. A paper presented at the 6th International Teletraffic Congress, Munich.

Jonin, G.L. and Sedol, Ya.Ya. (1970b) Investigation of telephone systems in presence of repeated calls. *Latvian Mathematical Yearbook*, **7**, Riga, 71–83 (in Russian).

Jonin, G.L. and Sedol, J.J. (1970c) *Tables of Probabilistic Characteristics of Fully Available Trunk Group in the Case of Repeated Calls*, Nauka, Moscow, (in Russian).

Jonin, G.L. (1971) A single-line system with repeated calls. In: *Scientific and Technical Conference for Problems of Information Networks and*

*Automatic Switching*, Moscow, 89–94 (in Russian).

Jonin, G.L. and Brezgunova, N.M. (1972) Single-line system with repeated calls in the case of Γ-distributed occupation time. *Latvian Math. Yearbook*, **11**, Riga, 65–71 (in Russian).

Jonin, G.L. and Sedol, J.J. (1973) Full-availability groups with repeated calls and time of advanced service. A paper presented at the 7th International Teletraffic Congress, Stockholm.

Jonin, G.L., Sedol, J.J. and Kibild, A.V. (1975) Generalized queueing model with repeated calls. In: *Information Networks and Automatic Switching*, 3rd All-Union Scientific and Technical Conference, Thesis of Reports, Nauka, Moscow, 38–40 (in Russian).

Jonin, G.L. (1980) An investigation of single-line system with repeated calls under independent discrete check of line state. *Latvian Math. Yearbook*, **24**, Riga, 204–209 (in Russian).

Jonin, G.L. (1982a) An investigation of a single-line system with repeated calls under service without interruption and with independent discrete check of channel state. In: *Models of Information Networks and Switching Systems*, Nauka, Moscow, 36–39 (in Russian).

Jonin, G.L. (1982b) Determination of probabilistic characteristics of single-line system with double connections and repeated calls. In: *Models of Systems of Distribution of Information and their Analysis*, Nauka, Moscow, 51–54 (in Russian).

Jonin, G.L. (1984) The systems with repeated calls: models, measurement, results. *Fundamentals of teletraffic theory. Proceedings of the Third International Seminar on Teletraffic Theory (Moscow, USSR, June 20-26, 1984)*, Moscow, 197–208.

Jonin, G.L. (1986) Systems of Information Transmission with Repeated Calls. Doctor of Technical Sciences Thesis, Moscow.

Jonin, G.L. and Sedol, Ya.Ya. (1988) Investigation of a system with repeated calls taking subscriber busyness into account. *Telecommunication and Radio Engineering*, No.11, 1–6.

Jonin, G.L. (1990) An approximate method for calculations of the probability characteristics of multilevel switching systems with repeated calls. *Telecommunications and Radio Engineering*, No.2, 26–29.

Kapyrin, V.A. (1977) A study of the stationary characteristics of a queueing system with recurring demand. *Cybernetics*, , **13**, 584–590.

Kerebel, R. (1969) Résultats d'observations des appels téléphoniques inefficaces dans le réseau de Paris. *Commutation et Electronique*, **26**, 95–112.

Kerebel, R. (1970) Nouveaux résultats d'observations concernant les appels téléphoniques inefficaces. *Commutation et Electronique*, **28**, 89–98.

Keilson, J., Cozzolino, J. and Young, H. (1968) A service system with unfilled requests repeated. *Operations Research*, **16**, 1126–1137.

Keilson, J. and Servi, L.D. (1993) The matrix $M/M/\infty$ system: Retrial models and Markov modulated sources. *Advances in Applied Probability*, **25**, No.2, 453–471.

Kelly, F.P. (1986) On autorepeat facilities and telephone network performance. *J. Roy. Statist. Soc.*, **B48**, No.1, 123–132.

Khalil, Z., Falin, G.I., Yang, T. (1992) Some analytical results for congestion in subscriber line modules. *Queueing Systems. Theory and Applications*, **10**, No.4, 381–402.

Khalil, Z. and Falin, G. (1994) Stochastic inequalities for $M/G/1$ retrial queues. *Operations Research Letters*, **16**, 285–290.

Kharkevich, A. *et al.* (1985) Approximate analysis of systems with repeated calls and multiphase service. *Teletraffic Issues in an Advanced Information Society. ITC-11. Proceedings of the 11 International Teletraffic Congress, Kyoto, September 4-11, 1985*, Pt. 2. (ed M.Akiyama), Elsevier Science Publishers, Amsterdam, 1029–1035.

Khomichkov, I.I. (1985) A model of a route of a circuit switching network with repeated calls. In: *Mathematical and Software Support for Systems of Automatic Design of Networks*, Mari State University, Ioshkar-Ola (in Russian).

Khomichkov, I.I. (1986) Calculation of characteristics of a model of a local area network with access protocol CSMA/CD. In: *Local Area Networks. Proceedings of the Conference of Scientists from Socialist Countries*, Riga, 150–154 (in Russian).

Khomichkov, I.I. (1987a) Generating functions of state probabilities of a single-line queue with repeated calls. *Vestnik Byelorussian Univ. Ser.1*, No.1, 51–55 (in Russian).

Khomichkov, I.I. (1987b) A model of local area computer network with random multiple access. *Automation and Computer Technics,* , No.1, 58–62 (in Russian).

Khomichkov, I.I. (1987c) Stationary queueing processes in systems with repeated calls. Ph.D. Thesis, Belorus. University, Minsk (in Russian).

Khomichkov, I.I. (1988a) Single-line queue with repeated calls and Cox input process of second order. *Vestnik Belorus. Univ. Ser.1*, No.1, 70–71 (in Russian).

Khomichkov, I.I. (1988b) Queueing system with repeated calls, preliminary service and blockings. *Izv. Akad. Nauk SSSR. Tekhn. Kybernet.*, No.1, 188–198 (in Russian).

Khomichkov, I.I. (1988c) Computing the parameters of a queueing system with repeated calls in case of collision. *Automation and Remote Control*, **49**, No.4, 458–463.

Khomichkov, I.I. (1988d) A queueing system with an unreliable instrument in the presence of repeated call. *Proceedings of Academy of Science of BSSR. Ser. Phys.-Math. Sciences*, No.5 (in Russian).

König, Hans-Joachim. (1982) Betrachtungen zur Wirkungwiederholter

Anrufversuche bei Ausfallsituationen. *Mitt. Inst. Post- und Fernmelde.*, No.2, 10–12.

Kornyshev, Yu.N. (1968) Calculation of coupled switching. *Trudy Utchebnih Institutov Svyasi*, **37**, 96–104 (in Russian).

Kornyshev, Yu.N. (1969) Calculation of a fully-available communication system with repeated calls. *Elektrosvyaz*, No.11, 65–72 (in Russian).

Kornyshev, Yu.N. (1974a) Repeated calls in a trunk-line. *Elektrosvyaz*, No.1, 35–41 (in Russian).

Kornyshev, Yu.N. (1974b) Waiting positions for overloading trunks.*Elektrosvyaz*, No.7, 32–39 (in Russian).

Kornyshev, Yu.N. (1977a) A system with repeated calls and finite number of sources.*Elektrosvyaz*, No.5, 58–62 (in Russian).

Kornyshev, Yu.N. (1977b) A single-line system with repeated orders and preliminary servicing. *Engineering Cybernetics*, No.2, 63–68.

Kornyshev, Yu.N. and Zelinskiy, A.M. (1980a) Analysis of subscriber's line states. In: *Information Networks and their Analysis*, Nauka, Moscow, 113–122 (in Russian).

Kornyshev, Yu.N. (1980b) A single-line system with heterogeneous repeated calls. In: *Teletraffic Theory and Networks with Controlled Elements*, Nauka, Moscow, 113–122 (in Russian).

Kornyshev, Yu.N. and Duz, V.I. (1990)The accuracy of evaluating call servicing quality characteristics in communications network. *Telecommunications and Radio Engineering*, No.4, 1–8.

Kosten, L. (1947) On the influence of repeated calls in the theory of probabilities of blocking. *De Ingenieur*, **59**, 1–25.

Kulkarni, V.G. (1982) Letter to the editor. *Journal of Applied Probability*, **19**, No.4, 901–904.

Kulkarni, V.G. (1983a) On queueing systems with retrials. *Journal of Applied Probability*, **20**, No.2, 380–389.

Kulkarni, V.G. (1983b) A game theoretic model for two types of customers competing for service. *Operations Research Letters*, **2**, No.3, 119–122.

Kulkarni, V.G. (1986) Expected waiting times in a multiclass batch arrival retrial queue. *Journal of Applied Probability*, **23**, No.1, 144–159.

Kulkarni, V.G. and Sethi, S.P. (1989) Deterministic retrial times are optimal in queues with forbidden states. *INFOR.*, **27**, No.3, 374–386.

Kulkarni, V.G. and Choi, B.D. (1990) Retrial queues with server subject to breakdowns and repairs. *Queueing Systems*, **7**, No.2, 191–208.

Kulkarni, V.G. and Liang, H.M. (1996) Retrial queues revisited. *Frontiers in queueing: models and applications in science and engineering. Recent developments, open problems and future research directions.* Edited by J.H.Dshalalow. CRC Press, Boca Raton, Part I, Chapter 2.

Langaris, C. and Moutzoukis, E. (1995) A retrial queue with structured batch arrivals, priorities and server vacations. *Queueing Systems*, **20**, 341–368.

Lederman, S. (1985) Congestion model for subscriber line modules. *GLOBECOM'85: IEEE Global Telecommunications Conference, New Orleans, Lousiana, December 2-5, 1985. Conference Record*, Vol.1, IEEE, New York, N.Y., 395–401.

Le Gall, P. (1968a) Les plans souples d'acheminement et l'influence des répétitions d'appels en période de saturation du réseau. *Actes de la Conférence O.T.A.N. sur "La survie des réseaux de télécommunications", île de Bendor, 17-21 Juin*.

Le Gall, P. (1968b) Sur l'écoulement dirigé du trafic dans les grands réseaux téléphoniques interurbains. *Commutation et Electronique*, **20**.

Le Gall, P. (1969) Sur une théorie des répétitions des appels téléphoniques. *Ann. Télécomm.*, **24**, 261–281.

Le Gall, P. (1970a) Sur le modèle du trafic téléphonique avec répétitions d'appels. *Commutation et Electronique*, **28**.

Le Gall, P. (1970b) Sur le qualité d'écoulement de trafic des réseaux téléphonique. *Commutation et Electronique*, **31**.

Le Gall, P. (1970c) Sur l'influence des répétitions d'appels dans l'écoulement du trafic téléphonique. A paper presented at the 6th International Teletraffic Congress, Munich.

Le Gall, P. (1971) Sur le taux d'efficacité et la stationnarité du trafic téléphonique. *Commutation et Electronique*, **35**, 7–36.

Le Gall, P. (1973) Sur l'utilisation et l'observation du taux d'efficacité du trafic téléphonique. A paper presented at the 7th International Teletraffic Congress, Stockholm.

Le Gall, P. (1976a) General telecommunications traffic without delay. A paper presented at the 8th International Teletraffic Congress, Melbourne.

Le Gall, P. (1976b) Trafic généraux de télécommunications sans attente. *Commutation et Electronique*, **55**, 5–24.

Le Gall, P. (1977a) Les appels répétés et les méthodes de calcul et d'observation du trafic. *Commutation et Electronique*, **56**, 105–116.

Le Gall, P. (1977b) La notion de qualité de service en téléphonie et les répétitions d'appels. *Commutation et Electronique*, **59**, 85–98.

Le Gall, P. (1984) The repeated call model and the queue with impatience. *Fundamentals of teletraffic theory. Proceedings of the Third International Seminar On Teletraffic Theory (Moscow, USSR, June 20-26, 1984)*. Moscow, pp. 278–289.

Li, H., Yang, T. (1995) A single-server retrial queue with server vacations and a finite number of input sources. *European Journal of Operational Research*, **85**, 149–160.

Liang, H.M. and Kulkarni, V.G. (1993a) Monotonicity properties of single-server retrial queues. *Communications in Statistics. Stochastic*

*Models*, **9**, No.3, 373–400.

Liang, H.M. and Kulkarni, V.G. (1993b) Stability condition for a single-server retrial queue. *Advances in Applied Probability*, **25**, No.3, 690–701.

Lind, G. (1976) Studies on the probability of a called subscriber being busy. A paper presented at the 8th International Teletraffic Congress, Melbourne.

Liu, K.S. (1980) Direct distance dialing: call completion and customer retrial behaviour. *The Bell System Technical Journal*, **59**, No.3, 295–311.

Lubacz, J. and Roberts, J. (1984) A new approach to the single server repeated attempt system with balking. *Fundamentals of Teletraffic Theory. Proceedings of the Third International Seminar on Teletraffic Theory*, Moscow, 290–293.

Lukashuk, L.I. (1990) Diffusion approximation and filtering for a queueing system with repeats. *Cybernetics*, No.2, 253–264.

Lupker, S.J., Fleet, G.J., Shelton, B.R. (1988) Callers' perceptions of post-dialing delays: the effects of a new signalling technology. *Behav. and Inf. Technol.*, **7**, No.3, 263–274.

Macfadyen, N.W. (1979) Statistical observation of repeated attempts in the arrival process. A paper presented at the 9th International Teletraffic Congress, Torremolinos.

Maheswaran, A. (1985) A real time simulation program for repeated calls. Report No.9, TRC, The University of Adelaide.

Martin, M. and Artalejo, J.R. (1995) Analysis of an $M/G/1$ queue with two types of impatient units. *Advances in Applied Probability*, **27**, No.3.

Meykshan, V.I. and Fidel'man, I.G. (1995) The design of communications networks with bypass routings when there are repeated calls and lines with undetected breakdowns. *Telecommunications and Radio Engineering*, **49**, No.7, 40–44.

Morrison, M., Fleming, P.Jr. (1976) Some blocking formulas for three-stage switching arrays, with multiple connection attempts. A paper presented at the 8th International Teletraffic Congress, Melbourne.

Moutzoukis, E., Langaris., C. (1996) Non-preemptive priorities and vacations in a multiclass retrial queueing system. *Stochastic Models*, **12**, No.3, 455–472.

de los Mozos, J.R., Buchheister, A. (1983) Blocking calculation method for digital switching networks with step-by-step hunting and retrials. A paper presented at the 10th International Teletraffic Congress, Montreal.

Myskja, A. (1971) A recording and processing system for accounting and traffic analysis on a large PABX. *IEEE Transactions on Communic. Technol.*, **5**, 693–699.

Myskja, A., Wallmann, O.O. (1972) An investigation of telephone user habits by means of computer technics. A paper presented at the 6th Intern. Symposium on Human Factors in Telecomm., Stockholm.

Myskja, A., Wallmann, O.O. (1973a) A statistical study of telephone traffic data with emphasis on subscriber behaviour. A paper presented at the 7th International Teletraffic Congress, Stockholm.

Myskja, A., Wallmann, O.O. (1973b) An investigation of telephone user habits by means of computer techniques. *IEEE Trans. on Communications*, **6**, 663–671.

Myskja, A., Aagesen, F.A. (1976) On the interaction between subscribers and a telephone system. A paper presented at the 8th International Teletraffic Congress, Melbourne.

Myskja, A., Aagesen F. A. (1977) On the interaction between subscribers and a telephone system. *Telektronik*, No.3, 271–282.

Myskja, A. (1979) Modelling of non-stationary traffic processes. A paper presented at the 9th International Teletraffic Congress, Torremolinos.

Nador, L. (1988) The effects of traffic overloads in automatic telephone networks. *Budavox Telecommun. Rev.*, No.1, 2–17.

Naumova, E.O. (1983) The loss probability measurement accuracy for the subscriber line with repeated calls. A paper presented at the International Teletraffic Congress, Montreal.

Naumova, E.O. and Shkolny, E.I. (1985) Fault lines' influence to the values of probabilistic characteristics and their measurement accuracy in system with expectation and repeated calls. *Teletraffic Issues in an Advanced Information Society. ITC-11. Proceedings of the 11 International Teletraffic Congress, Kyoto, September 4-11, 1985*, Pt. 2. (ed M.Akiyama), Elsevier Science Publishers, Amsterdam, 1024–1028.

Naumova, E.O. (1988) Formulas for variances of statistical estimations of characteristics of a single-line system with repeated calls. In: *Models and Methods of Investigation in Systems of Informatics*, Nauka, Moscow, 1988, 44–54 (in Russian).

Neuts, M.F. and Ramalhoto, M.F. (1984) A service model in which the server is required to search for customers. *Journal of Applied Probability*, **21**, No.1, 157–166.

Neuts, M.F. and Rao, B.M. (1990) Numerical investigation of multiserver retrial model. *Queueing Systems*, **7**, No.2, 169–190.

Noordegraaf, C. (1986) Herhaale oproepen: een studie naar abonneegedrag. *PTT-bedrijf*, **24**, No.1–4, 30–35.

Ohmura, H. (1985) An analysis of repeated call model with a finite number of sources. *Electronics and Communications in Japan*, Part 1, **68**, No.6, 112–121.

Paterok, M., Herzog, U., Bleisteiner, C. (1988) The influence of repeated calls on the performance measures of loss systems. *Teletraffic Science for New Cost-Effective Systems, Networks and Services. ITC-12. Pro-*

*ceedings of the 12 International Teletraffic Congress. Torino, Italy,
June 1–8, 1988*, Pt. 2. (ed M.Bonatti), Elsevier Science Publishers
B.V., Amsterdam, 1413–1419.

Pearce, C.E.M. (1985) The mathematical modelling of the problem of
repeated calls, Report No.4, Teletraffic Research Centre, The University of Adelaide.

Pearce, C.E.M.(1987) On the problem of re-attempted calls in teletraffic.
*Communications in Statistics-Stochastic Models*, **3**, No.3, 393–407.

Pearce, C.E.M. (1989) Extended continued fractions, recurrence relations and two dimensional Markov processes. *Advances in Applied
Probability*, **21**, 357–375.

Pellieux, G. (1973) Observation sur le comportement de l'abonné au
téléphone à fort trafic en cas d'échec. A paper presented at the 7th
International Teletraffic Congress, Stockholm.

Pellieux, G. (1973) Observation sur le comportement de l'abonne au
téléphone à fort trafic en cas d'échec. *Commutation & Electronique*,
**42**, 20–34.

Pellieux, G., Guerineau, J.P. (1974) Observation de comportement de
l'abonné en aval d'un goulet d'étranglement. *Commutation et Electronique*, **47**, 26–33.

Pevtsov, N.V. (1983) An estimate of the effect of subscriber behavior on
the load of a long-distance telephone network. *Telecommunications
and Radio Engineering*, No.4, 6–9.

Pourbabai, B. (1986) Stochastic modeling of a telecommunication system with repeated calls. A paper presented at the 18th JAACE Symposium on Stochastic Systems Theory and its Applications, Tokyo,
Japan.

Pourbabai, B. (1987) Analysis of a $G/M/K/0$ queueing loss system with
heterogeneous servers and retrials. *International Journal of System
Science*, **18**, No.5, 985–992.

Pourbabai, B. (1988a) A random access telecommunication system. *J.
Inf. Process. and Cyber. EIK.* , **24**, No.11–12, 613–625.

Pourbabai, B. (1988b) Asymptotic analysis of $G/G/K$ queueing-loss system with retrials and heterogeneous servers. *International Journal of
System Science*, **19**, No.6, 1047–1052.

Pourbabai, B. (1989) A finite capacity telecommunication system with
repeated calls. *J. Inf. Process. and Cyber. EIK*, **25**, No.8–9, 457–467.

Pourbabai, B. (1990) A note of $D/G/K$ loss system with retrials. *J.
Appl. Probab.*, **27**, No.2, 385–392.

Rasmussen, E. (1967) Subscriber reaction to unsuccessful calls.
*Teleteknik*, **11**, No.1.

Reeser, P.K. (1989) Simple approximation for blocking seen by peaked
traffic with delayed, correlated reattempts. *Teletraffic Science for New
Cost-Effective Systems, Networks and Services. ITC-12. Proceedings*

*of the 12 International Teletraffic Congress. Torino, Italy, June 1–8, 1988*, Pt. 2. (ed M.Bonatti), Elsevier Science Publishers B.V., Amsterdam, 1420–1426.

Ridout, G.E. (1984) A study of retrial queueing systems with buffers. M. A. Sc. Thesis, Department of Industrial Engineering, University of Toronto.

Riordan, J. (1962) *Stochastic Service Systems.* Wiley, New York.

Roberts, J.W. (1979) Recent observations of subscriber behaviour. A paper presented at the 9th International Teletraffic Congress, Torremolinos.

Robeva, R. (1988) A description of the subscribers set working regime in an idealised automatic telephone exchange by means of semi-Markov chain models. *International Seminar on Teletraffic Theory and Computer Modelling. Sofia, Bulgaria, March 21-26, 1988. Proceedings,* Sofia, 146–158.

Ruskov, P., Yanev, K., Dimitrov, B. and Boyanov, K. (1984) A model for investigation of local area computer networks. *Control Systems and Machines,* **5**, 37–40 (in Russian).

Ryan, F.A. and Johnson, T.C. (1972) The cost of getting engaged. *Post Off. Telecomm. J.,* **24**, No.1, 2–3.

Sabo, D.W. (1987) Closure methods for the single-server retrial queue. M.Sc. Thesis, Department of Mathematics, University of British Columbia, Vancouver.

Scavo, G.G., Miranda, G. (1985) Traffic offered grade of service and call completion ratio in a toll network versus subscriber retrial behaviour. *Teletraffic Issues in an Advanced Information Society. ITC-11. Proceedings of the 11 International Teletraffic Congress, Kyoto, September 4-11, 1985*, Pt. 2. (ed M.Akiyama), Elsevier Science Publishers, Amsterdam, 689–695.

Shkol'nyi, E.I. (1977) Estimates of the loss probabilities for a queueing system with repeated calls.*Engineering Cybernetics*, No.2, 74–78.

Shkol'nyy E.I., Naumova, E.O., Kharkevich, A.D. (1985) A method of determining the additional number of lines in overloaded groups taking repeated calls and traffic measurement errors into account. *Telecommunications and Radio Engineering*, No.2, 29–35.

Shneps-Shneppe, M.A. (1970) The effect of repeated calls on communication system. A paper presented at the 6th International Teletraffic Congress, Munich.

Shneps, M.A. (1974) *Numerical methods of teletraffic theory.* Moscow, Svyaz (in Russian).

Shneps, M.A. (1979) *Systems of Information Transmission. Methods of Calculation.* Moscow, Svyaz (in Russian).

Shneps-Shneppe, M.A., Ozols, A.A. (1983) Normalization of the load of long-distance telephone channels taking redialling into account.

*Telecommunications and Radio Engineering*, No.4, 9–12.

Shneps-Shneppe, M.A., Petrov, A.F. (1984) Optimal distribution of overall grade of service in hierarchical telephone network with repeated attempts. *Fundamentals of teletraffic theory. Proceedings of the Third International Seminar on Teletraffic Theory (Moscow, USSR, June 20-26, 1984)*. Moscow, 377–386.

Shneps-Shneppe, M.A., Petrov, A.F. (1986) Optimization of the quality of service of a toll telephone network. *Telecommunications and Radio Engineering*, No.6, 22–27.

Songhurst, D.J. (1984) Subscriber repeat attempts, congestion and the quality of service: a study based on network simulation. *Brit. Telecom. Technol. Journal*, **2**, 47–55.

Stastny, M. (1984) On the substitution of the basic retrial model for a complex loss model with retrials. *Fundamentals of teletraffic theory. Proceedings of the Third International Seminar on Teletraffic Theory (Moscow, USSR, June 20-26, 1984).*, Moscow, 395–399.

Stepanov, S.N. (1980) Integral equilibrium relations of non-full-access systems with repeated calls, and their applications. *Problems of Information Transmission*, **16**, No.4, 323–327.

Stepanov, S.N. (1981) Probabilistic characteristics of an incompletely accessible multi-phase service system with several types of repeated calls. *Problems of Control and Information Theory*, **10**, No.6, 387–401.

Stepanov, S.N. (1982) Approximate calculation of the probabilistic characteristics of a completely attainable bundle. *Engineering Cybernetics*, No.3, 93.

Stepanov, S.N. (1983a) *Numerical Methods of Calculation for Systems with Repeated Calls*, Nauka, Moscow (in Russian).

Stepanov, S.N. (1983b) Algorithms of approximate design of systems with repeated calls. *Automation and Remote Control*, **44**, No.1, part 2, 63–71.

Stepanov, S.N. (1983c) The design of a trunk group based on repeated calls and waiting. *Telecommunications and Radio Engineering*, No.6, 1–7.

Stepanov, S.N. (1983d) Properties of probability characteristics of a communication network with repeated calls. *Problems of Information Transmission*, **19**, No.1, 69–76.

Stepanov, S.N. (1983e) Asymptotic formulae and estimations for probabilistic characteristics of full-available group with absolutely persistent subscribers. *Problems of Control and Information Theory*, **12**, No.5, 361–369.

Stepanov, S.N. (1984a) Numerical calculation accuracy of communication models with repeated calls. *Problems of Control and Information Theory*, **13**, No.6, 371–381.

Stepanov, S.N. (1984b) Estimation of characteristics of multilinear systems with repeated calls. *Fundamentals of teletraffic theory. Proceedings of the Third International Seminar on Teletraffic Theory (Moscow, USSR, June 20-26, 1984)*, Moscow, 400–409.

Stepanov, S.N. (1985a) The construction of effective algorithms for numerical analysis of multilinear systems with repeated calls. *Teletraffic Issues in an Advanced Information Society. ITC-11. Proceedings of the 11 International Teletraffic Congress, Kyoto, September 4-11, 1985*, Pt. 2. (ed M.Akiyama), Elsevier Science Publishers, Amsterdam, 1151.

Stepanov, S.N. and Tsitovich, I.I. (1985b) The model of a full-available group with repeated calls and waiting positions in the case of extreme load. *Problems of Control and Information Theory*, **14**, No.1, 25–32.

Stepanov, S.N. (1986) Increasing the efficiency of numerical methods for models with repeated calls. *Problems of Information Transmission*, **22**, No.4, 313–326.

Stepanov, S.N. and Tsitovich, I.I. (1987) Qualitative methods of analysis of systems with repeated calls. *Problems of Information Transmission*, **23**, No.2, 156–173.

Stepanov, S.N. (1988) Optimal calculation of characteristics of model with repeated calls. *Teletraffic Science for New Cost-Effective Systems, Networks and Services. ITC-12. Proceedings of the 12 International Teletraffic Congress. Torino, Italy, June 1–8, 1988*, Pt. 2. (ed M.Bonatti), Elsevier Science Publishers B.V., Amsterdam, 1427–1433.

Stepanov, S.N. (1989a) Optimized algorithms for numerical calculation of the characteristics of multistream models with repeated calls. *Problems of Information Transmission*, **25**, No.2, 136–144.

Stepanov, S.N. and Tsitovich, I.I. (1989b) Equivalent definitions of the probabilistic characteristics of models with repeated calls and their application. *Problems of Information Transmission*, **25**, No.2, 145–153.

Stepanov, S.N. (1989c) Solution of simultaneous large-scale equilibrium equations. *Automation and Remote Control*, **50**, No.5, part 2, 647–655.

Stepanov, S.N. (1991) Asymptotic analysis models with repeated calls in case of extreme load. *Teletraffic and Datatraffic in a Period of Change. ITC-13. Proceedings of the 12 International Teletraffic Congress. Copenhagen, Denmark, June 19–26, 1991*, (eds A. Jensen and V. B. Iversen), Elsevier Science Publishers B.V., Amsterdam, 21–26.

Stepanov, S.N. (1993) Asymptotic analysis of models with repeated calls in the case of large losses. *Problems of Information Transmission*, **29**, No.3, 248–267.

Sukharev, Yu.I. (1984) Calculation of probabilistic characteristics of an $M/G/1/\infty$ system with repeated calls in the presence of network blocking. Paper #6258-84, All-Union Institute for Scientific and Technical Information, Moscow (in Russian).

Sukharev, Yu.I. (1987) On retrial queues with finite capacity of control unit. Paper #4236-B87, All-Union Institute for Scientific and Technical Information, Moscow (in Russian).

Sumita, U., Shanthikumar, J.G. (1985) APL software development for central control telephone switching systems via the row-continuous Markov chain procedure. *Teletraffic Issues in an Advanced Information Society. ITC-11. Proceedings of the 11 International Teletraffic Congress, Kyoto, September 4-11, 1985*, Pt. 1. (ed M.Akiyama), Elsevier Science Publishers, Amsterdam, 439–445.

Suprun, A.P. (1987) A stochastic model for estimating the influence of repeated calls in digital communication systems. *Moscow University Mathematics Bulletin*, **42**, No.3, 22–24.

Sutorikhin, N.B., Zaretsky, K.I., Fidelman, I.G. (1984) Reattempt influence on grade of service after eliminating the total failure state of an exchange route. *Fundamentals of teletraffic theory. Proceedings of the Third International Seminar on Teletraffic Theory (Moscow, USSR, June 20-26, 1984)*, Moscow, 420–429.

Syski, R.(1986) *Introduction to Congestion Theory in Telephone Systems*,2nd edn, North-Holland, Amsterdam.

Takemori, E., Usui, Y., Matsuda, J. (1985) Field data analysis for traffic engineering. *Teletraffic Issues in an Advanced Information Society. ITC-11. Proceedings of the 11 International Teletraffic Congress, Kyoto, September 4-11, 1985*, Pt. 1. (ed M.Akiyama), Elsevier Science Publishers, Amsterdam, 425–431.

Todorov, P., Poryazov, S. (1985) Basic dependences characterizing a model of an idealized (standard) subscriber automatic telephone exchange. *Teletraffic Issues in an Advanced Information Society. ITC-11. Proceedings of the 11 International Teletraffic Congress, Kyoto, September 4-11, 1985*, Pt. 2. (ed M.Akiyama), Elsevier Science Publishers, Amsterdam, 752–759.

Todorov, P. and Nicolaev, K. (1989) Distribution of the holding times in switching systems with losses without and with repeated calls. A paper presented at the International Seminar on Teletraffic Theory and Computer Modelling, Moscow.

Toledano, F., De los Mozos, I.R. (1973) An analytical model to describe the influence of the repeated call attempts. A paper presented at the 7th International Teletraffic Congress, Stockholm.

Tsankov, B. (1988) Blocking probability in digital link systems with step-by-step selection and restricted number of retrials. *International Seminar on Teletraffic Theory and Computer Modelling. Sofia, Bul-*

*garia, March 21-26, 1988. Proceedings*, Sofia, 205–215.

Udagawa, K., Miwa, E. (1965) A complete group of trunks and Poisson-type repeated calls which influence it. *Electronics and Communications in Japan*, **48**, No.10, 42–53.

Warfield, R., Foers, G. (1985) Application of bayesian teletraffic measurement to systems with queueing or repeated attempts. *Teletraffic Issues in an Advanced Information Society. ITC-11. Proceedings of the 11 International Teletraffic Congress, Kyoto, September 4-11, 1985*, Pt. 2. (ed M.Akiyama), Elsevier Science Publishers, Amsterdam, 1003–1009.

Wilkinson, R.I. (1956) Theories for toll traffic engineering in the USA. *Bell Syst. Techn. J.*, **35**, No.2, 421–507.

Wilkinson, R.I. and Radnik, R. (1968) Customers' retrials in toll circuit operation. A paper presented at the IEEE Int. Conf. on Communications.

Wilkinson, R.I., Radnik, R.C. (1967) The character and effect of customer retrials in intertoll circuit operation. A paper presented at the 5th International Teletraffic Congress, New York.

Wolff, R.W. (1989) *Stochastic Modeling and the Theory of Queues*, Prentice-Hall, Englewood Cliffs, NJ.

Yang, T. and Templeton, J.G.C. (1987) A survey on retrial queues. *Queueing Systems*, **2**, No.3, 201–233.

Yang, T., Posner, M.J.M. and Templeton, J.G.C. (1990a) The M/G/1 retrial queue with nonpersistent customers. *Queueing Systems*, **7**, No.2, 209–218.

Yang, T. (1990b) A broad class of retrial queues and the associated generalized recursive technique. Ph.D. Thesis, University of Toronto.

Yang, T., Posner, M.J.M., Templeton, J.G.C. (1992) The $C_a/M/S/m$ retrial queue: a computational approach.*ORSA Journal on Computing*, **4**, No.2, 182–191.

Yang, T. and Li, H. (1994a) The $M/G/1$ retrial queue with the server subject to starting failures. *Queueing Systems*, **16**, 83–96.

Yang, T., Posner, M.J.M., Templeton, J.G.C. and Li, H. (1994b) An approximation method for the $M/G/1$ retrial queue with general retrial times. *European Journal of Operational Research*, **76**, 552–562.

Yang, T. and Li, H. (1995) On the steady-state queue size distribution of the discrete-time $Geo/G/1$ queue with repeated customers. *Queueing Systems*, **21**, 199–215.

Yokoi, T. (1988) End-to-end blocking probability in telecommunication networks and repeated call attempts. *Rev. Elec. Commun. Lab.*, **36**, No.1, 23–28.

Zelinskiy, A.M. and Kornyshev, Yu. N. (1978) Two models of a system with repeated calls, *Telecommunications and Radio Engineering*, No.1, 43–47.

# Author index

# Subject index